STEAM TABLES

[WITH SI UNITS]

B.B. Arora

Department of Mechanical Engineering
Delhi Technological University (Formerly DCE)
Delhi

Akhilesh Arora

Department of Mechanical Engineering
Delhi Technological University (Formerly DCE)
Delhi

Vipin

Department of Production Engineering
Delhi Technological University (Formerly DCE)
Delhi

I.K. International Publishing House Pvt. Ltd.

NEW DELHI

Published by
I.K. International Publishing House Pvt. Ltd.
S-25, Green Park Extension
Uphaar Cinema Market
New Delhi–110 016 (India)
E-mail:info@ikinternational.com

ISBN: 978-93-84588-53-3

Published by Krishan Makhijani for I.K. International Publishing House Pvt. Ltd., S-25, Green Park Extension, Uphaar Cinema Market, New Delhi–110 016 and Printed by Rekha Printers Pvt. Ltd., Okhla Industrial Area, Phase II, New Delhi–110 020.

Preface

Performance analysis of boilers, steam turbines, plant process and other mechanical equipment connected with power generation require the thermodynamic data and transport properties of water and steam which are essential for the design, performance specifications.

With a view to present thermodynamic data catering to the need of Indian power industries, it gives us immense pleasure in presenting **"Steam Tables [With SI Units]"**. The Tables contain thermodynamic properties of water and steam which covers the whole practical range being practised in power plants and process industries.

The subject matter is intended for practising engineers in power plants and process industries and students pursuing undergraduate and postgraduate courses in Engineering. It also helps the students preparing for AMIE and competitive examinations such as GATE, IES, PSU and Civil Services, etc.

The book presents properties of water and steam in a simplified form. It is easy to understand and translate the properties for numericals and real-world problems. Hope this book will cater to the needs of faculty, students and practising engineers.

Any suggestion/criticism is welcome.

Authors

Steam Tables

Introduction

Water is converted into steam, which is used in the generation/process. During this process steam gets expansion and probably phase changes. Phase change may turn the steam into water, which further can be converted into steam by providing heat energy in the boiler or otherwise. This phenomenon is known as steam cycle.

To understand the steam cycle one needs to study the properties of water and steam at the important phases. The various terms associated with these properties and processes, and the use of steam tables is enlightened as under.

Basic thermodynamic terms

1. *Thermodynamic process*: It represents any process which changes the *state* of the working fluid.
2. *Thermodynamic cycle*: It represents a recurring series of thermodynamic processes through which an effect is produced by the transformation or redistribution of energy. In other words, it is a series of processes repeated over and over again in the same order.
3. *Specific volume* (v): It represents the space occupied per unit mass of a substance and is measured in m^3/kg. It is the mathematical inverse of density.
4. *Internal Energy* (u): It represents the energy content contained within the system and is measured as kJ/kg. It is the sum of the kinetic and potential energies of the particles that form the system while excluding the kinetic energy of motion of the system as a whole and the potential energy of the system as a whole due to external force fields
5. *Enthalpy* (h): It represents the total energy content of steam and is measured as kJ/kg. It expresses the internal energy and flow work, or the total potential energy and kinetic energy contained within a substance. .
6. *Entropy* (s): It represents the unavailability of energy and is measured in kJ/kg K.

Physical Parameter	**Synbol**	**Unit**
Temperature	t	°C
Pressure	p	bar
Specific Volume	v	(m^3/kg)
Internal Energy	u	(kJ/kg)
Enthalpy	h	(kJ/kg)
Entropy	s	(kJ/kg K)
Dryness fraction	x	
Suffix		
Feed water	f	
Evaporation	fg	
Saturation	g	

Property values at any state

(a) Wet Region

Specific Volume $v = xv_g + (1 - x)\, v_f$

Internal Energy $u = u_f + x\, u_{fg}$

Enthalpy $h = h_f + xh_{fg}$

(b) Saturation

Specific Volume $v = v_g$

Internal Energy $u = u_f + u_{fg}$

Enthalpy $h = h_f + h_{fg}$

Linear interpolation

- Consider a curve $y = f(x)$

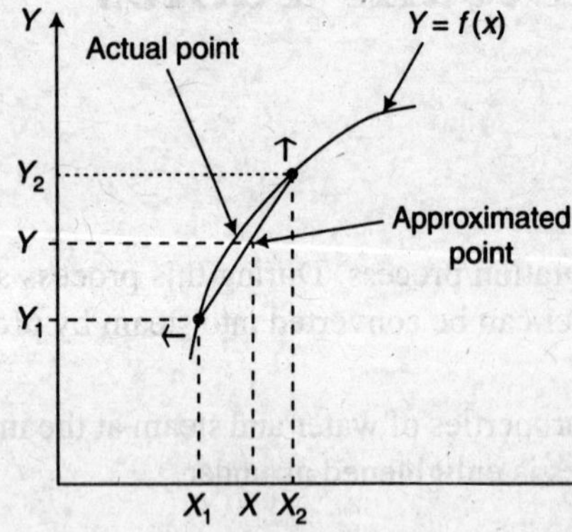

- Let X_1, X_2 be the X-values of two adjacent points, 1, 2 on the curve and let Y_1, Y_2 be the corresponding values of the dependent variable Y.
- Need to find the Y-value corresponding to a point, lying between 1 and 2, having the X-value X.
- Approximate portion of curve between 1 and 2 can be represented by a line (linear) segment as shown.
- Using the concept of similar triangles, $m = \dfrac{Y - Y_1}{(X - X_1)} = \dfrac{(Y_2 - Y_1)}{(X_2 - X_1)}$

Taking the two nearest points as line with constant slope m as explained above, one can proceed as under

$$y = mx + b$$

$$y_1 = mx_1 + b \quad y_2 = mx_2 + b$$

$$y_2 - y_1 = m\,(x_2 - x_1)$$

$$m = \frac{(y_2 - y_1)}{(x_2 - x_1)}$$

$$b = y_1 - mx_1$$

$$y = mx + y_1 - mx_1$$

$$y = y_1 + m\,(x - x_1)$$

$$y = y_1 + \left[\frac{(y_2 - y_1)}{(x_2 - x_1)}\right](x - x_1)$$

This is good approximation to determine the intermediate steam properties from Steam Tables such as Temperature, Pressure, Specific Volume, Internal Energy, Enthalpy and Entropy.

Tables

Steam Tables

Table 1 Saturation Temperatures and Properties of Water and Steam

Temp. t (°C)	Pressure P (bar)	Specific Volume $\left(\frac{m^3}{kg}\right)$		Internal Energy $\left(\frac{kJ}{kg}\right)$			Enthalpy $\left(\frac{kJ}{kg}\right)$			Entropy $\left(\frac{kJ}{kg\ K}\right)$		
		Sat. Liquid	Sat. Vapour	Sat. Liquid	Latent	Sat. Vapour	Sat. Liquid	Sat. Liquid	Sat. Vapour	Sat. Liquid	Latent	Sat. Vapour
		v_f	v_g	u_f	u_{fg}	u_g	h_f	h_{fg}	h_g	s_f	s_{fg}	s_g
0	0.006112	0.001000	206.1467	−0.04	2374.96	2374.91	− 0.04	2500.96	2500.91	−0.00015	9.15598	9.15582
0.01	0.006117	0.001000	206.0045	0.00	2374.93	2374.93	0.00	2500.93	2500.93	0.00000	9.15556	9.15555
1	0.006571	0.001000	192.4549	4.18	2372.11	2376.29	4.18	2498.57	2502.75	0.01526	9.11389	9.12915
2	0.007060	0.001000	179.7757	8.39	2369.28	2377.67	8.39	2496.19	2504.58	0.03061	9.07212	9.10273
3	0.007581	0.001000	168.0274	12.60	2366.44	2379.04	12.60	2493.81	2506.42	0.04589	9.03066	9.07655
4	0.008135	0.001000	157.1352	16.81	2363.60	2380.42	16.81	2491.44	2508.25	0.06110	8.98951	9.05061
5	0.008725	0.001000	147.0307	21.02	2360.77	2381.79	21.02	2489.06	2510.08	0.07625	8.94865	9.02491
6	0.009353	0.001000	137.6516	25.22	2357.94	2383.17	25.22	2486.69	2511.91	0.09134	8.90810	8.99944
7	0.010020	0.001000	128.9409	29.43	2355.11	2384.54	29.43	2484.31	2513.74	0.10637	8.86783	8.97420
8	0.010729	0.001000	120.8463	33.63	2352.29	2385.91	33.63	2481.94	2515.57	0.12133	8.82786	8.94919
9	0.011482	0.001000	113.3201	37.82	2349.46	2387.28	37.83	2479.57	2517.40	0.13624	8.78817	8.92441
10	0.012281	0.001000	106.3185	42.02	2346.63	2388.66	42.02	2477.20	2519.23	0.15109	8.74877	8.89986
11	0.013129	0.001000	99.8013	46.22	2343.81	2390.03	46.22	2474.84	2521.05	0.16588	8.70965	8.87552
12	0.014027	0.001001	93.7319	50.41	2340.99	2391.40	50.41	2472.47	2522.88	0.18061	8.67080	8.85141
13	0.014980	0.001001	88.0762	54.60	2338.17	2392.77	54.60	2470.10	2524.70	0.19528	8.63222	8.82751
14	0.015989	0.001001	82.8035	58.79	2335.34	2394.13	58.79	2467.73	2526.53	0.20990	8.59392	8.80382
15	0.017057	0.001001	77.8851	62.98	2332.52	2395.50	62.98	2465.37	2528.35	0.22447	8.55588	8.78034
16	0.018187	0.001001	73.2949	67.17	2329.70	2396.87	67.17	2463.00	2530.17	0.23898	8.51810	8.75708
17	0.019382	0.001001	69.0088	71.36	2326.88	2398.24	71.36	2460.63	2531.99	0.25343	8.48059	8.73402
18	0.020646	0.001001	65.0046	75.54	2324.06	2399.60	75.54	2458.27	2533.81	0.26784	8.44333	8.71116
19	0.021981	0.001002	61.2619	79.73	2321.24	2400.97	79.73	2455.90	2535.63	0.28219	8.40632	8.68851
20	0.023392	0.001002	57.7619	83.91	2318.42	2402.33	83.91	2453.53	2537.45	0.29649	8.36957	8.66606
21	0.024881	0.001002	54.4872	88.10	2315.60	2403.69	88.10	2451.16	2539.26	0.31073	8.33306	8.64380
22	0.026452	0.001002	51.4218	92.28	2312.78	2405.06	92.28	2448.80	2541.08	0.32493	8.29681	8.62174
23	0.028110	0.001003	48.5510	96.46	2309.95	2406.42	96.47	2446.43	2542.89	0.33908	8.26079	8.59987
24	0.029857	0.001003	45.8611	100.65	2307.13	2407.78	100.65	2444.06	2544.70	0.35318	8.22501	8.57819
25	0.031698	0.001003	43.3395	104.83	2304.31	2409.14	104.83	2441.69	2546.52	0.36723	8.18947	8.55670
26	0.033638	0.001003	40.9747	109.01	2301.49	2410.50	109.01	2439.31	2548.33	0.38123	8.15417	8.53540
27	0.035680	0.001004	38.7558	113.19	2298.66	2411.85	113.19	2436.94	2550.14	0.39518	8.11909	8.51428
28	0.037830	0.001004	36.6729	117.37	2295.84	2413.21	117.37	2434.57	2551.94	0.40909	8.08425	8.49334
29	0.040091	0.001004	34.7167	121.55	2293.02	2414.57	121.55	2432.19	2553.75	0.42295	8.04963	8.47258
30	0.042469	0.001004	32.8788	125.73	2290.19	2415.92	125.74	2429.82	2555.55	0.43676	8.01524	8.45200
31	0.044969	0.001005	31.1512	129.91	2287.36	2417.27	129.92	2427.44	2557.36	0.45053	7.98107	8.43160
32	0.047596	0.001005	29.5265	134.09	2284.54	2418.63	134.10	2425.06	2559.16	0.46425	7.94712	8.41137
33	0.050354	0.001005	27.9981	138.27	2281.71	2419.98	138.28	2422.68	2560.96	0.47792	7.91339	8.39131
34	0.053251	0.001006	26.5595	142.45	2278.88	2421.33	142.46	2420.30	2562.76	0.49155	7.87987	8.37142
35	0.056291	0.001006	25.2049	146.63	2276.05	2422.68	146.64	2417.92	2564.56	0.50514	7.84657	8.35170
36	0.059480	0.001006	23.9289	150.81	2273.21	2424.02	150.82	2415.54	2566.35	0.51868	7.81347	8.33215
37	0.062824	0.001007	22.7264	154.99	2270.38	2425.37	155.00	2413.15	2568.15	0.53217	7.78059	8.31276
38	0.066330	0.001007	21.5927	159.17	2267.55	2426.71	159.17	2410.76	2569.94	0.54563	7.74791	8.29353
39	0.070003	0.001008	20.5234	163.35	2264.71	2428.06	163.35	2408.37	2571.73	0.55904	7.71543	8.27447

Temp. t (°C)	Pressure P (bar)	Specific Volume $\left(\frac{m^3}{kg}\right)$		Internal Energy $\left(\frac{kJ}{kg}\right)$			Enthalpy $\left(\frac{kJ}{kg}\right)$			Entropy $\left(\frac{kJ}{kg\,K}\right)$		
		Sat. Liquid	Sat. Vapour	Sat. Liquid	Latent	Sat. Vapour	Sat. Liquid	Sat. Liquid	Sat. Vapour	Sat. Liquid	Latent	Sat. Vapour
		v_f	v_g	u_f	u_{fg}	u_g	h_f	h_{fg}	h_g	s_f	s_{fg}	s_g
40	0.073851	0.001008	19.5145	167.53	2261.87	2429.40	167.53	2405.98	2573.52	0.57241	7.68315	8.25556
41	0.077880	0.001008	18.5622	171.71	2259.03	2430.74	171.72	2403.59	2575.30	0.58573	7.65108	8.23681
42	0.082098	0.001009	17.6629	175.89	2256.19	2432.08	175.90	2401.19	2577.09	0.59902	7.61920	8.21822
43	0.086511	0.001009	16.8133	180.07	2253.35	2433.42	180.08	2398.80	2578.87	0.61226	7.58751	8.19977
44	0.091127	0.001010	16.0104	184.25	2250.51	2434.76	184.26	2396.40	2580.65	0.62546	7.55602	8.18149
45	0.095953	0.001010	15.2514	188.43	2247.66	2436.09	188.44	2394.00	2582.43	0.63862	7.52472	8.16335
46	0.100998	0.001010	14.5336	192.61	2244.82	2437.42	192.62	2391.59	2584.21	0.65174	7.49361	8.14535
47	0.106269	0.001011	13.8544	196.79	2241.97	2438.76	196.80	2389.19	2585.98	0.66482	7.46269	8.12751
48	0.111775	0.001011	13.2115	200.97	2239.12	2440.09	200.98	2386.78	2587.76	0.67786	7.43195	8.10981
49	0.117524	0.001012	12.6029	205.15	2236.26	2441.41	205.16	2384.37	2589.53	0.69086	7.40140	8.09226
50	0.123525	0.001012	12.0264	209.33	2233.41	2442.74	209.34	2381.95	2591.30	0.70382	7.37103	8.07484
51	0.129786	0.001013	11.4801	213.51	2230.55	2444.07	213.53	2379.54	2593.06	0.71674	7.34083	8.05757
52	0.136318	0.001013	10.9624	217.70	2227.69	2445.39	217.71	2377.12	2594.83	0.72962	7.31081	8.04043
53	0.143129	0.001014	10.4714	221.88	2224.83	2446.71	221.89	2374.70	2596.59	0.74246	7.28097	8.02344
54	0.150229	0.001014	10.0058	226.06	2221.97	2448.03	226.08	2372.27	2598.35	0.75527	7.25131	8.00657
55	0.157628	0.001015	9.5639	230.24	2219.11	2449.35	230.26	2369.84	2600.10	0.76803	7.22181	7.98985
56	0.165337	0.001015	9.1445	234.43	2216.24	2450.67	234.44	2367.41	2601.86	0.78076	7.19249	7.97325
57	0.173365	0.001016	8.7463	238.61	2213.37	2451.98	238.63	2364.98	2603.61	0.79345	7.16333	7.95679
58	0.181723	0.001016	8.3680	242.79	2210.50	2453.29	242.81	2362.54	2605.36	0.80611	7.13434	7.94045
59	0.190422	0.001017	8.0086	246.98	2207.62	2454.60	247.00	2360.11	2607.10	0.81872	7.10552	7.92424
60	0.199474	0.001017	7.6670	251.16	2204.75	2455.91	251.18	2357.66	2608.85	0.83130	7.07686	7.90816
61	0.208889	0.001018	7.3422	255.35	2201.87	2457.22	255.37	2355.22	2610.59	0.84385	7.04836	7.89221
62	0.218680	0.001018	7.0333	259.53	2198.98	2458.52	259.56	2352.77	2612.32	0.85635	7.02002	7.87638
63	0.228859	0.001019	6.7395	263.72	2196.10	2459.82	263.74	2350.32	2614.06	0.86883	6.99184	7.86067
64	0.239437	0.001019	6.4598	267.91	2193.21	2461.12	267.93	2347.86	2615.79	0.88126	6.96382	7.84508
65	0.250427	0.001020	6.1935	272.09	2190.32	2462.42	272.12	2345.40	2617.52	0.89366	6.93595	7.82962
66	0.261843	0.001020	5.9399	276.28	2187.43	2463.71	276.31	2342.94	2619.24	0.90603	6.90824	7.81427
67	0.273697	0.001021	5.6984	280.47	2184.53	2465.00	280.50	2340.47	2620.97	0.91836	6.88068	7.79904
68	0.286002	0.001022	5.4682	284.66	2181.64	2466.29	284.69	2338.00	2622.69	0.93066	6.85327	7.78392
69	0.298772	0.001022	5.2488	288.85	2178.73	2467.58	288.88	2335.52	2624.40	0.94292	6.82600	7.76892
70	0.312022	0.001023	5.0396	293.04	2175.83	2468.87	293.07	2333.04	2626.11	0.95514	6.79889	7.75403
71	0.325765	0.001023	4.8401	297.23	2172.92	2470.15	297.26	2330.56	2627.82	0.96734	6.77192	7.73926
72	0.340016	0.001024	4.6497	301.42	2170.01	2471.43	301.45	2328.07	2629.53	0.97950	6.74509	7.72459
73	0.354790	0.001025	4.4681	305.61	2167.10	2472.71	305.65	2325.58	2631.23	0.99162	6.71841	7.71003
74	0.370102	0.001025	4.2947	309.80	2164.18	2473.98	309.84	2323.09	2632.93	1.00372	6.69187	7.69559
75	0.385967	0.001026	4.1291	313.99	2161.26	2475.25	314.03	2320.59	2634.62	1.01578	6.66547	7.68125
76	0.402402	0.001026	3.9709	318.19	2158.33	2476.52	318.23	2318.08	2636.31	1.02780	6.63921	7.66701
77	0.419422	0.001027	3.8198	322.38	2155.41	2477.79	322.42	2315.58	2638.00	1.03980	6.61308	7.65288
78	0.437043	0.001028	3.6754	326.58	2152.48	2479.05	326.62	2313.06	2639.68	1.05176	6.58709	7.63885
79	0.455283	0.001028	3.5374	330.77	2149.54	2480.31	330.82	2310.54	2641.36	1.06369	6.56124	7.62493
80	0.474158	0.001029	3.4053	334.97	2146.60	2481.57	335.02	2308.02	2643.04	1.07559	6.53552	7.61111
81	0.493686	0.001030	3.2791	339.16	2143.66	2482.83	339.22	2305.49	2644.71	1.08746	6.50993	7.59738

Temp. t (°C)	Pressure P (bar)	Specific Volume $\left(\frac{m^3}{kg}\right)$		Internal Energy $\left(\frac{kJ}{kg}\right)$			Enthalpy $\left(\frac{kJ}{kg}\right)$			Entropy $\left(\frac{kJ}{kg\,K}\right)$		
		Sat. Liquid	Sat. Vapour	Sat. Liquid	Latent	Sat. Vapour	Sat. Liquid	Sat. Liquid	Sat. Vapour	Sat. Liquid	Latent	Sat. Vapour
		v_f	v_g	u_f	u_{fg}	u_g	h_f	h_{fg}	h_g	s_f	s_{fg}	s_g
82	0.513884	0.001030	3.1583	343.36	2140.72	2484.08	343.41	2302.96	2646.38	1.09929	6.48447	7.58376
83	0.534770	0.001031	3.0427	347.56	2137.77	2485.33	347.62	2300.42	2648.04	1.11110	6.45914	7.57023
84	0.556363	0.001032	2.9320	351.76	2134.81	2486.57	351.82	2297.88	2649.70	1.12287	6.43393	7.55680
85	0.578681	0.001032	2.8260	355.96	2131.86	2487.81	356.02	2295.33	2651.35	1.13461	6.40885	7.54347
86	0.601744	0.001033	2.7246	360.16	2128.89	2489.05	360.22	2292.78	2653.00	1.14633	6.38390	7.53023
87	0.625570	0.001034	2.6273	364.36	2125.93	2490.29	364.43	2290.22	2654.65	1.15801	6.35907	7.51708
88	0.650179	0.001035	2.5342	368.56	2122.96	2491.52	368.63	2287.66	2656.29	1.16966	6.33437	7.50403
89	0.675591	0.001035	2.4449	372.77	2119.98	2492.75	372.84	2285.09	2657.93	1.18128	6.30978	7.49106
90	0.701827	0.001036	2.3593	376.97	2117.01	2493.98	377.04	2282.51	2659.56	1.19288	6.28532	7.47819
91	0.728906	0.001037	2.2772	381.18	2114.02	2495.20	381.25	2279.93	2661.18	1.20444	6.26097	7.46541
92	0.756851	0.001037	2.1984	385.38	2111.04	2496.42	385.46	2277.35	2662.81	1.21597	6.23674	7.45271
93	0.785682	0.001038	2.1229	389.59	2108.04	2497.63	389.67	2274.75	2664.42	1.22748	6.21263	7.44010
94	0.815421	0.001039	2.0504	393.80	2105.05	2498.84	393.88	2272.15	2666.04	1.23895	6.18863	7.42758
95	0.846090	0.001040	1.9808	398.00	2102.05	2500.05	398.09	2269.55	2667.64	1.25040	6.16474	7.41514
96	0.877711	0.001040	1.9139	402.21	2099.04	2501.26	402.31	2266.94	2669.24	1.26182	6.14097	7.40279
97	0.910308	0.001041	1.8498	406.42	2096.03	2502.46	406.52	2264.32	2670.84	1.27321	6.11731	7.39052
98	0.943902	0.001042	1.7881	410.64	2093.02	2503.65	410.73	2261.70	2672.43	1.28458	6.09376	7.37833
99	0.978518	0.001043	1.7289	414.85	2089.99	2504.84	414.95	2259.07	2674.02	1.29591	6.07032	7.36623
100	1.01325	0.001043	1.6720	419.06	2086.97	2506.03	419.17	2256.43	2675.60	1.30722	6.04698	7.35420
101	1.05091	0.001044	1.6172	423.28	2083.94	2507.22	423.39	2253.79	2677.18	1.31850	6.02376	7.34226
102	1.08873	0.001045	1.5646	427.49	2080.90	2508.40	427.61	2251.14	2678.74	1.32975	6.00063	7.33039
103	1.12768	0.001046	1.5140	431.71	2077.86	2509.57	431.83	2248.48	2680.31	1.34098	5.97762	7.31860
104	1.16776	0.001047	1.4654	435.93	2074.82	2510.75	436.05	2245.82	2681.87	1.35218	5.95470	7.30688
105	1.20902	0.001047	1.4186	440.15	2071.76	2511.91	440.28	2243.14	2683.42	1.36335	5.93189	7.29525
106	1.25147	0.001048	1.3735	444.37	2068.71	2513.08	444.50	2240.46	2684.97	1.37450	5.90918	7.28368
107	1.29515	0.001049	1.3301	448.59	2065.64	2514.24	448.73	2237.78	2686.51	1.38562	5.88657	7.27220
108	1.34007	0.001050	1.2884	452.82	2062.57	2515.39	452.96	2235.09	2688.04	1.39672	5.86406	7.26078
109	1.38627	0.001051	1.2482	457.04	2059.50	2516.54	457.19	2232.38	2689.57	1.40779	5.84165	7.24944
110	1.43377	0.001052	1.2094	461.27	2056.42	2517.68	461.42	2229.68	2691.09	1.41883	5.81933	7.23817
111	1.48260	0.001052	1.1721	465.49	2053.33	2518.83	465.65	2226.96	2692.61	1.42985	5.79711	7.22697
112	1.53279	0.001053	1.1362	469.72	2050.24	2519.96	469.88	2224.23	2694.12	1.44084	5.77499	7.21583
113	1.58437	0.001054	1.1016	473.95	2047.14	2521.09	474.12	2221.50	2695.62	1.45181	5.75296	7.20477
114	1.63736	0.001055	1.0682	478.18	2044.04	2522.22	478.36	2218.76	2697.12	1.46276	5.73102	7.19378
115	1.69180	0.001056	1.0360	482.42	2040.93	2523.34	482.59	2216.01	2698.61	1.47368	5.70918	7.18285
116	1.74771	0.001057	1.0049	486.65	2037.81	2524.46	486.84	2213.26	2700.09	1.48457	5.68742	7.17199
117	1.80513	0.001058	0.9750	490.89	2034.68	2525.57	491.08	2210.49	2701.57	1.49544	5.66576	7.16120
118	1.86408	0.001059	0.9461	495.12	2031.55	2526.68	495.32	2207.72	2703.04	1.50629	5.64418	7.15047
119	1.92460	0.001059	0.9182	499.36	2028.42	2527.78	499.57	2204.94	2704.50	1.51711	5.62269	7.13980
120	1.98671	0.001060	0.8913	503.60	2025.27	2528.88	503.81	2202.14	2705.96	1.52791	5.60129	7.12920
121	2.05045	0.001061	0.8654	507.84	2022.12	2529.97	508.06	2199.34	2707.41	1.53869	5.57997	7.11866
122	2.11585	0.001062	0.8403	512.09	2018.97	2531.05	512.31	2196.53	2708.85	1.54944	5.55875	7.10819
123	2.18295	0.001063	0.8161	516.33	2015.80	2532.13	516.56	2193.72	2710.28	1.56017	5.53760	7.09777

Temp. t (°C)	Pressure P (bar)	Specific Volume $\left(\frac{m^3}{kg}\right)$		Internal Energy $\left(\frac{kJ}{kg}\right)$			Enthalpy $\left(\frac{kJ}{kg}\right)$			Entropy $\left(\frac{kJ}{kg\ K}\right)$		
		Sat. Liquid	Sat. Vapour	Sat. Liquid	Latent	Sat. Vapour	Sat. Liquid	Sat. Liquid	Sat. Vapour	Sat. Liquid	Latent	Sat. Vapour
		v_f	v_g	u_f	u_{fg}	u_g	h_f	h_{fg}	h_g	s_f	s_{fg}	s_g
124	2.25176	0.001064	0.7927	520.58	2012.63	2533.21	520.82	2190.89	2711.71	1.57088	5.51654	7.08741
125	2.32234	0.001065	0.7701	524.83	2009.45	2534.28	525.07	2188.05	2713.13	1.58156	5.49556	7.07712
126	2.39470	0.001066	0.7483	529.08	2006.27	2535.34	529.33	2185.21	2714.54	1.59222	5.47466	7.06688
127	2.46889	0.001067	0.7272	533.33	2003.07	2536.40	533.59	2182.35	2715.94	1.60286	5.45384	7.05670
128	2.54494	0.001068	0.7068	537.58	1999.87	2537.46	537.85	2179.49	2717.34	1.61348	5.43310	7.04658
129	2.62288	0.001069	0.6871	541.84	1996.66	2538.50	542.12	2176.61	2718.73	1.62407	5.41244	7.03652
130	2.70275	0.001070	0.6681	546.10	1993.45	2539.54	546.38	2173.73	2720.11	1.63465	5.39186	7.02651
131	2.78458	0.001071	0.6497	550.35	1990.23	2540.58	550.65	2170.83	2721.48	1.64520	5.37136	7.01656
132	2.86841	0.001072	0.6318	554.61	1986.99	2541.61	554.92	2167.93	2722.85	1.65573	5.35093	7.00666
133	2.95427	0.001073	0.6146	558.88	1983.76	2542.63	559.19	2165.01	2724.20	1.66624	5.33058	6.99682
134	3.04220	0.001074	0.5979	563.14	1980.51	2543.65	563.47	2162.08	2725.55	1.67673	5.31030	6.98703
135	3.13224	0.001075	0.5818	567.41	1977.25	2544.66	567.75	2159.15	2726.89	1.68719	5.29009	6.97729
136	3.22442	0.001076	0.5662	571.68	1973.99	2545.67	572.02	2156.20	2728.23	1.69764	5.26996	6.96760
137	3.31878	0.001077	0.5510	575.95	1970.72	2546.67	576.31	2153.24	2729.55	1.70807	5.24990	6.95797
138	3.41536	0.001078	0.5364	580.22	1967.44	2547.66	580.59	2150.28	2730.86	1.71847	5.22992	6.94839
139	3.51420	0.001079	0.5222	584.50	1964.15	2548.65	584.87	2147.30	2732.17	1.72886	5.21000	6.93885
140	3.61533	0.001080	0.5085	588.77	1960.86	2549.63	589.16	2144.31	2733.47	1.73922	5.19015	6.92937
141	3.71880	0.001081	0.4952	593.05	1957.55	2550.60	593.45	2141.30	2734.76	1.74957	5.17037	6.91994
142	3.82463	0.001082	0.4823	597.33	1954.24	2551.57	597.75	2138.29	2736.04	1.75989	5.15066	6.91055
143	3.93289	0.001083	0.4698	601.61	1950.92	2552.53	602.04	2135.27	2737.31	1.77020	5.13101	6.90121
144	4.04359	0.001084	0.4577	605.90	1947.58	2553.48	606.34	2132.23	2738.57	1.78048	5.11144	6.89192
145	4.15678	0.001085	0.4460	610.19	1944.24	2554.43	610.64	2129.18	2739.82	1.79075	5.09192	6.88267
146	4.27251	0.001086	0.4346	614.48	1940.89	2555.37	614.94	2126.13	2741.07	1.80100	5.07248	6.87348
147	4.39081	0.001087	0.4236	618.77	1937.54	2556.31	619.25	2123.05	2742.30	1.81123	5.05309	6.86432
148	4.51173	0.001088	0.4129	623.06	1934.17	2557.23	623.56	2119.97	2743.53	1.82144	5.03377	6.85521
149	4.63531	0.001089	0.4025	627.36	1930.79	2558.15	627.87	2116.88	2744.74	1.83163	5.01452	6.84615
150	4.76158	0.001091	0.3925	631.66	1927.40	2559.06	632.18	2113.77	2745.95	1.84180	4.99532	6.83713
151	4.89060	0.001092	0.3827	635.96	1924.01	2559.97	636.50	2110.65	2747.14	1.85196	4.97619	6.82815
152	5.02240	0.001093	0.3733	640.27	1920.60	2560.87	640.82	2107.51	2748.33	1.86210	4.95711	6.81921
153	5.15703	0.001094	0.3641	644.57	1917.19	2561.76	645.14	2104.37	2749.51	1.87222	4.93810	6.81032
154	5.29453	0.001095	0.3551	648.88	1913.76	2562.64	649.46	2101.21	2750.67	1.88232	4.91915	6.80146
155	5.43494	0.001096	0.3465	653.19	1910.33	2563.52	653.79	2098.04	2751.83	1.89240	4.90025	6.79265
156	5.57831	0.001097	0.3381	657.51	1906.88	2564.39	658.12	2094.86	2752.98	1.90247	4.88141	6.78388
157	5.72469	0.001098	0.3299	661.83	1903.42	2565.25	662.45	2091.66	2754.11	1.91252	4.86263	6.77515
158	5.87412	0.001100	0.3220	666.15	1899.96	2566.10	666.79	2088.45	2755.24	1.92255	4.84390	6.76645
159	6.02663	0.001101	0.3143	670.47	1896.48	2566.95	671.13	2085.22	2756.35	1.93257	4.82523	6.75780
160	6.18229	0.001102	0.3068	674.79	1893.00	2567.79	675.47	2081.99	2757.46	1.94256	4.80662	6.74918
161	6.34114	0.001103	0.2995	679.12	1889.50	2568.62	679.82	2078.73	2758.55	1.95255	4.78806	6.74060
162	6.50321	0.001104	0.2925	683.45	1885.99	2569.44	684.17	2075.47	2759.64	1.96251	4.76955	6.73206
163	6.66857	0.001106	0.2856	687.79	1882.47	2570.26	688.52	2072.19	2760.71	1.97246	4.75110	6.72356
164	6.83725	0.001107	0.2789	692.12	1878.94	2571.06	692.88	2068.90	2761.77	1.98239	4.73269	6.71509
165	7.00930	0.001108	0.2724	696.46	1875.40	2571.86	697.24	2065.59	2762.83	1.99231	4.71434	6.70665

Temp. t (°C)	Pressure P (bar)	Specific Volume $\left(\frac{m^3}{kg}\right)$		Internal Energy $\left(\frac{kJ}{kg}\right)$			Enthalpy $\left(\frac{kJ}{kg}\right)$			Entropy $\left(\frac{kJ}{kg\,K}\right)$		
		Sat. Liquid	Sat. Vapour	Sat. Liquid	Latent	Sat. Vapour	Sat. Liquid	Sat. Liquid	Sat. Vapour	Sat. Liquid	Latent	Sat. Vapour
		v_f	v_g	u_f	u_{fg}	u_g	h_f	h_{fg}	h_g	s_f	s_{fg}	s_g
166	7.18477	0.001109	0.2661	700.80	1871.85	2572.65	701.60	2062.27	2763.87	2.00221	4.69604	6.69826
167	7.36372	0.001110	0.2600	705.15	1868.29	2573.44	705.97	2058.93	2764.90	2.01210	4.67779	6.68989
168	7.54617	0.001112	0.2540	709.50	1864.72	2574.21	710.34	2055.58	2765.91	2.02197	4.65959	6.68156
169	7.73220	0.001113	0.2482	713.85	1861.13	2574.98	714.71	2052.21	2766.92	2.03182	4.64144	6.67326
170	7.92184	0.001114	0.2426	718.20	1857.53	2575.74	719.08	2048.83	2767.92	2.04166	4.62334	6.66500
171	8.11514	0.001116	0.2371	722.56	1853.93	2576.49	723.46	2045.44	2768.90	2.05148	4.60529	6.65677
172	8.31216	0.001117	0.2318	726.92	1850.31	2577.23	727.85	2042.03	2769.87	2.06129	4.58728	6.64857
173	8.51295	0.001118	0.2266	731.28	1846.67	2577.96	732.24	2038.60	2770.83	2.07109	4.56932	6.64040
174	8.71754	0.001119	0.2215	735.65	1843.03	2578.68	736.63	2035.16	2771.78	2.08087	4.55140	6.63227
175	8.92601	0.001121	0.2166	740.02	1839.38	2579.40	741.02	2031.70	2772.72	2.09063	4.53353	6.62416
176	9.13839	0.001122	0.2118	744.39	1835.71	2580.10	745.42	2028.23	2773.65	2.10038	4.51570	6.61609
177	9.35473	0.001123	0.2071	748.77	1832.03	2580.80	749.82	2024.74	2774.56	2.11012	4.49792	6.60804
178	9.57510	0.001125	0.2026	753.15	1828.34	2581.49	754.23	2021.23	2775.46	2.11984	4.48018	6.60002
179	9.79954	0.001126	0.1982	757.54	1824.63	2582.17	758.64	2017.71	2776.35	2.12955	4.46248	6.59204
180	10.02811	0.001127	0.1938	761.92	1820.91	2582.84	763.05	2014.17	2777.23	2.13925	4.44483	6.58408
181	10.2608	0.001129	0.1896	766.31	1817.18	2583.50	767.47	2010.62	2778.09	2.14893	4.42722	6.57615
182	10.4978	0.001130	0.1856	770.71	1813.44	2584.15	771.89	2007.05	2778.94	2.15860	4.40964	6.56824
183	10.7390	0.001132	0.1816	775.11	1809.69	2584.79	776.32	2003.46	2779.78	2.16825	4.39211	6.56037
184	10.9846	0.001133	0.1777	779.51	1805.92	2585.43	780.75	1999.86	2780.61	2.17790	4.37462	6.55252
185	11.2346	0.001134	0.1739	783.91	1802.14	2586.05	785.19	1996.24	2781.42	2.18752	4.35717	6.54469
186	11.4891	0.001136	0.1702	788.32	1798.34	2586.66	789.63	1992.60	2782.22	2.19714	4.33975	6.53689
187	11.7480	0.001137	0.1666	792.73	1794.53	2587.27	794.07	1988.94	2783.01	2.20674	4.32238	6.52912
188	12.0115	0.001139	0.1631	797.15	1790.71	2587.86	798.52	1985.27	2783.79	2.21634	4.30504	6.52137
189	12.2796	0.001140	0.1597	801.57	1786.88	2588.45	802.97	1981.58	2784.55	2.22592	4.28773	6.51365
190	12.5524	0.001141	0.1564	806.00	1783.03	2589.02	807.43	1977.87	2785.30	2.23548	4.27047	6.50595
191	12.8299	0.001143	0.1531	810.42	1779.16	2589.59	811.89	1974.14	2786.03	2.24504	4.25324	6.49827
192	13.1121	0.001144	0.1499	814.86	1775.29	2590.14	816.36	1970.39	2786.75	2.25458	4.23604	6.49062
193	13.3993	0.001146	0.1469	819.29	1771.40	2590.69	820.83	1966.63	2787.46	2.26411	4.21888	6.48299
194	13.6913	0.001147	0.1438	823.73	1767.49	2591.22	825.30	1962.85	2788.15	2.27363	4.20175	6.47538
195	13.9882	0.001149	0.1409	828.18	1763.57	2591.75	829.78	1959.05	2788.83	2.28314	4.18466	6.46780
196	14.2902	0.001150	0.1380	832.63	1759.64	2592.26	834.27	1955.23	2789.50	2.29264	4.16759	6.46023
197	14.5972	0.001152	0.1352	837.08	1755.69	2592.77	838.76	1951.39	2790.15	2.30212	4.15056	6.45269
198	14.9094	0.001153	0.1325	841.54	1751.72	2593.26	843.26	1947.53	2790.79	2.31160	4.13357	6.44516
199	15.2268	0.001155	0.1298	846.00	1747.74	2593.74	847.76	1943.65	2791.41	2.32106	4.11660	6.43766
200	15.5493	0.001157	0.1272	850.46	1743.75	2594.21	852.26	1939.76	2792.02	2.33052	4.09966	6.43018
201	15.8772	0.001158	0.1247	854.93	1739.74	2594.67	856.77	1935.84	2792.61	2.33996	4.08275	6.42271
202	16.2105	0.001160	0.1222	859.41	1735.72	2595.12	861.29	1931.90	2793.19	2.34939	4.06588	6.41527
203	16.5492	0.001161	0.1198	863.89	1731.68	2595.56	865.81	1927.95	2793.76	2.35882	4.04903	6.40784
204	16.8933	0.001163	0.1174	868.37	1727.62	2595.99	870.34	1923.97	2794.31	2.36823	4.03221	6.40044
205	17.2430	0.001164	0.1151	872.86	1723.55	2596.41	874.87	1919.97	2794.84	2.37763	4.01541	6.39305
206	17.5984	0.001166	0.1128	877.35	1719.46	2596.82	879.41	1915.95	2795.36	2.38703	3.99865	6.38567
207	17.9593	0.001168	0.1106	881.85	1715.36	2597.21	883.95	1911.91	2795.86	2.39641	3.98191	6.37832

Temp. t (°C)	Pressure P (bar)	Specific Volume $\left(\frac{m^3}{kg}\right)$		Internal Energy $\left(\frac{kJ}{kg}\right)$			Enthalpy $\left(\frac{kJ}{kg}\right)$			Entropy $\left(\frac{kJ}{kg\,K}\right)$		
		Sat. Liquid	Sat. Vapour	Sat. Liquid	Latent	Sat. Vapour	Sat. Liquid	Sat. Liquid	Sat. Vapour	Sat. Liquid	Latent	Sat. Vapour
		v_f	v_g	u_f	u_{fg}	u_g	h_f	h_{fg}	h_g	s_f	s_{fg}	s_g
208	18.3260	0.001169	0.1085	886.36	1711.24	2597.59	888.50	1907.85	2796.35	2.40578	3.96519	6.37098
209	18.6985	0.001171	0.1064	890.86	1707.10	2597.96	893.05	1903.77	2796.83	2.41515	3.94850	6.36365
210	19.0769	0.001173	0.1043	895.38	1702.95	2598.32	897.61	1899.67	2797.28	2.42451	3.93184	6.35635
211	19.4611	0.001174	0.1023	899.89	1698.78	2598.67	902.18	1895.54	2797.72	2.43385	3.91520	6.34905
212	19.8514	0.001176	0.1003	904.42	1694.59	2599.01	906.75	1891.40	2798.15	2.44319	3.89858	6.34177
213	20.2476	0.001178	0.0984	908.94	1690.39	2599.33	911.33	1887.23	2798.56	2.45252	3.88199	6.33451
214	20.6500	0.001180	0.0965	913.48	1686.17	2599.64	915.91	1883.04	2798.95	2.46185	3.86542	6.32726
215	21.0586	0.001181	0.0947	918.02	1681.93	2599.95	920.50	1878.83	2799.33	2.47116	3.84887	6.32003
216	21.4733	0.001183	0.0929	922.56	1677.67	2600.23	925.10	1874.59	2799.69	2.48046	3.83234	6.31280
217	21.8944	0.001185	0.0911	927.11	1673.40	2600.51	929.70	1870.33	2800.03	2.48976	3.81583	6.30559
218	22.3218	0.001187	0.0894	931.66	1669.11	2600.77	934.31	1866.05	2800.36	2.49905	3.79934	6.29840
219	22.7557	0.001188	0.0877	936.22	1664.80	2601.02	938.93	1861.74	2800.67	2.50834	3.78288	6.29121
220	23.1960	0.001190	0.0861	940.79	1660.47	2601.26	943.55	1857.41	2800.96	2.51761	3.76643	6.28404
221	23.6429	0.001192	0.0845	945.36	1656.13	2601.48	948.18	1853.06	2801.24	2.52688	3.75000	6.27688
222	24.0965	0.001194	0.0829	949.93	1651.76	2601.70	952.81	1848.69	2801.50	2.53614	3.73358	6.26972
223	24.5567	0.001196	0.0814	954.52	1647.38	2601.89	957.45	1844.28	2801.74	2.54539	3.71719	6.26258
224	25.0237	0.001197	0.0799	959.10	1642.97	2602.08	962.10	1839.86	2801.96	2.55464	3.70081	6.25545
225	25.4974	0.001199	0.0784	963.70	1638.55	2602.25	966.76	1835.41	2802.16	2.56388	3.68445	6.24833
226	25.9781	0.001201	0.0770	968.30	1634.11	2602.41	971.42	1830.93	2802.35	2.57312	3.66810	6.24122
227	26.4658	0.001203	0.0756	972.90	1629.65	2602.56	976.09	1826.43	2802.52	2.58235	3.65177	6.23411
228	26.9604	0.001205	0.0742	977.52	1625.17	2602.69	980.77	1821.91	2802.67	2.59157	3.63545	6.22702
229	27.4622	0.001207	0.0728	982.14	1620.67	2602.80	985.45	1817.35	2802.80	2.60079	3.61914	6.21993
230	27.9711	0.001209	0.0715	986.76	1616.15	2602.91	990.14	1812.77	2802.92	2.61000	3.60285	6.21285
231	28.4873	0.001211	0.0702	991.39	1611.61	2603.00	994.84	1808.17	2803.01	2.61920	3.58657	6.20577
232	29.0107	0.001213	0.0689	996.03	1607.04	2603.07	999.55	1803.54	2803.09	2.62840	3.57030	6.19870
233	29.5415	0.001215	0.0677	1000.67	1602.46	2603.13	1004.26	1798.88	2803.14	2.63760	3.55404	6.19164
234	30.0798	0.001217	0.0665	1005.32	1597.86	2603.18	1008.98	1794.20	2803.18	2.64679	3.53780	6.18459
235	30.6255	0.001219	0.0653	1009.98	1593.23	2603.21	1013.71	1789.48	2803.20	2.65598	3.52156	6.17754
236	31.1789	0.001221	0.0641	1014.64	1588.58	2603.23	1018.45	1784.74	2803.19	2.66516	3.50534	6.17049
237	31.7399	0.001223	0.0630	1019.31	1583.92	2603.23	1023.20	1779.97	2803.17	2.67433	3.48912	6.16345
238	32.3086	0.001225	0.0619	1023.99	1579.22	2603.21	1027.95	1775.18	2803.13	2.68351	3.47291	6.15641
239	32.8851	0.001227	0.0608	1028.68	1574.51	2603.19	1032.71	1770.35	2803.06	2.69268	3.45670	6.14938
240	33.4695	0.001229	0.0597	1033.37	1569.77	2603.14	1037.48	1765.50	2802.98	2.70184	3.44051	6.14235
241	34.0618	0.001231	0.0587	1038.07	1565.02	2603.08	1042.26	1760.61	2802.87	2.71100	3.42432	6.13532
242	34.6622	0.001234	0.0576	1042.77	1560.23	2603.01	1047.05	1755.70	2802.75	2.72016	3.40813	6.12830
243	35.2706	0.001236	0.0566	1047.49	1555.43	2602.91	1051.84	1750.76	2802.60	2.72932	3.39195	6.12127
244	35.8871	0.001238	0.0556	1052.21	1550.60	2602.81	1056.65	1745.78	2802.43	2.73847	3.37578	6.11425
245	36.5119	0.001240	0.0547	1056.93	1545.75	2602.68	1061.46	1740.78	2802.24	2.74762	3.35960	6.10723
246	37.1450	0.001242	0.0537	1061.67	1540.87	2602.54	1066.29	1735.74	2802.03	2.75677	3.34343	6.10021
247	37.7865	0.001245	0.0528	1066.42	1535.97	2602.38	1071.12	1730.68	2801.80	2.76592	3.32727	6.09318
248	38.4365	0.001247	0.0519	1071.17	1531.04	2602.21	1075.96	1725.58	2801.54	2.77506	3.31110	6.08616
249	39.0949	0.001249	0.0510	1075.93	1526.09	2602.02	1080.81	1720.45	2801.26	2.78420	3.29493	6.07914

Temp. t (°C)	Pressure P (bar)	Specific Volume $\left(\frac{m^3}{kg}\right)$		Internal Energy $\left(\frac{kJ}{kg}\right)$			Enthalpy $\left(\frac{kJ}{kg}\right)$			Entropy $\left(\frac{kJ}{kg\,K}\right)$		
		Sat. Liquid	Sat. Vapour	Sat. Liquid	Latent	Sat. Vapour	Sat. Liquid	Sat. Liquid	Sat. Vapour	Sat. Liquid	Latent	Sat. Vapour
		v_f	v_g	u_f	u_{fg}	u_g	h_f	h_{fg}	h_g	s_f	s_{fg}	s_g
250	39.7620	0.001252	0.0501	1080.70	1521.12	2601.81	1085.67	1715.29	2800.96	2.79334	3.27877	6.07211
251	40.4377	0.001254	0.0492	1085.47	1516.11	2601.58	1090.54	1710.09	2800.63	2.80248	3.26260	6.06508
252	41.1222	0.001256	0.0484	1090.26	1511.08	2601.34	1095.42	1704.86	2800.29	2.81162	3.24643	6.05805
253	41.8156	0.001259	0.0476	1095.05	1506.03	2601.08	1100.31	1699.60	2799.92	2.82076	3.23026	6.05102
254	42.5178	0.001261	0.0467	1099.85	1500.95	2600.80	1105.21	1694.31	2799.52	2.82990	3.21409	6.04398
255	43.2291	0.001263	0.0459	1104.66	1495.84	2600.50	1110.12	1688.98	2799.10	2.83903	3.19791	6.03694
256	43.9494	0.001266	0.0452	1109.48	1490.71	2600.19	1115.04	1683.61	2798.66	2.84817	3.18173	6.02990
257	44.6788	0.001268	0.0444	1114.31	1485.54	2599.85	1119.98	1678.21	2798.19	2.85731	3.16554	6.02285
258	45.4175	0.001271	0.0436	1119.15	1480.35	2599.50	1124.92	1672.78	2797.69	2.86644	3.14935	6.01579
259	46.1654	0.001273	0.0429	1123.99	1475.13	2599.12	1129.87	1667.30	2797.17	2.87558	3.13315	6.00873
260	46.9228	0.001276	0.0422	1128.85	1469.89	2598.73	1134.83	1661.80	2796.63	2.88472	3.11694	6.00166
261	47.6896	0.001278	0.0415	1133.71	1464.61	2598.32	1139.81	1656.25	2796.06	2.89386	3.10072	5.99458
262	48.4659	0.001281	0.0408	1138.59	1459.30	2597.89	1144.80	1650.67	2795.46	2.90300	3.08450	5.98750
263	49.2519	0.001284	0.0401	1143.47	1453.97	2597.44	1149.79	1645.05	2794.84	2.91214	3.06826	5.98040
264	50.0476	0.001286	0.0394	1148.37	1448.60	2596.96	1154.80	1639.39	2794.19	2.92129	3.05201	5.97330
265	50.8530	0.001289	0.0387	1153.27	1443.20	2596.47	1159.82	1633.69	2793.52	2.93043	3.03575	5.96619
266	51.6684	0.001292	0.0381	1158.18	1437.77	2595.96	1164.86	1627.95	2792.81	2.93958	3.01948	5.95907
267	52.4936	0.001294	0.0375	1163.11	1432.31	2595.42	1169.90	1622.18	2792.08	2.94874	3.00320	5.95193
268	53.3289	0.001297	0.0368	1168.04	1426.82	2594.87	1174.96	1616.36	2791.32	2.95789	2.98690	5.94479
269	54.1743	0.001300	0.0362	1172.99	1421.30	2594.29	1180.03	1610.50	2790.53	2.96705	2.97058	5.93763
270	55.0299	0.001303	0.0356	1177.95	1415.74	2593.69	1185.12	1604.60	2789.72	2.97622	2.95425	5.93047
271	55.8958	0.001306	0.0350	1182.92	1410.15	2593.06	1190.21	1598.66	2788.87	2.98538	2.93790	5.92328
272	56.7721	0.001308	0.0344	1187.90	1404.52	2592.42	1195.32	1592.67	2788.00	2.99456	2.92153	5.91609
273	57.6588	0.001311	0.0339	1192.89	1398.87	2591.75	1200.45	1586.64	2787.09	3.00373	2.90514	5.90888
274	58.5560	0.001314	0.0333	1197.89	1393.17	2591.06	1205.58	1580.57	2786.16	3.01292	2.88874	5.90165
275	59.4639	0.001317	0.0328	1202.90	1387.44	2590.34	1210.73	1574.45	2785.19	3.02210	2.87231	5.89441
276	60.3825	0.001320	0.0322	1207.93	1381.68	2589.61	1215.90	1568.29	2784.19	3.03130	2.85586	5.88715
277	61.3119	0.001323	0.0317	1212.96	1375.88	2588.84	1221.08	1562.08	2783.16	3.04050	2.83938	5.87988
278	62.2522	0.001326	0.0312	1218.02	1370.04	2588.05	1226.27	1555.83	2782.10	3.04970	2.82288	5.87258
279	63.2034	0.001329	0.0307	1223.08	1364.16	2587.24	1231.48	1549.53	2781.01	3.05892	2.80635	5.86527
280	64.1657	0.001333	0.0302	1228.15	1358.25	2586.40	1236.70	1543.18	2779.88	3.06814	2.78980	5.85794
281	65.1392	0.001336	0.0297	1233.24	1352.30	2585.54	1241.94	1536.78	2778.72	3.07736	2.77322	5.85058
282	66.1239	0.001339	0.0292	1238.34	1346.30	2584.65	1247.20	1530.33	2777.53	3.08660	2.75661	5.84321
283	67.1200	0.001342	0.0287	1243.46	1340.27	2583.73	1252.47	1523.83	2776.30	3.09585	2.73997	5.83581
284	68.1274	0.001345	0.0282	1248.59	1334.20	2582.79	1257.75	1517.28	2775.04	3.10510	2.72330	5.82840
285	69.1464	0.001349	0.0278	1253.73	1328.08	2581.81	1263.06	1510.68	2773.74	3.11436	2.70659	5.82095
286	70.1770	0.001352	0.0273	1258.89	1321.93	2580.82	1268.38	1504.03	2772.41	3.12364	2.68985	5.81349
287	71.2194	0.001356	0.0269	1264.06	1315.73	2579.79	1273.71	1497.32	2771.03	3.13292	2.67308	5.80600
288	72.2735	0.001359	0.0264	1269.24	1309.49	2578.73	1279.07	1490.56	2769.63	3.14221	2.65626	5.79848
289	73.3395	0.001362	0.0260	1274.45	1303.20	2577.65	1284.44	1483.74	2768.18	3.15152	2.63941	5.79093
290	74.4175	0.001366	0.0256	1279.66	1296.87	2576.53	1289.83	1476.87	2766.70	3.16084	2.62252	5.78336
291	75.5076	0.001370	0.0251	1284.89	1290.49	2575.38	1295.23	1469.94	2765.18	3.17017	2.60559	5.77576

Temp. t (°C)	Pressure P (bar)	Specific Volume $\left(\frac{m^3}{kg}\right)$		Internal Energy $\left(\frac{kJ}{kg}\right)$			Enthalpy $\left(\frac{kJ}{kg}\right)$			Entropy $\left(\frac{kJ}{kg\,K}\right)$		
		Sat. Liquid	Sat. Vapour	Sat. Liquid	Latent	Sat. Vapour	Sat. Liquid	Sat. Liquid	Sat. Vapour	Sat. Liquid	Latent	Sat. Vapour
		v_f	v_g	u_f	u_{fg}	u_g	h_f	h_{fg}	h_g	s_f	s_{fg}	s_g
292	76.6099	0.001373	0.0247	1290.14	1284.07	2574.21	1300.66	1462.95	2763.61	3.17951	2.58861	5.76813
293	77.7244	0.001377	0.0243	1295.40	1277.59	2573.00	1306.11	1455.91	2762.01	3.18887	2.57159	5.76046
294	78.8514	0.001381	0.0239	1300.69	1271.07	2571.76	1311.57	1448.80	2760.37	3.19824	2.55453	5.75277
295	79.9908	0.001384	0.0235	1305.98	1264.50	2570.49	1317.06	1441.63	2758.69	3.20762	2.53742	5.74504
296	81.1428	0.001388	0.0231	1311.30	1257.88	2569.18	1322.56	1434.40	2756.96	3.21702	2.52025	5.73728
297	82.3075	0.001392	0.0228	1316.63	1251.21	2567.84	1328.09	1427.11	2755.19	3.22644	2.50304	5.72948
298	83.4849	0.001396	0.0224	1321.98	1244.49	2566.47	1333.63	1419.75	2753.38	3.23587	2.48577	5.72164
299	84.6753	0.001400	0.0220	1327.35	1237.71	2565.06	1339.20	1412.32	2751.52	3.24532	2.46845	5.71377
300	85.8786	0.001404	0.0217	1332.73	1230.88	2563.62	1344.79	1404.83	2749.62	3.25479	2.45107	5.70586
301	87.0951	0.001408	0.0213	1338.14	1224.00	2562.14	1350.40	1397.27	2747.67	3.26427	2.43364	5.69791
302	88.3247	0.001412	0.0210	1343.56	1217.06	2560.62	1356.03	1389.64	2745.68	3.27378	2.41614	5.68992
303	89.5677	0.001416	0.0206	1349.01	1210.06	2559.06	1361.69	1381.94	2743.63	3.28330	2.39858	5.68188
304	90.8240	0.001421	0.0203	1354.47	1203.00	2557.47	1367.37	1374.17	2741.54	3.29284	2.38096	5.67380
305	92.0939	0.001425	0.0199	1359.96	1195.88	2555.84	1373.08	1366.32	2739.40	3.30241	2.36326	5.66567
306	93.3774	0.001429	0.0196	1365.47	1188.70	2554.17	1378.81	1358.40	2737.21	3.31200	2.34550	5.65750
307	94.6747	0.001434	0.0193	1370.99	1181.46	2552.45	1384.57	1350.40	2734.96	3.32161	2.32767	5.64928
308	95.9858	0.001438	0.0190	1376.55	1174.15	2550.70	1390.35	1342.32	2732.67	3.33125	2.30976	5.64101
309	97.3109	0.001443	0.0186	1382.12	1166.78	2548.90	1396.16	1334.16	2730.32	3.34091	2.29178	5.63268
310	98.6501	0.001447	0.0183	1387.72	1159.34	2547.06	1402.00	1325.92	2727.91	3.35059	2.27372	5.62431
311	100.0035	0.001452	0.0180	1393.34	1151.83	2545.17	1407.86	1317.59	2725.45	3.36030	2.25557	5.61587
312	101.3712	0.001457	0.0177	1398.98	1144.25	2543.24	1413.75	1309.18	2722.93	3.37004	2.23734	5.60738
313	102.7533	0.001462	0.0174	1404.66	1136.60	2541.26	1419.68	1300.68	2720.36	3.37981	2.21902	5.59883
314	104.1500	0.001467	0.0171	1410.35	1128.88	2539.23	1425.63	1292.09	2717.72	3.38961	2.20061	5.59022
315	105.5614	0.001472	0.0168	1416.08	1121.08	2537.15	1431.61	1283.41	2715.02	3.39944	2.18211	5.58155
316	106.9875	0.001477	0.0166	1421.83	1113.20	2535.03	1437.63	1274.63	2712.26	3.40931	2.16351	5.57281
317	108.4286	0.001482	0.0163	1427.61	1105.24	2532.85	1443.68	1265.75	2709.43	3.41920	2.14480	5.56401
318	109.8848	0.001488	0.0160	1433.41	1097.21	2530.62	1449.76	1256.78	2706.54	3.42914	2.12600	5.55513
319	111.3561	0.001493	0.0157	1439.25	1089.08	2528.34	1455.88	1247.71	2703.58	3.43910	2.10708	5.54618
320	112.8427	0.001499	0.0155	1445.12	1080.88	2525.99	1462.03	1238.53	2700.56	3.44911	2.08805	5.53716
321	114.3448	0.001504	0.0152	1451.02	1072.58	2523.60	1468.22	1229.24	2697.46	3.45915	2.06891	5.52806
322	115.8624	0.001510	0.0149	1456.95	1064.19	2521.14	1474.44	1219.84	2694.29	3.46924	2.04964	5.51888
323	117.3958	0.001516	0.0147	1462.91	1055.71	2518.62	1480.71	1210.33	2691.04	3.47937	2.03025	5.50962
324	118.9450	0.001522	0.0144	1468.91	1047.13	2516.05	1487.01	1200.70	2687.72	3.48954	2.01073	5.50027
325	120.5102	0.001528	0.0142	1474.95	1038.46	2513.40	1493.36	1190.96	2684.32	3.49976	1.99107	5.49083
326	122.0914	0.001534	0.0139	1481.02	1029.68	2510.70	1499.75	1181.09	2680.84	3.51003	1.97127	5.48130
327	123.6890	0.001541	0.0137	1487.13	1020.79	2507.92	1506.18	1171.09	2677.27	3.52035	1.95133	5.47167
328	125.3030	0.001547	0.0135	1493.28	1011.80	2505.08	1512.66	1160.96	2673.62	3.53072	1.93123	5.46195
329	126.9335	0.001554	0.0132	1499.46	1002.69	2502.16	1519.18	1150.69	2669.88	3.54114	1.91098	5.45212
330	128.5807	0.001560	0.0130	1505.69	993.47	2499.17	1525.76	1140.29	2666.05	3.55162	1.89056	5.44218
331	130.2448	0.001567	0.0127	1511.97	984.13	2496.10	1532.38	1129.74	2662.12	3.56216	1.86997	5.43213
332	131.9259	0.001574	0.0125	1518.29	974.66	2492.95	1539.06	1119.04	2658.10	3.57277	1.84920	5.42197
333	133.6242	0.001582	0.0123	1524.65	965.07	2489.72	1545.79	1108.19	2653.97	3.58344	1.82824	5.41168

Temp. t (°C)	Pressure P (bar)	Specific Volume $\left(\frac{m^3}{kg}\right)$		Internal Energy $\left(\frac{kJ}{kg}\right)$			Enthalpy $\left(\frac{kJ}{kg}\right)$			Entropy $\left(\frac{kJ}{kg\,K}\right)$		
		Sat. Liquid	Sat. Vapour	Sat. Liquid	Latent	Sat. Vapour	Sat. Liquid	Sat. Liquid	Sat. Vapour	Sat. Liquid	Latent	Sat. Vapour
		v_f	v_g	u_f	u_{fg}	u_g	h_f	h_{fg}	h_g	s_f	s_{fg}	s_g
334	135.3398	0.001589	0.0121	1531.06	955.34	2486.40	1552.57	1097.17	2649.74	3.59417	1.80709	5.40126
335	137.0730	0.001597	0.0118	1537.53	945.47	2483.00	1559.41	1085.99	2645.41	3.60498	1.78573	5.39072
336	138.8238	0.001604	0.0116	1544.05	935.46	2479.50	1566.32	1074.64	2640.96	3.61587	1.76417	5.38003
337	140.5924	0.001612	0.0114	1550.62	925.29	2475.91	1573.29	1063.11	2636.40	3.62683	1.74237	5.36920
338	142.3790	0.001621	0.0112	1557.25	914.97	2472.22	1580.32	1051.39	2631.71	3.63787	1.72035	5.35822
339	144.1839	0.001629	0.0110	1563.93	904.49	2468.43	1587.42	1039.48	2626.90	3.64901	1.69808	5.34709
340	146.0071	0.001638	0.0108	1570.68	893.84	2464.53	1594.60	1027.36	2621.96	3.66023	1.67555	5.33578
341	147.8488	0.001647	0.0106	1577.50	883.01	2460.51	1601.85	1015.04	2616.88	3.67155	1.65275	5.32430
342	149.7094	0.001656	0.0104	1584.38	871.99	2456.38	1609.17	1002.49	2611.66	3.68297	1.62967	5.31264
343	151.5889	0.001665	0.0102	1591.34	860.78	2452.13	1616.58	989.71	2606.30	3.69450	1.60629	5.30079
344	153.4875	0.001675	0.0100	1598.37	849.37	2447.74	1624.08	976.69	2600.77	3.70614	1.58258	5.28873
345	155.4055	0.001685	0.0098	1605.48	837.74	2443.22	1631.67	963.42	2595.08	3.71791	1.55855	5.27645
346	157.3430	0.001695	0.0096	1612.68	825.88	2438.56	1639.35	949.87	2589.23	3.72980	1.53416	5.26395
347	159.3003	0.001706	0.0094	1619.96	813.79	2433.75	1647.14	936.05	2583.18	3.74183	1.50939	5.25121
348	161.2777	0.001717	0.0092	1627.34	801.44	2428.78	1655.03	921.92	2576.95	3.75400	1.48422	5.23822
349	163.2753	0.001729	0.0090	1634.81	788.82	2423.64	1663.04	907.48	2570.52	3.76633	1.45862	5.22494
350	165.2933	0.001741	0.0088	1642.40	775.92	2418.32	1671.17	892.70	2563.87	3.77882	1.43256	5.21138
351	167.3321	0.001753	0.0086	1650.09	762.71	2412.81	1679.43	877.57	2557.00	3.79149	1.40602	5.19751
352	169.3919	0.001766	0.0084	1657.91	749.18	2407.09	1687.83	862.05	2549.88	3.80435	1.37895	5.18331
353	171.4730	0.001780	0.0082	1665.86	735.31	2401.16	1696.37	846.13	2542.50	3.81742	1.35132	5.16874
354	173.5755	0.001794	0.0081	1673.94	721.06	2395.00	1705.08	829.77	2534.84	3.83071	1.32308	5.15379
355	175.6999	0.001808	0.0079	1682.18	706.40	2388.58	1713.95	812.93	2526.88	3.84425	1.29417	5.13842
356	177.8464	0.001824	0.0077	1690.58	691.31	2381.89	1723.02	795.59	2518.60	3.85804	1.26454	5.12259
357	180.0153	0.001840	0.0075	1699.16	675.75	2374.91	1732.29	777.68	2509.97	3.87213	1.23413	5.10626
358	182.2069	0.001857	0.0073	1707.93	659.66	2367.59	1741.78	759.17	2500.95	3.88654	1.20284	5.08938
359	184.4217	0.001876	0.0071	1716.92	643.00	2359.92	1751.52	739.99	2491.51	3.90130	1.17059	5.07190
360	186.6599	0.001895	0.0070	1726.16	625.70	2351.86	1761.53	720.07	2481.60	3.91646	1.13728	5.05374
361	188.9220	0.001916	0.0068	1735.66	607.69	2343.35	1771.85	699.32	2471.16	3.93207	1.10276	5.03483
362	191.2084	0.001938	0.0066	1745.47	588.88	2334.35	1782.51	677.63	2460.15	3.94818	1.06688	5.01507
363	193.5195	0.001961	0.0064	1755.62	569.15	2324.78	1793.58	654.89	2448.47	3.96488	1.02945	4.99433
364	195.8558	0.001987	0.0062	1766.18	548.38	2314.56	1805.10	630.92	2436.02	3.98225	0.99022	4.97248
365	198.2179	0.002015	0.0060	1777.22	526.36	2303.59	1817.16	605.53	2422.69	4.00043	0.94888	4.94931
366	200.6063	0.002046	0.0058	1788.83	502.89	2291.72	1829.87	578.44	2408.32	4.01957	0.90502	4.92459
367	203.0217	0.002080	0.0056	1801.14	477.63	2278.77	1843.37	549.30	2392.67	4.03990	0.85808	4.89798
368	205.4649	0.002119	0.0054	1814.32	450.15	2264.47	1857.87	517.60	2375.46	4.06172	0.80729	4.86901
369	207.9367	0.002164	0.0052	1828.65	419.79	2248.44	1873.65	482.59	2356.24	4.08549	0.75152	4.83701
370	210.4382	0.002217	0.0050	1844.53	385.55	2230.08	1891.19	443.11	2334.30	4.11193	0.68897	4.80090
371	212.9708	0.002283	0.0047	1862.67	345.67	2208.34	1911.29	397.16	2308.44	4.14228	0.61656	4.75883
372	215.5364	0.002370	0.0044	1884.52	296.57	2181.09	1935.60	340.62	2276.22	4.17906	0.52798	4.70703
373	218.1382	0.002504	0.0041	1914.17	228.30	2142.47	1968.79	262.07	2230.86	4.22949	0.40559	4.63508
374	220.5000	0.002880	0.0034	1985.90	79.50	2065.40	2049.40	91.00	2140.40	4.35290	0.14060	4.49350
374.14	220.9000	0.003155	0.0032	2029.60	0.00	2029.60	2099.30	0.00	2099.30	4.42980	0.00000	4.42980

Table 2 Saturation Pressures and Properties of Water and Steam

Temp. t (°C)	Pressure P (bar)	Specific Volume $\left(\frac{m^3}{kg}\right)$		Internal Energy $\left(\frac{kJ}{kg}\right)$			Enthalpy $\left(\frac{kJ}{kg}\right)$			Entropy $\left(\frac{kJ}{kg\,K}\right)$		
		Sat. Liquid	Sat. Vapour	Sat. Liquid	Latent	Sat. Vapour	Sat. Liquid	Sat. Liquid	Sat. Vapour	Sat. Liquid	Latent	Sat. Vapour
		v_f	v_g	u_f	u_{fg}	u_g	h_f	h_{fg}	h_g	s_f	s_{fg}	s_g
0.10	0.006113	0.001000	206.1188	–0.03	2374.95	2374.92	–0.03	2500.95	2500.92	–0.00013	9.15590	9.15577
1.88	0.007	0.001000	181.2310	7.89	2369.61	2377.50	7.89	2496.47	2504.37	0.02879	9.07707	9.10585
3.76	0.008	0.001000	159.6533	15.81	2364.28	2380.09	15.81	2492.00	2507.81	0.05749	8.99927	9.05676
5.45	0.009	0.001000	142.7691	22.89	2359.51	2382.40	22.89	2488.00	2510.90	0.08298	8.93056	9.01354
6.97	0.010	0.001000	129.1890	29.30	2355.20	2384.50	29.30	2484.38	2513.69	0.10593	8.86902	8.97494
8.37	0.011	0.001000	118.0238	35.16	2351.25	2386.41	35.17	2481.07	2516.24	0.12681	8.81328	8.94008
9.65	0.012	0.001000	108.6781	40.57	2347.61	2388.18	40.57	2478.02	2518.60	0.14597	8.76234	8.90831
10.85	0.013	0.001000	100.7379	45.59	2344.23	2389.82	45.59	2475.19	2520.78	0.16369	8.71543	8.87911
11.97	0.014	0.001001	93.9062	50.28	2341.07	2391.36	50.29	2472.54	2522.82	0.18017	8.67195	8.85212
13.02	0.015	0.001001	87.9644	54.69	2338.11	2392.79	54.69	2470.05	2524.74	0.19558	8.63144	8.82702
14.01	0.016	0.001001	82.7482	58.84	2335.31	2394.15	58.84	2467.71	2526.55	0.21006	8.59350	8.80356
14.95	0.017	0.001001	78.1314	62.76	2332.67	2395.43	62.77	2465.49	2528.25	0.22371	8.55784	8.78155
15.84	0.018	0.001001	74.0155	66.49	2330.16	2396.65	66.49	2463.38	2529.88	0.23663	8.52419	8.76082
16.69	0.019	0.001001	70.3226	70.04	2327.77	2397.81	70.04	2461.38	2531.42	0.24890	8.49233	8.74123
17.50	0.020	0.001001	66.9902	73.43	2325.48	2398.91	73.43	2459.46	2532.89	0.26058	8.46209	8.72267
18.27	0.021	0.001002	63.9676	76.67	2323.30	2399.97	76.68	2457.63	2534.30	0.27172	8.43330	8.70502
19.01	0.022	0.001002	61.2132	79.78	2321.20	2400.98	79.79	2455.87	2535.65	0.28238	8.40583	8.68821
19.73	0.023	0.001002	58.6925	82.77	2319.19	2401.96	82.77	2454.18	2536.95	0.29260	8.37956	8.67215
20.41	0.024	0.001002	56.3768	85.65	2317.25	2402.90	85.65	2452.55	2538.20	0.30240	8.35439	8.65680
21.08	0.025	0.001002	54.2418	88.42	2315.38	2403.80	88.42	2450.98	2539.40	0.31184	8.33024	8.64208
21.72	0.026	0.001002	52.2670	91.10	2313.57	2404.67	91.10	2449.46	2540.57	0.32093	8.30702	8.62795
22.34	0.027	0.001002	50.4349	93.69	2311.83	2405.51	93.69	2448.00	2541.69	0.32970	8.28466	8.61436
22.94	0.028	0.001002	48.7305	96.19	2310.14	2406.33	96.20	2446.58	2542.78	0.33817	8.26310	8.60127
23.52	0.029	0.001003	47.1407	98.62	2308.50	2407.12	98.62	2445.20	2543.83	0.34636	8.24229	8.58865
24.08	0.030	0.001003	45.6543	100.98	2306.91	2407.89	100.98	2443.87	2544.85	0.35430	8.22217	8.57647
24.63	0.031	0.001003	44.2613	103.27	2305.36	2408.63	103.27	2442.57	2545.84	0.36199	8.20270	8.56470
25.16	0.032	0.001003	42.9533	105.49	2303.86	2409.35	105.50	2441.31	2546.80	0.36946	8.18384	8.55330
25.68	0.033	0.001003	41.7225	107.66	2302.40	2410.06	107.66	2440.08	2547.74	0.37671	8.16555	8.54226
26.18	0.034	0.001003	40.5624	109.77	2300.98	2410.74	109.77	2438.89	2548.65	0.38376	8.14780	8.53156
26.67	0.035	0.001003	39.4668	111.82	2299.59	2411.41	111.82	2437.72	2549.54	0.39062	8.13055	8.52117
27.15	0.036	0.001004	38.4306	113.83	2298.23	2412.06	113.83	2436.58	2550.41	0.39730	8.11378	8.51108
27.62	0.037	0.001004	37.4490	115.78	2296.91	2412.69	115.79	2435.47	2551.26	0.40381	8.09747	8.50127
28.08	0.038	0.001004	36.5177	117.69	2295.62	2413.31	117.70	2434.39	2552.08	0.41016	8.08158	8.49173
28.52	0.039	0.001004	35.6330	119.56	2294.36	2413.92	119.56	2433.33	2552.89	0.41635	8.06609	8.48244
28.96	0.040	0.001004	34.7914	121.39	2293.13	2414.51	121.39	2432.29	2553.68	0.42240	8.05099	8.47340
29.39	0.041	0.001004	33.9898	123.17	2291.92	2415.09	123.18	2431.27	2554.45	0.42831	8.03626	8.46457
29.81	0.042	0.001004	33.2254	124.92	2290.74	2415.66	124.93	2430.28	2555.21	0.43409	8.02188	8.45597
30.22	0.043	0.001004	32.4956	126.64	2289.58	2416.21	126.64	2429.30	2555.94	0.43974	8.00783	8.44757
30.62	0.044	0.001005	31.7981	128.32	2288.44	2416.76	128.32	2428.35	2556.67	0.44528	7.99409	8.43937
31.01	0.045	0.001005	31.1309	129.96	2287.33	2417.29	129.97	2427.41	2557.38	0.45069	7.98066	8.43135
31.40	0.046	0.001005	30.4919	131.58	2286.24	2417.81	131.58	2426.49	2558.08	0.45600	7.96752	8.42352

Temp. t (°C)	Pressure P (bar)	Specific Volume $\left(\frac{m^3}{kg}\right)$		Internal Energy $\left(\frac{kJ}{kg}\right)$			Enthalpy $\left(\frac{kJ}{kg}\right)$			Entropy $\left(\frac{kJ}{kg\,K}\right)$		
		Sat. Liquid	Sat. Vapour	Sat. Liquid	Latent	Sat. Vapour	Sat. Liquid	Sat. Liquid	Sat. Vapour	Sat. Liquid	Latent	Sat. Vapour
		v_f	v_g	u_f	u_{fg}	u_g	h_f	h_{fg}	h_g	s_f	s_{fg}	s_g
31.78	0.047	0.001005	29.8794	133.16	2285.16	2418.33	133.17	2425.59	2558.76	0.46120	7.95466	8.41585
32.15	0.048	0.001005	29.2918	134.72	2284.11	2418.83	134.72	2424.71	2559.43	0.46630	7.94206	8.40836
32.52	0.049	0.001005	28.7275	136.24	2283.08	2419.32	136.25	2423.84	2560.09	0.47130	7.92972	8.40101
32.87	0.050	0.001005	28.1852	137.75	2282.06	2419.81	137.75	2422.98	2560.73	0.47620	7.91762	8.39382
34.58	0.055	0.001006	25.7622	144.88	2277.23	2422.11	144.89	2418.92	2563.80	0.49945	7.86049	8.35994
36.16	0.060	0.001006	23.7334	151.47	2272.76	2424.24	151.48	2415.16	2566.64	0.52082	7.80824	8.32906
37.63	0.065	0.001007	22.0087	157.61	2268.61	2426.21	157.61	2411.66	2569.27	0.54061	7.76010	8.30070
39.00	0.070	0.001008	20.5243	163.34	2264.71	2428.06	163.35	2408.38	2571.73	0.55903	7.71546	8.27448
40.29	0.075	0.001008	19.2328	168.74	2261.05	2429.79	168.75	2405.29	2574.04	0.57628	7.67384	8.25011
41.51	0.080	0.001008	18.0987	173.83	2257.59	2431.42	173.84	2402.37	2576.21	0.59249	7.63485	8.22734
42.66	0.085	0.001009	17.0945	178.66	2254.31	2432.97	178.67	2399.61	2578.27	0.60780	7.59818	8.20598
43.76	0.090	0.001009	16.1991	183.24	2251.19	2434.43	183.25	2396.97	2580.23	0.62230	7.56357	8.18586
44.81	0.095	0.001010	15.3954	187.62	2248.22	2435.83	187.63	2394.46	2582.09	0.63607	7.53078	8.16685
45.81	0.100	0.001010	14.6700	191.79	2245.37	2437.16	191.81	2392.06	2583.86	0.64919	7.49965	8.14884
47.68	0.110	0.001011	13.4120	199.64	2240.02	2439.66	199.65	2387.54	2587.19	0.67372	7.44170	8.11542
49.42	0.120	0.001012	12.3582	206.90	2235.07	2441.97	206.91	2383.36	2590.27	0.69628	7.38869	8.08496
51.03	0.130	0.001013	11.4624	213.65	2230.46	2444.11	213.67	2379.46	2593.12	0.71717	7.33983	8.05699
52.55	0.140	0.001013	10.6912	219.98	2226.13	2446.11	219.99	2375.80	2595.79	0.73663	7.29451	8.03114
53.97	0.150	0.001014	10.0201	225.93	2222.06	2447.99	225.94	2372.35	2598.29	0.75486	7.25224	8.00711
55.31	0.160	0.001015	9.4307	231.55	2218.21	2449.76	231.56	2369.09	2600.65	0.77201	7.21265	7.98465
56.59	0.170	0.001015	8.9087	236.88	2214.56	2451.44	236.89	2365.99	2602.88	0.78820	7.17539	7.96359
57.80	0.180	0.001016	8.4432	241.95	2211.08	2453.03	241.96	2363.04	2605.00	0.80354	7.14021	7.94375
58.95	0.190	0.001017	8.0253	246.78	2207.76	2454.54	246.80	2360.22	2607.02	0.81812	7.10689	7.92501
60.06	0.200	0.001017	7.6481	251.40	2204.58	2455.98	251.42	2357.52	2608.95	0.83202	7.07523	7.90725
61.12	0.210	0.001018	7.3058	255.83	2201.53	2457.37	255.85	2354.93	2610.79	0.84529	7.04508	7.89038
62.13	0.220	0.001018	6.9938	260.09	2198.60	2458.69	260.11	2352.44	2612.55	0.85800	7.01630	7.87430
63.11	0.230	0.001019	6.7081	264.18	2195.78	2459.96	264.20	2350.05	2614.25	0.87019	6.98876	7.85895
64.05	0.240	0.001019	6.4455	268.13	2193.06	2461.19	268.15	2347.73	2615.88	0.88191	6.96236	7.84427
64.96	0.250	0.001020	6.2034	271.93	2190.43	2462.37	271.96	2345.49	2617.45	0.89319	6.93702	7.83021
65.84	0.260	0.001020	5.9794	275.62	2187.89	2463.51	275.64	2343.33	2618.97	0.90407	6.91263	7.81670
66.69	0.270	0.001021	5.7715	279.18	2185.43	2464.61	279.21	2341.23	2620.44	0.91457	6.88915	7.80371
67.52	0.280	0.001021	5.5780	282.63	2183.04	2465.67	282.66	2339.19	2621.86	0.92472	6.86649	7.79121
68.32	0.290	0.001022	5.3975	285.99	2180.72	2466.70	286.02	2337.21	2623.23	0.93455	6.84461	7.77915
69.09	0.300	0.001022	5.2287	289.24	2178.46	2467.70	289.27	2335.29	2624.56	0.94407	6.82344	7.76751
70.58	0.320	0.001023	4.9218	295.49	2174.13	2469.62	295.52	2331.59	2627.11	0.96228	6.78310	7.74538
72.00	0.340	0.001024	4.6499	301.41	2170.01	2471.43	301.45	2328.08	2629.53	0.97948	6.74512	7.72461
73.34	0.360	0.001025	4.4075	307.05	2166.09	2473.15	307.09	2324.72	2631.81	0.99579	6.70926	7.70505
74.63	0.380	0.001026	4.1898	312.43	2162.35	2474.78	312.47	2321.52	2633.99	1.01129	6.67528	7.68657
75.86	0.400	0.001026	3.9933	317.58	2158.76	2476.34	317.62	2318.45	2636.07	1.02607	6.64298	7.66905
77.03	0.420	0.001027	3.8149	322.52	2155.31	2477.83	322.56	2315.49	2638.06	1.04020	6.61221	7.65241
78.16	0.440	0.001028	3.6523	327.27	2151.99	2479.26	327.31	2312.65	2639.96	1.05373	6.58283	7.63656
79.25	0.460	0.001029	3.5034	331.83	2148.80	2480.63	331.88	2309.91	2641.79	1.06671	6.55472	7.62142

Temp. t (°C)	Pressure P (bar)	Specific Volume $\left(\frac{m^3}{kg}\right)$		Internal Energy $\left(\frac{kJ}{kg}\right)$			Enthalpy $\left(\frac{kJ}{kg}\right)$			Entropy $\left(\frac{kJ}{kg\,K}\right)$		
		Sat. Liquid	Sat. Vapour	Sat. Liquid	Latent	Sat. Vapour	Sat. Liquid	Sat. Liquid	Sat. Vapour	Sat. Liquid	Latent	Sat. Vapour
		v_f	v_g	u_f	u_{fg}	u_g	h_f	h_{fg}	h_g	s_f	s_{fg}	s_g
80.30	0.480	0.001029	3.3665	336.24	2145.71	2481.95	336.29	2307.26	2643.54	1.07919	6.52776	7.60694
81.32	0.500	0.001030	3.2403	340.49	2142.73	2483.22	340.54	2304.69	2645.24	1.09120	6.50186	7.59306
83.71	0.550	0.001032	2.9637	350.54	2135.67	2486.21	350.59	2298.62	2649.22	1.11945	6.44126	7.56070
85.93	0.600	0.001033	2.7320	359.85	2129.11	2488.96	359.91	2292.97	2652.88	1.14546	6.38576	7.53121
87.99	0.650	0.001035	2.5348	368.53	2122.98	2491.51	368.60	2287.68	2656.28	1.16958	6.33454	7.50412
89.93	0.700	0.001036	2.3650	376.68	2117.21	2493.89	376.75	2282.69	2659.45	1.19208	6.28699	7.47907
91.76	0.750	0.001037	2.2172	384.36	2111.76	2496.12	384.44	2277.97	2662.41	1.21318	6.24260	7.45578
93.49	0.800	0.001039	2.0873	391.63	2106.59	2498.22	391.71	2273.49	2665.21	1.23305	6.20096	7.43401
95.13	0.850	0.001040	1.9722	398.53	2101.67	2500.20	398.62	2269.22	2667.84	1.25183	6.16176	7.41359
96.69	0.900	0.001041	1.8696	405.11	2096.97	2502.08	405.20	2265.14	2670.34	1.26965	6.12470	7.39435
98.18	0.950	0.001042	1.7774	411.39	2092.48	2503.86	411.49	2261.23	2672.72	1.28660	6.08957	7.37617
99.61	1.000	0.001043	1.6941	417.40	2088.16	2505.56	417.51	2257.47	2674.98	1.30277	6.05617	7.35893
100.98	1.050	0.001044	1.6186	423.18	2084.01	2507.19	423.28	2253.85	2677.14	1.31823	6.02432	7.34255
102.29	1.100	0.001045	1.5496	428.73	2080.01	2508.74	428.84	2250.36	2679.20	1.33304	5.99390	7.32693
103.56	1.150	0.001046	1.4865	434.08	2076.16	2510.23	434.20	2246.99	2681.18	1.34726	5.96477	7.31202
104.78	1.200	0.001047	1.4285	439.24	2072.42	2511.66	439.36	2243.72	2683.08	1.36094	5.93682	7.29776
105.97	1.250	0.001048	1.3750	444.23	2068.81	2513.04	444.36	2240.56	2684.91	1.37412	5.90996	7.28408
107.11	1.300	0.001049	1.3255	449.05	2065.31	2514.36	449.19	2237.48	2686.68	1.38684	5.88411	7.27094
108.22	1.350	0.001050	1.2795	453.73	2061.91	2515.64	453.88	2234.50	2688.37	1.39913	5.85919	7.25831
109.29	1.400	0.001051	1.2367	458.27	2058.60	2516.87	458.42	2231.59	2690.02	1.41101	5.83513	7.24614
110.34	1.450	0.001052	1.1968	462.68	2055.38	2518.07	462.84	2228.77	2691.60	1.42253	5.81187	7.23440
111.35	1.500	0.001053	1.1594	466.97	2052.25	2519.22	467.13	2226.01	2693.14	1.43370	5.78937	7.22306
112.34	1.550	0.001054	1.1244	471.15	2049.20	2520.34	471.31	2223.32	2694.63	1.44454	5.76756	7.21210
113.30	1.600	0.001054	1.0915	475.21	2046.22	2521.43	475.38	2220.69	2696.07	1.45507	5.74642	7.20149
114.23	1.650	0.001055	1.0605	479.18	2043.31	2522.48	479.35	2218.12	2697.47	1.46532	5.72589	7.19121
115.15	1.700	0.001056	1.0313	483.04	2040.46	2523.51	483.22	2215.61	2698.83	1.47529	5.70594	7.18124
116.04	1.750	0.001057	1.0037	486.82	2037.68	2524.50	487.01	2213.15	2700.15	1.48501	5.68655	7.17156
116.91	1.800	0.001058	0.9776	490.51	2034.96	2525.47	490.70	2210.74	2701.44	1.49448	5.66767	7.16215
117.76	1.850	0.001058	0.9528	494.12	2032.30	2526.42	494.32	2208.37	2702.69	1.50373	5.64927	7.15300
118.60	1.900	0.001059	0.9293	497.65	2029.68	2527.34	497.85	2206.06	2703.91	1.51275	5.63135	7.14410
119.41	1.950	0.001060	0.9070	501.11	2027.12	2528.23	501.32	2203.79	2705.10	1.52157	5.61386	7.13543
120.21	2.000	0.001061	0.8858	504.50	2024.61	2529.11	504.71	2201.55	2706.26	1.53018	5.59679	7.12698
120.99	2.050	0.001061	0.8655	507.81	2022.15	2529.96	508.03	2199.36	2707.40	1.53861	5.58013	7.11874
121.76	2.100	0.001062	0.8462	511.07	2019.72	2530.79	511.29	2197.21	2708.50	1.54686	5.56384	7.11070
122.51	2.150	0.001063	0.8278	514.26	2017.35	2531.61	514.49	2195.09	2709.58	1.55494	5.54791	7.10284
123.25	2.200	0.001063	0.8102	517.39	2015.01	2532.40	517.63	2193.01	2710.64	1.56285	5.53232	7.09517
123.97	2.250	0.001064	0.7933	520.47	2012.71	2533.18	520.71	2190.96	2711.67	1.57061	5.51707	7.08768
124.69	2.300	0.001065	0.7771	523.49	2010.45	2533.94	523.74	2188.94	2712.68	1.57821	5.50213	7.08034
125.39	2.350	0.001065	0.7616	526.46	2008.23	2534.69	526.71	2186.96	2713.67	1.58567	5.48750	7.07317
126.07	2.400	0.001066	0.7468	529.38	2006.04	2535.42	529.64	2185.00	2714.64	1.59299	5.47315	7.06615
126.75	2.450	0.001067	0.7325	532.26	2003.88	2536.14	532.52	2183.07	2715.59	1.60018	5.45909	7.05927
127.41	2.500	0.001067	0.7187	535.08	2001.76	2536.84	535.35	2181.17	2716.52	1.60724	5.44529	7.05253

Temp. t (°C)	Pressure P (bar)	Specific Volume $\left(\frac{m^3}{kg}\right)$		Internal Energy $\left(\frac{kJ}{kg}\right)$			Enthalpy $\left(\frac{kJ}{kg}\right)$			Entropy $\left(\frac{kJ}{kg\,K}\right)$		
		Sat. Liquid	Sat. Vapour	Sat. Liquid	Latent	Sat. Vapour	Sat. Liquid	Sat. Liquid	Sat. Vapour	Sat. Liquid	Latent	Sat. Vapour
		v_f	v_g	u_f	u_{fg}	u_g	h_f	h_{fg}	h_g	s_f	s_{fg}	s_g
128.07	2.550	0.001068	0.7055	537.86	1999.66	2537.52	538.13	2179.30	2717.43	1.61418	5.43175	7.04592
128.71	2.600	0.001068	0.6928	540.60	1997.60	2538.20	540.88	2177.45	2718.33	1.62099	5.41845	7.03944
129.34	2.650	0.001069	0.6805	543.29	1995.57	2538.86	543.58	2175.62	2719.20	1.62770	5.40539	7.03309
129.97	2.700	0.001070	0.6687	545.95	1993.56	2539.51	546.24	2173.82	2720.06	1.63429	5.39256	7.02685
130.58	2.750	0.001070	0.6573	548.57	1991.58	2540.15	548.86	2172.05	2720.91	1.64077	5.37995	7.02073
131.19	2.800	0.001071	0.6463	551.15	1989.63	2540.77	551.44	2170.29	2721.74	1.64716	5.36756	7.01472
131.78	2.850	0.001071	0.6357	553.69	1987.70	2541.39	553.99	2168.56	2722.55	1.65344	5.35537	7.00881
132.37	2.900	0.001072	0.6254	556.19	1985.80	2541.99	556.51	2166.85	2723.35	1.65963	5.34338	7.00301
132.95	2.950	0.001073	0.6154	558.67	1983.92	2542.58	558.98	2165.15	2724.14	1.66572	5.33158	6.99730
133.52	3.000	0.001073	0.6058	561.11	1982.06	2543.17	561.43	2163.48	2724.91	1.67173	5.31996	6.99169
134.09	3.050	0.001074	0.5965	563.52	1980.22	2543.74	563.84	2161.83	2725.67	1.67764	5.30853	6.98617
134.64	3.100	0.001074	0.5875	565.89	1978.41	2544.30	566.23	2160.19	2726.42	1.68348	5.29727	6.98074
135.19	3.150	0.001075	0.5787	568.24	1976.62	2544.86	568.58	2158.58	2727.15	1.68923	5.28617	6.97540
135.74	3.200	0.001075	0.5702	570.56	1974.85	2545.41	570.90	2156.98	2727.88	1.69490	5.27524	6.97014
136.27	3.250	0.001076	0.5620	572.84	1973.10	2545.94	573.19	2155.39	2728.59	1.70049	5.26447	6.96496
136.80	3.300	0.001076	0.5540	575.11	1971.37	2546.47	575.46	2153.83	2729.29	1.70601	5.25385	6.95987
137.33	3.350	0.001077	0.5462	577.34	1969.65	2546.99	577.70	2152.28	2729.98	1.71146	5.24339	6.95484
137.84	3.400	0.001078	0.5387	579.55	1967.96	2547.51	579.91	2150.74	2730.66	1.71683	5.23306	6.94989
138.35	3.450	0.001078	0.5314	581.73	1966.28	2548.01	582.10	2149.23	2731.33	1.72214	5.22288	6.94502
138.86	3.500	0.001079	0.5242	583.89	1964.62	2548.51	584.26	2147.72	2731.99	1.72738	5.21283	6.94021
139.36	3.550	0.001079	0.5173	586.02	1962.98	2549.00	586.40	2146.23	2732.63	1.73256	5.20291	6.93547
139.85	3.600	0.001080	0.5105	588.13	1961.35	2549.48	588.52	2144.76	2733.27	1.73767	5.19313	6.93079
140.34	3.650	0.001080	0.5040	590.22	1959.74	2549.96	590.61	2143.29	2733.90	1.74272	5.18346	6.92618
140.82	3.700	0.001081	0.4976	592.28	1958.15	2550.43	592.68	2141.85	2734.53	1.74771	5.17393	6.92163
141.30	3.750	0.001081	0.4913	594.32	1956.57	2550.89	594.73	2140.41	2735.14	1.75264	5.16451	6.91714
141.77	3.800	0.001082	0.4853	596.34	1955.00	2551.35	596.75	2138.99	2735.74	1.75751	5.15520	6.91271
142.24	3.850	0.001082	0.4793	598.34	1953.45	2551.80	598.76	2137.58	2736.34	1.76233	5.14601	6.90834
142.70	3.900	0.001083	0.4736	600.32	1951.92	2552.24	600.75	2136.18	2736.93	1.76709	5.13693	6.90402
143.16	3.950	0.001083	0.4679	602.28	1950.40	2552.68	602.71	2134.79	2737.51	1.77180	5.12795	6.89976
143.61	4.000	0.001084	0.4624	604.22	1948.89	2553.11	604.66	2133.42	2738.08	1.77646	5.11909	6.89555
144.06	4.050	0.001084	0.4570	606.15	1947.39	2553.54	606.58	2132.06	2738.64	1.78107	5.11032	6.89139
144.50	4.100	0.001085	0.4518	608.05	1945.91	2553.96	608.49	2130.71	2739.20	1.78563	5.10165	6.88728
144.94	4.150	0.001085	0.4467	609.93	1944.44	2554.38	610.38	2129.37	2739.75	1.79014	5.09308	6.88322
145.38	4.200	0.001085	0.4417	611.80	1942.99	2554.79	612.26	2128.04	2740.29	1.79461	5.08460	6.87921
145.81	4.250	0.001086	0.4368	613.65	1941.54	2555.19	614.11	2126.72	2740.83	1.79903	5.07622	6.87525
146.23	4.300	0.001086	0.4320	615.48	1940.11	2555.59	615.95	2125.41	2741.36	1.80340	5.06793	6.87133
146.66	4.350	0.001087	0.4273	617.30	1938.69	2555.99	617.77	2124.11	2741.88	1.80773	5.05973	6.86745
147.08	4.400	0.001087	0.4228	619.10	1937.28	2556.38	619.58	2122.82	2742.40	1.81201	5.05161	6.86362
147.49	4.450	0.001088	0.4183	620.88	1935.88	2556.76	621.37	2121.54	2742.91	1.81626	5.04358	6.85983
147.90	4.500	0.001088	0.4139	622.65	1934.49	2557.14	623.14	2120.27	2743.41	1.82046	5.03563	6.85609
148.31	4.550	0.001089	0.4096	624.40	1933.11	2557.52	624.90	2119.01	2743.91	1.82462	5.02776	6.85238
148.72	4.600	0.001089	0.4054	626.14	1931.75	2557.89	626.64	2117.75	2744.40	1.82874	5.01997	6.84871

Temp. t (°C)	Pressure P (bar)	Specific Volume $\left(\frac{m^3}{kg}\right)$		Internal Energy $\left(\frac{kJ}{kg}\right)$			Enthalpy $\left(\frac{kJ}{kg}\right)$			Entropy $\left(\frac{kJ}{kg\ K}\right)$		
		Sat. Liquid	Sat. Vapour	Sat. Liquid	Latent	Sat. Vapour	Sat. Liquid	Sat. Liquid	Sat. Vapour	Sat. Liquid	Latent	Sat. Vapour
		v_f	v_g	u_f	u_{fg}	u_g	h_f	h_{fg}	h_g	s_f	s_{fg}	s_g
149.12	4.650	0.001090	0.4013	627.87	1930.39	2558.26	628.37	2116.51	2744.88	1.83283	5.01226	6.84509
149.51	4.700	0.001090	0.3973	629.58	1929.05	2558.62	630.09	2115.28	2745.36	1.83687	5.00462	6.84150
149.91	4.750	0.001090	0.3934	631.27	1927.71	2558.98	631.79	2114.05	2745.84	1.84088	4.99706	6.83794
150.30	4.800	0.001091	0.3895	632.95	1926.39	2559.34	633.47	2112.83	2746.31	1.84485	4.98958	6.83443
150.69	4.850	0.001091	0.3857	634.62	1925.07	2559.69	635.15	2111.62	2746.77	1.84879	4.98216	6.83095
151.07	4.90	0.001092	0.3820	636.27	1923.76	2560.03	636.81	2110.42	2747.23	1.85269	4.97481	6.82750
151.45	4.95	0.001092	0.3784	637.91	1922.46	2560.38	638.45	2109.23	2747.68	1.85656	4.96753	6.82409
151.83	5.00	0.001093	0.3748	639.54	1921.18	2560.72	640.09	2108.04	2748.13	1.86039	4.96032	6.82071
152.58	5.10	0.001093	0.3679	642.76	1918.63	2561.39	643.32	2105.69	2749.01	1.86796	4.94610	6.81406
153.31	5.20	0.001094	0.3612	645.93	1916.11	2562.04	646.50	2103.38	2749.87	1.87540	4.93213	6.80753
154.04	5.30	0.001095	0.3548	649.05	1913.63	2562.68	649.63	2101.09	2750.72	1.88272	4.91840	6.80112
154.75	5.40	0.001096	0.3486	652.13	1911.17	2563.30	652.72	2098.82	2751.55	1.88991	4.90491	6.79482
155.46	5.50	0.001097	0.3426	655.16	1908.75	2563.92	655.77	2096.59	2752.35	1.89700	4.89165	6.78864
156.15	5.60	0.001097	0.3368	658.15	1906.36	2564.52	658.77	2094.38	2753.15	1.90397	4.87860	6.78257
156.83	5.70	0.001098	0.3313	661.10	1904.00	2565.11	661.73	2092.19	2753.92	1.91084	4.86577	6.77661
157.51	5.80	0.001099	0.3259	664.01	1901.67	2565.68	664.65	2090.03	2754.68	1.91760	4.85314	6.77074
158.17	5.90	0.001100	0.3206	666.89	1899.36	2566.25	667.53	2087.90	2755.43	1.92427	4.84071	6.76497
158.83	6.00	0.001101	0.3156	669.72	1897.09	2566.80	670.38	2085.78	2756.16	1.93083	4.82846	6.75930
159.47	6.10	0.001101	0.3107	672.52	1894.83	2567.35	673.19	2083.69	2756.88	1.93731	4.81641	6.75371
160.11	6.20	0.001102	0.3060	675.28	1892.60	2567.88	675.96	2081.62	2757.58	1.94369	4.80453	6.74822
160.74	6.30	0.001103	0.3014	678.01	1890.40	2568.41	678.70	2079.57	2758.27	1.94998	4.79283	6.74281
161.37	6.40	0.001104	0.2969	680.70	1888.22	2568.92	681.41	2077.54	2758.95	1.95619	4.78129	6.73748
161.98	6.50	0.001104	0.2926	683.37	1886.06	2569.43	684.08	2075.53	2759.62	1.96232	4.76992	6.73223
162.59	6.60	0.001105	0.2884	686.00	1883.92	2569.92	686.73	2073.54	2760.27	1.96836	4.75870	6.72706
163.19	6.70	0.001106	0.2843	688.60	1881.81	2570.41	689.34	2071.57	2760.91	1.97433	4.74764	6.72196
163.78	6.80	0.001107	0.2804	691.17	1879.72	2570.89	691.92	2069.62	2761.54	1.98022	4.73672	6.71694
164.37	6.90	0.001107	0.2765	693.71	1877.65	2571.36	694.48	2067.68	2762.16	1.98603	4.72595	6.71199
164.95	7.00	0.001108	0.2728	696.23	1875.59	2571.82	697.00	2065.77	2762.77	1.99178	4.71533	6.70711
165.52	7.10	0.001109	0.2691	698.72	1873.56	2572.28	699.50	2063.86	2763.37	1.99745	4.70484	6.70229
166.09	7.20	0.001109	0.2656	701.18	1871.55	2572.72	701.97	2061.98	2763.96	2.00306	4.69448	6.69754
166.65	7.30	0.001110	0.2622	703.61	1869.55	2573.16	704.42	2060.11	2764.53	2.00860	4.68425	6.69285
167.20	7.40	0.001111	0.2588	706.02	1867.57	2573.59	706.84	2058.26	2765.10	2.01408	4.67414	6.68822
167.75	7.50	0.001111	0.2555	708.40	1865.61	2574.02	709.24	2056.42	2765.66	2.01949	4.66416	6.68365
168.29	7.60	0.001112	0.2523	710.76	1863.67	2574.44	711.61	2054.60	2766.21	2.02484	4.65430	6.67914
168.83	7.70	0.001113	0.2492	713.10	1861.75	2574.85	713.96	2052.79	2766.75	2.03013	4.64456	6.67469
169.36	7.80	0.001113	0.2462	715.41	1859.84	2575.25	716.28	2051.00	2767.28	2.03536	4.63493	6.67029
169.89	7.90	0.001114	0.2432	717.70	1857.95	2575.65	718.58	2049.22	2767.80	2.04054	4.62541	6.66594
170.41	8.00	0.001115	0.2403	719.97	1856.07	2576.04	720.87	2047.45	2768.32	2.04566	4.61599	6.66165
170.92	8.10	0.001115	0.2375	722.22	1854.21	2576.43	723.12	2045.70	2768.82	2.05072	4.60669	6.65741
171.43	8.20	0.001116	0.2348	724.45	1852.36	2576.81	725.36	2043.96	2769.32	2.05573	4.59748	6.65321
171.94	8.30	0.001117	0.2321	726.65	1850.53	2577.18	727.58	2042.23	2769.81	2.06069	4.58838	6.64907
172.44	8.40	0.001117	0.2295	728.84	1848.71	2577.55	729.78	2040.52	2770.30	2.06560	4.57937	6.64497

Temp. t (°C)	Pressure P (bar)	Specific Volume $\left(\frac{m^3}{kg}\right)$		Internal Energy $\left(\frac{kJ}{kg}\right)$			Enthalpy $\left(\frac{kJ}{kg}\right)$			Entropy $\left(\frac{kJ}{kg\,K}\right)$		
		Sat. Liquid	Sat. Vapour	Sat. Liquid	Latent	Sat. Vapour	Sat. Liquid	Sat. Liquid	Sat. Vapour	Sat. Liquid	Latent	Sat. Vapour
		v_f	v_g	u_f	u_{fg}	u_g	h_f	h_{fg}	h_g	s_f	s_{fg}	s_g
172.94	8.50	0.001118	0.2269	731.00	1846.91	2577.91	731.95	2038.82	2770.77	2.07046	4.57046	6.64092
173.43	8.60	0.001119	0.2244	733.15	1845.12	2578.27	734.11	2037.13	2771.24	2.07527	4.56165	6.63692
173.91	8.70	0.001119	0.2219	735.28	1843.34	2578.62	736.25	2035.45	2771.70	2.08004	4.55292	6.63296
174.40	8.80	0.001120	0.2195	737.39	1841.58	2578.97	738.37	2033.78	2772.16	2.08475	4.54429	6.62904
174.88	8.90	0.001121	0.2172	739.48	1839.83	2579.31	740.48	2032.13	2772.61	2.08943	4.53574	6.62516
175.35	9.00	0.001121	0.2149	741.55	1838.09	2579.65	742.56	2030.48	2773.05	2.09405	4.52728	6.62133
175.82	9.10	0.001122	0.2126	743.61	1836.37	2579.98	744.63	2028.85	2773.48	2.09864	4.51890	6.61753
176.29	9.20	0.001122	0.2104	745.65	1834.65	2580.30	746.68	2027.23	2773.91	2.10318	4.51060	6.61378
176.75	9.30	0.001123	0.2083	747.67	1832.95	2580.63	748.72	2025.62	2774.33	2.10768	4.50238	6.61006
177.21	9.40	0.001124	0.2062	749.68	1831.27	2580.94	750.73	2024.01	2774.75	2.11213	4.49425	6.60638
177.66	9.50	0.001124	0.2041	751.67	1829.59	2581.26	752.74	2022.42	2775.16	2.11655	4.48619	6.60274
178.11	9.60	0.001125	0.2021	753.64	1827.92	2581.56	754.72	2020.84	2775.56	2.12093	4.47820	6.59913
178.56	9.70	0.001125	0.2001	755.60	1826.27	2581.87	756.69	2019.27	2775.96	2.12527	4.47029	6.59556
179.00	9.80	0.001126	0.1981	757.54	1824.62	2582.17	758.65	2017.70	2776.35	2.12957	4.46245	6.59202
179.44	9.90	0.001127	0.1962	759.47	1822.99	2582.46	760.59	2016.15	2776.74	2.13384	4.45468	6.58852
179.88	10.00	0.001127	0.1944	761.39	1821.37	2582.76	762.51	2014.61	2777.12	2.13807	4.44698	6.58505
180.74	10.20	0.001128	0.1907	765.17	1818.15	2583.33	766.32	2011.54	2777.87	2.14642	4.43179	6.57820
181.59	10.40	0.001130	0.1872	768.90	1814.98	2583.88	770.08	2008.52	2778.60	2.15463	4.41686	6.57148
182.43	10.60	0.001131	0.1838	772.58	1811.85	2584.42	773.78	2005.52	2779.30	2.16271	4.40218	6.56489
183.25	10.80	0.001132	0.1806	776.20	1808.75	2584.95	777.43	2002.56	2779.99	2.17066	4.38774	6.55840
184.06	11.00	0.001133	0.1775	779.78	1805.68	2585.46	781.03	1999.63	2780.66	2.17849	4.37354	6.55203
184.86	11.20	0.001134	0.1744	783.31	1802.66	2585.96	784.58	1996.74	2781.31	2.18620	4.35957	6.54577
185.65	11.40	0.001135	0.1715	786.79	1799.66	2586.45	788.08	1993.87	2781.95	2.19379	4.34581	6.53961
186.43	11.60	0.001136	0.1687	790.22	1796.70	2586.92	791.54	1991.03	2782.56	2.20128	4.33227	6.53354
187.20	11.80	0.001137	0.1659	793.61	1793.78	2587.39	794.95	1988.21	2783.17	2.20865	4.31893	6.52758
187.96	12.00	0.001138	0.1633	796.96	1790.88	2587.84	798.33	1985.43	2783.75	2.21592	4.30579	6.52171
188.70	12.20	0.001140	0.1607	800.27	1788.01	2588.28	801.66	1982.67	2784.32	2.22309	4.29284	6.51593
189.44	12.40	0.001141	0.1582	803.53	1785.17	2588.70	804.95	1979.93	2784.88	2.23016	4.28007	6.51023
190.17	12.60	0.001142	0.1558	806.76	1782.36	2589.12	808.20	1977.22	2785.42	2.23713	4.26749	6.50462
190.89	12.80	0.001143	0.1535	809.95	1779.58	2589.53	811.41	1974.54	2785.95	2.24402	4.25508	6.49909
191.60	13.00	0.001144	0.1512	813.10	1776.82	2589.92	814.59	1971.88	2786.47	2.25081	4.24284	6.49364
192.31	13.20	0.001145	0.1490	816.22	1774.09	2590.31	817.73	1969.24	2786.97	2.25751	4.23076	6.48827
193.00	13.40	0.001146	0.1468	819.30	1771.39	2590.69	820.84	1966.62	2787.46	2.26413	4.21884	6.48297
193.69	13.60	0.001147	0.1448	822.35	1768.71	2591.06	823.91	1964.03	2787.94	2.27067	4.20707	6.47774
194.37	13.80	0.001148	0.1427	825.37	1766.05	2591.42	826.95	1961.45	2788.40	2.27713	4.19546	6.47259
195.04	14.00	0.001149	0.1408	828.35	1763.42	2591.77	829.96	1958.90	2788.86	2.28351	4.18399	6.46750
195.70	14.20	0.001150	0.1389	831.30	1760.81	2592.11	832.94	1956.36	2789.30	2.28982	4.17266	6.46248
196.36	14.40	0.001151	0.1370	834.23	1758.22	2592.44	835.88	1953.85	2789.73	2.29605	4.16147	6.45752
197.01	14.60	0.001152	0.1352	837.12	1755.65	2592.77	838.80	1951.35	2790.15	2.30221	4.15042	6.45262
197.65	14.80	0.001153	0.1334	839.98	1753.11	2593.09	841.69	1948.88	2790.57	2.30830	4.13949	6.44779
198.29	15.00	0.001154	0.1317	842.82	1750.58	2593.40	844.55	1946.42	2790.97	2.31432	4.12869	6.44301
199.85	15.50	0.001156	0.1276	849.78	1744.36	2594.14	851.58	1940.35	2791.93	2.32908	4.10224	6.43132
201.37	16.00	0.001159	0.1237	856.59	1738.25	2594.84	858.44	1934.38	2792.83	2.34345	4.07651	6.41996

Temp. t (°C)	Pressure P (bar)	Specific Volume $\left(\frac{m^3}{kg}\right)$		Internal Energy $\left(\frac{kJ}{kg}\right)$			Enthalpy $\left(\frac{kJ}{kg}\right)$			Entropy $\left(\frac{kJ}{kg\,K}\right)$		
		Sat. Liquid	Sat. Vapour	Sat. Liquid	Latent	Sat. Vapour	Sat. Liquid	Sat. Liquid	Sat. Vapour	Sat. Liquid	Latent	Sat. Vapour
		v_f	v_g	u_f	u_{fg}	u_g	h_f	h_{fg}	h_g	s_f	s_{fg}	s_g
202.86	16.50	0.001161	0.1201	863.24	1732.26	2595.50	865.16	1928.52	2793.68	2.35745	4.05147	6.40892
204.31	17.00	0.001163	0.1167	869.75	1726.37	2596.12	871.72	1922.75	2794.47	2.37111	4.02706	6.39817
205.72	17.50	0.001166	0.1134	876.12	1720.59	2596.70	878.16	1917.06	2795.22	2.38444	4.00327	6.38770
207.11	18.00	0.001168	0.1104	882.35	1714.90	2597.25	884.46	1911.46	2795.92	2.39745	3.98005	6.37750
208.47	18.50	0.001170	0.1075	888.47	1709.30	2597.77	890.63	1905.94	2796.58	2.41017	3.95737	6.36754
209.80	19.00	0.001172	0.1047	894.46	1703.79	2598.25	896.69	1900.50	2797.19	2.42261	3.93521	6.35782
211.10	19.50	0.001175	0.1021	900.35	1698.36	2598.71	902.64	1895.13	2797.77	2.43479	3.91354	6.34832
212.38	20.00	0.001177	0.0996	906.12	1693.01	2599.13	908.47	1889.83	2798.31	2.44671	3.89233	6.33904
213.63	20.50	0.001179	0.0972	911.79	1687.74	2599.53	914.21	1884.60	2798.81	2.45839	3.87157	6.32995
214.86	21.00	0.001181	0.0949	917.37	1682.54	2599.90	919.85	1879.43	2799.28	2.46983	3.85123	6.32106
216.06	21.50	0.001183	0.0928	922.85	1677.40	2600.25	925.39	1874.32	2799.71	2.48106	3.83129	6.31235
217.25	22.00	0.001185	0.0907	928.24	1672.34	2600.57	930.85	1869.27	2800.12	2.49207	3.81174	6.30381
218.41	22.50	0.001187	0.0887	933.54	1667.33	2600.88	936.21	1864.28	2800.49	2.50288	3.79255	6.29543
219.56	23.00	0.001189	0.0868	938.76	1662.39	2601.16	941.50	1859.34	2800.83	2.51350	3.77372	6.28722
220.68	23.50	0.001191	0.0850	943.90	1657.51	2601.41	946.70	1854.45	2801.15	2.52393	3.75523	6.27916
221.79	24.00	0.001193	0.0832	948.97	1652.69	2601.65	951.83	1849.61	2801.44	2.53418	3.73706	6.27124
222.88	24.50	0.001195	0.0816	953.95	1647.92	2601.87	956.88	1844.82	2801.71	2.54426	3.71920	6.26346
223.95	25.0	0.001197	0.0800	958.87	1643.20	2602.07	961.87	1840.08	2801.95	2.55418	3.70164	6.25581
226.05	26.0	0.001201	0.0769	968.51	1633.91	2602.42	971.63	1830.73	2802.36	2.57353	3.66736	6.24090
228.08	27.0	0.001205	0.0741	977.88	1624.81	2602.70	981.14	1821.55	2802.68	2.59230	3.63415	6.22645
230.06	28.0	0.001209	0.0714	987.02	1615.89	2602.91	990.41	1812.52	2802.92	2.61051	3.60193	6.21245
231.98	29.0	0.001213	0.0690	995.93	1607.14	2603.07	999.45	1803.63	2803.08	2.62822	3.57063	6.19885
233.85	30.0	0.001217	0.0667	1004.64	1598.54	2603.17	1008.29	1794.89	2803.17	2.64543	3.54019	6.18563
235.68	31.0	0.001220	0.0645	1013.14	1590.08	2603.22	1016.92	1786.27	2803.20	2.66220	3.51056	6.17276
237.46	32.0	0.001224	0.0625	1021.46	1581.76	2603.22	1025.38	1777.78	2803.15	2.67855	3.48168	6.16022
239.20	33.0	0.001228	0.0606	1029.60	1573.58	2603.18	1033.65	1769.40	2803.05	2.69449	3.45350	6.14799
240.90	34.0	0.001231	0.0588	1037.58	1565.51	2603.09	1041.76	1761.12	2802.89	2.71005	3.42600	6.13605
242.56	35.0	0.001235	0.0571	1045.40	1557.56	2602.96	1049.72	1752.95	2802.67	2.72526	3.39912	6.12438
244.18	36.0	0.001238	0.0554	1053.06	1549.72	2602.78	1057.52	1744.88	2802.40	2.74013	3.37284	6.11297
245.77	37.0	0.001242	0.0539	1060.59	1541.98	2602.57	1065.19	1736.90	2802.08	2.75469	3.34712	6.10181
247.33	38.0	0.001245	0.0525	1067.98	1534.35	2602.33	1072.71	1729.00	2801.71	2.76893	3.32193	6.09087
248.86	39.0	0.001249	0.0511	1075.24	1526.80	2602.05	1080.11	1721.19	2801.30	2.78289	3.29725	6.08014
250.35	40.0	0.001252	0.0498	1082.38	1519.35	2601.73	1087.39	1713.45	2800.85	2.79658	3.27305	6.06963
251.82	41.0	0.001256	0.0485	1089.41	1511.98	2601.39	1094.55	1705.80	2800.35	2.81000	3.24931	6.05930
253.26	42.0	0.001259	0.0473	1096.31	1504.69	2601.01	1101.60	1698.21	2799.81	2.82317	3.22600	6.04916
254.68	43.0	0.001263	0.0462	1103.12	1497.48	2600.60	1108.55	1690.69	2799.24	2.83610	3.20310	6.03920
256.07	44.0	0.001266	0.0451	1109.82	1490.35	2600.16	1115.39	1683.24	2798.62	2.84881	3.18060	6.02941
257.44	45.0	0.001269	0.0441	1116.42	1483.28	2599.70	1122.13	1675.85	2797.97	2.86129	3.15848	6.01977
258.78	46.0	0.001273	0.0431	1122.92	1476.28	2599.21	1128.78	1668.51	2797.29	2.87357	3.13672	6.01028
260.10	47.0	0.001276	0.0421	1129.34	1469.35	2598.69	1135.34	1661.24	2796.57	2.88564	3.11530	6.00094
261.40	48.0	0.001280	0.0412	1135.67	1462.48	2598.15	1141.81	1654.02	2795.82	2.89753	3.09421	5.99174
262.68	49.0	0.001283	0.0403	1141.91	1455.67	2597.58	1148.20	1646.85	2795.04	2.90922	3.07345	5.98267
263.94	50.0	0.001286	0.0394	1148.07	1448.92	2596.99	1154.50	1639.73	2794.23	2.92074	3.05298	5.97372

Temp. t (°C)	Pressure P (bar)	Specific Volume $\left(\frac{m^3}{kg}\right)$		Internal Energy $\left(\frac{kJ}{kg}\right)$			Enthalpy $\left(\frac{kJ}{kg}\right)$			Entropy $\left(\frac{kJ}{kg\,K}\right)$		
		Sat. Liquid	Sat. Vapour	Sat. Liquid	Latent	Sat. Vapour	Sat. Liquid	Sat. Liquid	Sat. Vapour	Sat. Liquid	Latent	Sat. Vapour
		v_f	v_g	u_f	u_{fg}	u_g	h_f	h_{fg}	h_g	s_f	s_{fg}	s_g
266.40	52.0	0.001293	0.0378	1160.17	1435.58	2595.74	1166.89	1625.63	2792.52	2.94328	3.01292	5.95619
268.79	54.0	0.001299	0.0364	1171.97	1422.43	2594.41	1178.99	1611.71	2790.70	2.96517	2.97393	5.93910
271.12	56.0	0.001306	0.0350	1183.51	1409.48	2592.99	1190.82	1597.95	2788.77	2.98648	2.93595	5.92243
273.38	58.0	0.001312	0.0337	1194.79	1396.70	2591.49	1202.41	1584.33	2786.74	3.00724	2.89889	5.90612
275.59	60.0	0.001319	0.0324	1205.84	1384.08	2589.92	1213.75	1570.85	2784.61	3.02748	2.86269	5.89017
277.73	62.0	0.001326	0.0313	1216.67	1371.60	2588.27	1224.88	1557.50	2782.39	3.04724	2.82729	5.87453
279.83	64.0	0.001332	0.0302	1227.28	1359.27	2586.55	1235.81	1544.27	2780.08	3.06655	2.79264	5.85920
281.87	66.0	0.001339	0.0292	1237.70	1347.06	2584.76	1246.54	1531.14	2777.68	3.08544	2.75869	5.84414
283.87	68.0	0.001345	0.0283	1247.94	1334.97	2582.91	1257.09	1518.11	2775.20	3.10393	2.72540	5.82933
285.83	70.0	0.001352	0.0274	1258.00	1322.98	2580.99	1267.46	1505.17	2772.64	3.12205	2.69272	5.81477
287.74	72.0	0.001358	0.0265	1267.90	1311.10	2579.01	1277.68	1492.31	2769.99	3.13981	2.66061	5.80042
289.61	74.0	0.001365	0.0257	1277.65	1299.32	2576.96	1287.74	1479.53	2767.28	3.15724	2.62905	5.78629
291.45	76.0	0.001371	0.0250	1287.24	1287.62	2574.86	1297.66	1466.82	2764.48	3.17435	2.59799	5.77234
293.25	78.0	0.001378	0.0242	1296.70	1276.00	2572.70	1307.45	1454.17	2761.61	3.19117	2.56741	5.75858
295.01	80.0	0.001384	0.0235	1306.02	1264.45	2570.48	1317.10	1441.57	2758.67	3.20770	2.53728	5.74498
296.74	82.0	0.001391	0.0229	1315.22	1252.97	2568.20	1326.63	1429.03	2755.66	3.22396	2.50757	5.73153
298.43	84.0	0.001398	0.0222	1324.31	1241.56	2565.86	1336.04	1416.53	2752.58	3.23997	2.47826	5.71823
300.10	86.0	0.001404	0.0216	1333.27	1230.20	2563.47	1345.35	1404.08	2749.43	3.25574	2.44933	5.70507
301.74	88.0	0.001411	0.0210	1342.13	1218.89	2561.02	1354.55	1391.66	2746.21	3.27127	2.42075	5.69202
303.35	90.0	0.001418	0.0205	1350.89	1207.63	2558.52	1363.65	1379.26	2742.92	3.28659	2.39250	5.67909
304.93	92.0	0.001425	0.0200	1359.55	1196.41	2555.96	1372.66	1366.90	2739.56	3.30171	2.36457	5.66627
306.48	94.0	0.001431	0.0194	1368.12	1185.22	2553.35	1381.58	1354.56	2736.13	3.31662	2.33693	5.65355
308.01	96.0	0.001438	0.0190	1376.61	1174.07	2550.68	1390.41	1342.23	2732.64	3.33135	2.30957	5.64092
309.52	98.0	0.001445	0.0185	1385.01	1162.95	2547.95	1399.17	1329.91	2729.08	3.34590	2.28247	5.62837
311.00	100.0	0.001452	0.0180	1393.33	1151.85	2545.17	1407.85	1317.61	2725.46	3.36028	2.25561	5.61589
312.46	102.0	0.001459	0.0176	1401.57	1140.77	2542.34	1416.45	1305.31	2721.76	3.37450	2.22899	5.60349
313.89	104.0	0.001466	0.0172	1409.74	1129.71	2539.45	1424.99	1293.01	2718.00	3.38856	2.20258	5.59115
315.31	106.0	0.001474	0.0168	1417.85	1118.66	2536.50	1433.47	1280.71	2714.17	3.40249	2.17638	5.57886
316.70	108.0	0.001481	0.0164	1425.89	1107.61	2533.50	1441.88	1268.40	2710.28	3.41627	2.15036	5.56662
318.08	110.0	0.001488	0.0160	1433.87	1096.57	2530.44	1450.24	1256.07	2706.31	3.42992	2.12451	5.55443
319.43	112.0	0.001496	0.0156	1441.80	1085.53	2527.33	1458.55	1243.73	2702.28	3.44345	2.09883	5.54227
320.77	114.0	0.001503	0.0153	1449.67	1074.48	2524.15	1466.80	1231.37	2698.17	3.45685	2.07330	5.53015
322.09	116.0	0.001511	0.0149	1457.49	1063.43	2520.92	1475.01	1218.99	2694.00	3.47015	2.04790	5.51805
323.39	118.0	0.001518	0.0146	1465.26	1052.37	2517.62	1483.17	1206.58	2689.75	3.48334	2.02263	5.50597
324.68	120.0	0.001526	0.0143	1472.98	1041.29	2514.27	1491.29	1194.14	2685.43	3.49644	1.99747	5.49391
325.94	122.0	0.001534	0.0140	1480.67	1030.19	2510.85	1499.38	1181.66	2681.04	3.50944	1.97242	5.48185
327.19	124.0	0.001542	0.0136	1488.31	1019.06	2507.38	1507.43	1169.14	2676.57	3.52235	1.94745	5.46980
328.43	126.0	0.001550	0.0133	1495.92	1007.91	2503.83	1515.45	1156.58	2672.03	3.53518	1.92257	5.45775
329.65	128.0	0.001558	0.0131	1503.50	996.73	2500.23	1523.44	1143.96	2667.40	3.54793	1.89775	5.44569
330.85	130.0	0.001566	0.0128	1511.05	985.51	2496.55	1531.41	1131.29	2662.70	3.56062	1.87299	5.43361
332.04	132.0	0.001575	0.0125	1518.56	974.25	2492.81	1539.35	1118.57	2657.92	3.57323	1.84828	5.42152
333.22	134.0	0.001583	0.0122	1526.06	962.94	2489.00	1547.27	1105.78	2653.05	3.58579	1.82361	5.40940
334.38	136.0	0.001592	0.0120	1533.53	951.58	2485.11	1555.18	1092.92	2648.10	3.59830	1.79896	5.39725

Temp. t (°C)	Pressure P (bar)	Specific Volume $\left(\frac{m^3}{kg}\right)$		Internal Energy $\left(\frac{kJ}{kg}\right)$			Enthalpy $\left(\frac{kJ}{kg}\right)$			Entropy $\left(\frac{kJ}{kg\,K}\right)$		
		Sat. Liquid	Sat. Vapour	Sat. Liquid	Latent	Sat. Vapour	Sat. Liquid	Sat. Liquid	Sat. Vapour	Sat. Liquid	Latent	Sat. Vapour
		v_f	v_g	u_f	u_{fg}	u_g	h_f	h_{fg}	h_g	s_f	s_{fg}	s_g
335.53	138.0	0.001601	0.0117	1540.98	940.17	2481.15	1563.07	1079.99	2643.06	3.61075	1.77431	5.38506
336.67	140.0	0.001610	0.0115	1548.42	928.70	2477.12	1570.95	1066.98	2637.93	3.62316	1.74968	5.37284
337.79	142.0	0.001619	0.0112	1555.84	917.17	2473.01	1578.83	1053.88	2632.71	3.63553	1.72502	5.36056
338.90	144.0	0.001628	0.0110	1563.25	905.56	2468.82	1586.70	1040.70	2627.40	3.64787	1.70035	5.34822
340.00	146.0	0.001638	0.0108	1570.66	893.88	2464.54	1594.57	1027.41	2621.98	3.66019	1.67564	5.33583
341.08	148.0	0.001647	0.0106	1578.06	882.12	2460.18	1602.44	1014.02	2616.46	3.67248	1.65088	5.32336
342.16	150.0	0.001657	0.0103	1585.46	870.27	2455.73	1610.32	1000.52	2610.84	3.68475	1.62606	5.31081
343.22	152.0	0.001667	0.0101	1592.86	858.32	2451.18	1618.21	986.90	2605.11	3.69702	1.60116	5.29818
344.27	154.0	0.001678	0.0099	1600.27	846.27	2446.54	1626.11	973.16	2599.26	3.70929	1.57617	5.28546
345.31	156.0	0.001688	0.0097	1607.69	834.11	2441.80	1634.03	959.27	2593.30	3.72156	1.55106	5.27262
346.34	158.0	0.001699	0.0095	1615.12	821.83	2436.95	1641.97	945.24	2587.21	3.73384	1.52584	5.25968
347.36	160.0	0.001710	0.0093	1622.57	809.43	2432.00	1649.93	931.06	2580.99	3.74614	1.50049	5.24662
348.36	162.0	0.001721	0.0091	1630.04	796.89	2426.93	1657.93	916.71	2574.64	3.75846	1.47497	5.23343
349.36	164.0	0.001733	0.0089	1637.54	784.20	2421.74	1665.96	902.19	2568.15	3.77081	1.44927	5.22009
350.35	166.0	0.001745	0.0087	1645.06	771.36	2416.42	1674.03	887.47	2561.50	3.78321	1.42338	5.20659
351.33	168.0	0.001757	0.0086	1652.62	758.34	2410.97	1682.15	872.56	2554.70	3.79566	1.39726	5.19292
352.29	170.0	0.001770	0.0084	1660.23	745.15	2405.38	1690.32	857.42	2547.74	3.80817	1.37090	5.17907
353.25	172.0	0.001783	0.0082	1667.88	731.75	2399.63	1698.55	842.05	2540.60	3.82075	1.34427	5.16501
354.20	174.0	0.001797	0.0080	1675.58	718.15	2393.73	1706.84	826.42	2533.27	3.83341	1.31732	5.15073
355.14	176.0	0.001811	0.0078	1683.35	704.31	2387.66	1715.22	810.52	2525.74	3.84617	1.29005	5.13622
356.07	178.0	0.001825	0.0077	1691.18	690.22	2381.40	1723.67	794.33	2518.00	3.85904	1.26240	5.12144
356.99	180.0	0.001840	0.0075	1699.10	675.86	2374.95	1732.22	777.81	2510.03	3.87203	1.23434	5.10637
357.91	182.0	0.001856	0.0073	1707.10	661.19	2368.29	1740.88	760.94	2501.81	3.88517	1.20582	5.09099
358.81	184.0	0.001872	0.0072	1715.20	646.20	2361.41	1749.65	743.68	2493.33	3.89848	1.17678	5.07526
359.71	186.0	0.001889	0.0070	1723.42	630.85	2354.27	1758.56	726.00	2484.56	3.91197	1.14718	5.05915
360.59	188.0	0.001907	0.0068	1731.76	615.10	2346.86	1767.62	707.85	2475.47	3.92567	1.11694	5.04261
361.47	190.0	0.001926	0.0067	1740.26	598.90	2339.16	1776.85	689.18	2466.03	3.93962	1.08598	5.02560
362.34	192.0	0.001946	0.0065	1748.92	582.21	2331.12	1786.27	669.94	2456.21	3.95385	1.05420	5.00805
363.21	194.0	0.001966	0.0064	1757.77	564.95	2322.72	1795.92	650.04	2445.96	3.96841	1.02150	4.98991
364.06	196.0	0.001989	0.0062	1766.85	547.06	2313.91	1805.83	629.41	2435.23	3.98335	0.98775	4.97109
364.91	198.0	0.002012	0.0060	1776.19	528.44	2304.63	1816.03	607.93	2423.96	3.99873	0.95277	4.95150
365.75	200.0	0.002038	0.0059	1785.84	508.98	2294.81	1826.59	585.47	2412.06	4.01463	0.91637	4.93100
366.58	202.0	0.002065	0.0057	1795.85	488.52	2284.37	1837.57	561.87	2399.44	4.03117	0.87829	4.90945
367.40	204.0	0.002095	0.0055	1806.32	466.88	2273.20	1849.06	536.91	2385.97	4.04847	0.83819	4.88666
368.22	206.0	0.002128	0.0054	1817.33	443.82	2261.15	1861.18	510.30	2371.47	4.06670	0.79564	4.86234
369.03	208.0	0.002165	0.0052	1829.03	418.97	2248.01	1874.07	481.65	2355.72	4.08613	0.75002	4.83615
369.83	210.0	0.002207	0.0050	1841.62	391.86	2233.49	1887.97	450.39	2338.36	4.10709	0.70047	4.80756
370.62	212.0	0.002256	0.0048	1855.41	361.73	2217.14	1903.23	415.66	2318.89	4.13011	0.64567	4.77578
371.40	214.0	0.002314	0.0046	1870.90	327.32	2198.22	1920.43	376.02	2296.45	4.15609	0.58339	4.73948
372.18	216.0	0.002389	0.0044	1889.05	286.26	2175.31	1940.65	328.75	2269.40	4.18672	0.50943	4.69615
372.95	218.0	0.002495	0.0041	1912.24	232.79	2145.03	1966.62	267.24	2233.86	4.22618	0.41362	4.63980
373.71	220.0	0.002703	0.0036	1951.65	140.77	2092.42	2011.12	161.48	2172.60	4.29421	0.24964	4.54385
374.14	220.0	0.003155	0.0032	20.00	18.95	2025.44	2073.67	21.72	2095.39	4.39060	0.03357	4.42417

Table 3(1) Superheated Steam Properties (Pressure Base) Pressure = 11.0 bar and saturation temperature, Tsat = 184.062°C

t (°C)	v $\left(\frac{m^3}{kg}\right)$	u $\left(\frac{kJ}{kg}\right)$	h $\left(\frac{kJ}{kg}\right)$	s $\left(\frac{kJ}{kg\,K}\right)$
Sat Vapour (g)	0.1775	2585.46	2780.66	6.552
10	0.1829	2606.03	2807.22	6.610
20	0.1881	2625.40	2832.35	6.663
30	0.1932	2644.04	2856.59	6.713
40	0.1982	2662.16	2880.20	6.761
50	0.2031	2679.90	2903.31	6.807
60	0.2079	2697.34	2926.05	6.851
70	0.2127	2714.54	2948.48	6.894
80	0.2174	2731.55	2970.66	6.936
90	0.2220	2748.41	2992.64	6.977
100	0.2266	2765.16	3014.47	7.016
110	0.2312	2781.82	3036.16	7.055
120	0.2358	2798.42	3057.76	7.092
130	0.2403	2814.96	3079.27	7.129
140	0.2448	2831.47	3100.73	7.166
150	0.2493	2847.95	3122.13	7.201
160	0.2537	2864.42	3143.51	7.236
170	0.2582	2880.89	3164.86	7.270
180	0.2626	2897.37	3186.20	7.304
190	0.2670	2913.85	3207.53	7.337
200	0.2714	2930.35	3228.87	7.370
210	0.2758	2946.88	3250.21	7.402
220	0.2801	2963.43	3271.57	7.434
230	0.2845	2980.01	3292.95	7.465
240	0.2889	2996.62	3314.36	7.496
250	0.2932	3013.28	3335.79	7.527
260	0.2975	3029.97	3357.25	7.557
270	0.3019	3046.71	3378.75	7.587
280	0.3062	3063.49	3400.29	7.616
290	0.3105	3080.31	3421.87	7.645
300	0.3148	3097.19	3443.48	7.674
310	0.3191	3114.12	3465.15	7.702
320	0.3234	3131.10	3486.86	7.731
330	0.3277	3148.13	3508.62	7.758
340	0.3320	3165.22	3530.42	7.786
350	0.3363	3182.36	3552.28	7.813
360	0.3406	3199.56	3574.20	7.840
370	0.3449	3216.82	3596.17	7.867
380	0.3491	3234.14	3618.19	7.893
390	0.3534	3251.52	3640.27	7.920
400	0.3577	3268.96	3662.40	7.945
410	0.3619	3286.46	3684.60	7.971
420	0.3662	3304.03	3706.85	7.997
430	0.3705	3321.65	3729.17	8.022
440	0.3747	3339.34	3751.54	8.047

t (°C)	v $\left(\frac{m^3}{kg}\right)$	u $\left(\frac{kJ}{kg}\right)$	h $\left(\frac{kJ}{kg}\right)$	s $\left(\frac{kJ}{kg\,K}\right)$
Sat Vapour (g)	0.1775	2585.46	2780.66	6.552
450	0.3790	3357.10	3773.98	8.072
460	0.3832	3374.92	3796.48	8.097
470	0.3875	3392.80	3819.04	8.121
480	0.3917	3410.75	3841.66	8.145
490	0.3960	3428.77	3864.35	8.169
500	0.4002	3446.85	3887.10	8.193
510	0.4045	3464.99	3909.92	8.217
520	0.4087	3483.21	3932.80	8.241
530	0.4130	3501.49	3955.75	8.264
540	0.4172	3519.84	3978.76	8.287
550	0.4214	3538.25	4001.84	8.310
560	0.4257	3556.73	4024.98	8.333
570	0.4299	3575.28	4048.19	8.356
580	0.4342	3593.90	4071.46	8.378
590	0.4384	3612.58	4094.80	8.401
600	0.4426	3631.33	4118.21	8.423
610	0.4468	3650.15	4141.68	8.445
620	0.4511	3669.04	4165.22	8.467
630	0.4553	3687.99	4188.83	8.489
640	0.4595	3707.01	4212.50	8.510
650	0.4638	3726.10	4236.23	8.532
660	0.4680	3745.25	4260.04	8.553
670	0.4722	3764.47	4283.91	8.575
680	0.4764	3783.76	4307.84	8.596
690	0.4807	3803.12	4331.84	8.617
700	0.4849	3822.54	4355.91	8.638
710	0.4891	3842.03	4380.04	8.658
720	0.4933	3861.58	4404.24	8.679
730	0.4975	3881.20	4428.50	8.700
740	0.5018	3900.89	4452.83	8.720
750	0.5060	3920.64	4477.23	8.740
760	0.5102	3940.46	4501.68	8.761
770	0.5144	3960.34	4526.20	8.781
780	0.5186	3980.28	4550.79	8.801
790	0.5229	4000.29	4575.44	8.820
800	0.5271	4020.37	4600.15	8.840
810	0.5313	4040.51	4624.93	8.860
820	0.5355	4060.71	4649.77	8.879
830	0.5397	4080.97	4674.67	8.899
840	0.5439	4101.30	4699.63	8.918
850	0.5482	4121.69	4724.66	8.937
860	0.5524	4142.15	4749.75	8.956
870	0.5566	4162.66	4774.90	8.975

t (°C)	v $\left(\frac{m^3}{kg}\right)$	u $\left(\frac{kJ}{kg}\right)$	h $\left(\frac{kJ}{kg}\right)$	s $\left(\frac{kJ}{kg\,K}\right)$
Sat Vapour (g)	0.1775	2585.46	2780.66	6.552
880	0.5608	4183.24	4800.11	8.994
890	0.5650	4203.87	4825.38	9.013
900	0.5692	4224.57	4850.71	9.032
910	0.5734	4245.33	4876.10	9.050
920	0.5776	4266.15	4901.55	9.069
930	0.5819	4287.03	4927.06	9.087
940	0.5861	4307.97	4952.63	9.106
950	0.5903	4328.96	4978.26	9.124
960	0.5945	4350.02	5003.95	9.142
970	0.5987	4371.13	5029.69	9.160
980	0.6029	4392.30	5055.49	9.178
990	0.6071	4413.53	5081.35	9.196
1000	0.6113	4434.81	5107.26	9.214
1010	0.6155	4456.16	5133.23	9.232
1020	0.6197	4477.55	5159.26	9.250
1030	0.6239	4499.01	5185.34	9.267
1040	0.6282	4520.52	5211.48	9.285
1050	0.6324	4542.08	5237.67	9.302
1060	0.6366	4563.70	5263.92	9.320
1070	0.6408	4585.37	5290.22	9.337
1080	0.6450	4607.10	5316.57	9.354
1090	0.6492	4628.88	5342.98	9.371
1100	0.6534	4650.71	5369.44	9.388
1110	0.6576	4672.59	5395.95	9.405
1120	0.6618	4694.53	5422.51	9.422
1130	0.6660	4716.52	5449.13	9.439
1140	0.6702	4738.56	5475.79	9.456
1150	0.6744	4760.65	5502.51	9.472
1160	0.6786	4782.79	5529.28	9.489
1170	0.6828	4804.98	5556.09	9.505
1180	0.6870	4827.22	5582.96	9.522
1190	0.6912	4849.51	5609.88	9.538
1200	0.6954	4871.85	5636.84	9.555
1210	0.6996	4894.24	5663.85	9.571
1220	0.7039	4916.67	5690.91	9.587
1230	0.7081	4939.16	5718.02	9.603
1240	0.7123	4961.69	5745.17	9.619
1250	0.7165	4984.27	5772.37	9.635
1260	0.7207	5006.89	5799.62	9.651
1270	0.7249	5029.56	5826.92	9.667
1280	0.7291	5052.28	5854.26	9.683
1290	0.7333	5075.04	5881.64	9.698
1300	0.7375	5097.84	5909.07	9.714

Table 3(2) Superheated Steam Properties (Pressure Base) Pressure = 12.0 bar and saturation temperature, Tsat = 187.957°C

t (°C)	$v\left(\frac{m^3}{kg}\right)$	$u\left(\frac{kJ}{kg}\right)$	$h\left(\frac{kJ}{kg}\right)$	$s\left(\frac{kJ}{kg\,K}\right)$
Sat Vapour (g)	*0.1633*	*2587.84*	*2783.75*	*6.522*
10	0.1683	2608.77	2810.76	6.580
20	0.1732	2628.42	2836.25	6.633
30	0.1779	2647.30	2860.80	6.684
40	0.1825	2665.63	2884.67	6.732
50	0.1871	2683.56	2908.02	6.778
60	0.1915	2701.16	2930.97	6.822
70	0.1959	2718.51	2953.58	6.865
80	0.2002	2735.66	2975.94	6.907
90	0.2045	2752.65	2998.08	6.948
100	0.2088	2769.51	3020.05	6.987
110	0.2130	2786.28	3041.88	7.026
120	0.2172	2802.97	3063.60	7.063
130	0.2214	2819.61	3085.24	7.100
140	0.2255	2836.21	3106.80	7.137
150	0.2296	2852.78	3128.31	7.172
160	0.2337	2869.33	3149.79	7.207
170	0.2378	2885.87	3171.23	7.241
180	0.2419	2902.42	3192.66	7.275
190	0.2459	2918.97	3214.08	7.308
200	0.2500	2935.54	3235.50	7.341
210	0.2540	2952.13	3256.92	7.373
220	0.2580	2968.74	3278.36	7.404
230	0.2620	2985.38	3299.81	7.436
240	0.2660	3002.05	3321.28	7.467
250	0.2700	3018.76	3342.78	7.497
260	0.2740	3035.51	3364.31	7.527
270	0.2780	3052.30	3385.88	7.557
280	0.2820	3069.13	3407.47	7.586
290	0.2859	3086.01	3429.11	7.615
300	0.2899	3102.94	3450.79	7.644
310	0.2938	3119.91	3472.51	7.672
320	0.2978	3136.94	3494.28	7.700
330	0.3017	3154.02	3516.09	7.728
340	0.3057	3171.15	3537.95	7.755
350	0.3096	3188.34	3559.86	7.783
360	0.3135	3205.59	3581.83	7.810
370	0.3175	3222.89	3603.85	7.836
380	0.3214	3240.25	3625.92	7.863
390	0.3253	3257.67	3648.05	7.889
400	0.3292	3275.15	3670.23	7.915
410	0.3331	3292.70	3692.47	7.940
420	0.3371	3310.30	3714.77	7.966
430	0.3410	3327.97	3737.13	7.991
440	0.3449	3345.70	3759.55	8.016

t (°C)	$v\left(\frac{m^3}{kg}\right)$	$u\left(\frac{kJ}{kg}\right)$	$h\left(\frac{kJ}{kg}\right)$	$s\left(\frac{kJ}{kg\,K}\right)$
Sat Vapour (g)	*0.1633*	*2587.84*	*2783.75*	*6.522*
450	0.3488	3363.49	3782.04	8.041
460	0.3527	3381.35	3804.58	8.065
470	0.3566	3399.27	3827.18	8.090
480	0.3605	3417.26	3849.85	8.114
490	0.3644	3435.32	3872.58	8.138
500	0.3683	3453.43	3895.37	8.162
510	0.3722	3471.62	3918.23	8.186
520	0.3761	3489.87	3941.15	8.209
530	0.3800	3508.19	3964.14	8.232
540	0.3839	3526.57	3987.19	8.256
550	0.3877	3545.02	4010.31	8.279
560	0.3916	3563.54	4033.49	8.301
570	0.3955	3582.12	4056.74	8.324
580	0.3994	3600.77	4080.05	8.346
590	0.4033	3619.49	4103.43	8.369
600	0.4072	3638.28	4126.87	8.391
610	0.4110	3657.13	4150.38	8.413
620	0.4149	3676.05	4173.96	8.435
630	0.4188	3695.03	4197.60	8.457
640	0.4227	3714.09	4221.31	8.478
650	0.4266	3733.21	4245.08	8.500
660	0.4304	3752.40	4268.92	8.521
670	0.4343	3771.65	4292.82	8.542
680	0.4382	3790.97	4316.79	8.564
690	0.4421	3810.36	4340.83	8.585
700	0.4459	3829.81	4364.93	8.605
710	0.4498	3849.33	4389.10	8.626
720	0.4537	3868.92	4413.33	8.647
730	0.4575	3888.57	4437.62	8.667
740	0.4614	3908.29	4461.99	8.688
750	0.4653	3928.07	4486.41	8.708
760	0.4692	3947.91	4510.90	8.728
770	0.4730	3967.83	4535.45	8.748
780	0.4769	3987.80	4560.07	8.768
790	0.4808	4007.84	4584.75	8.788
800	0.4846	4027.95	4609.50	8.807
810	0.4885	4048.11	4634.30	8.827
820	0.4924	4068.35	4659.17	8.846
830	0.4962	4088.64	4684.11	8.866
840	0.5001	4109.00	4709.10	8.885
850	0.5040	4129.41	4734.16	8.904
860	0.5078	4149.90	4759.27	8.923
870	0.5117	4170.44	4784.45	8.942

t (°C)	$v\left(\frac{m^3}{kg}\right)$	$u\left(\frac{kJ}{kg}\right)$	$h\left(\frac{kJ}{kg}\right)$	$s\left(\frac{kJ}{kg\,K}\right)$
Sat Vapour (g)	*0.1633*	*2587.84*	*2783.75*	*6.522*
880	0.5155	4191.04	4809.69	8.961
890	0.5194	4211.71	4834.99	8.980
900	0.5233	4232.43	4860.35	8.999
910	0.5271	4253.22	4885.77	9.017
920	0.5310	4274.06	4911.25	9.036
930	0.5349	4294.97	4936.79	9.054
940	0.5387	4315.93	4962.39	9.073
950	0.5426	4336.95	4988.04	9.091
960	0.5464	4358.03	5013.75	9.109
970	0.5503	4379.17	5039.52	9.127
980	0.5542	4400.37	5065.35	9.145
990	0.5580	4421.62	5091.24	9.163
1000	0.5619	4442.93	5117.18	9.181
1010	0.5657	4464.30	5143.17	9.199
1020	0.5696	4485.72	5169.23	9.216
1030	0.5734	4507.20	5195.33	9.234
1040	0.5773	4528.73	5221.50	9.251
1050	0.5812	4550.32	5247.71	9.269
1060	0.5850	4571.96	5273.98	9.286
1070	0.5889	4593.66	5300.31	9.303
1080	0.5927	4615.40	5326.69	9.320
1090	0.5966	4637.21	5353.12	9.338
1100	0.6004	4659.06	5379.60	9.355
1110	0.6043	4680.97	5406.13	9.372
1120	0.6082	4702.93	5432.72	9.388
1130	0.6120	4724.94	5459.36	9.405
1140	0.6159	4747.00	5486.05	9.422
1150	0.6197	4769.11	5512.79	9.439
1160	0.6236	4791.28	5539.58	9.455
1170	0.6274	4813.49	5566.41	9.472
1180	0.6313	4835.75	5593.30	9.488
1190	0.6351	4858.06	5620.24	9.504
1200	0.6390	4880.42	5647.22	9.521
1210	0.6429	4902.83	5674.26	9.537
1220	0.6467	4925.29	5701.34	9.553
1230	0.6506	4947.79	5728.47	9.569
1240	0.6544	4970.34	5755.64	9.585
1250	0.6583	4992.94	5782.87	9.601
1260	0.6621	5015.58	5810.13	9.617
1270	0.6660	5038.27	5837.45	9.633
1280	0.6698	5061.01	5864.81	9.649
1290	0.6737	5083.79	5892.21	9.664
1300	0.6775	5106.61	5919.66	9.680

Table 3(3) Superheated Steam Properties (Pressure Base) Pressure = 13.0 bar and saturation temperature, Tsat = 191.605°C

t (°C)	$v\left(\frac{m^3}{kg}\right)$	$u\left(\frac{kJ}{kg}\right)$	$h\left(\frac{kJ}{kg}\right)$	$s\left(\frac{kJ}{kg\,K}\right)$
Sat Vapour (g)	*0.1512*	*2589.92*	*2786.47*	*6.494*
10	0.1559	2611.21	2813.92	6.552
20	0.1605	2631.14	2839.76	6.606
30	0.1649	2650.25	2864.60	6.657
40	0.1692	2668.79	2888.73	6.705
50	0.1734	2686.89	2912.31	6.751
60	0.1775	2704.66	2935.46	6.796
70	0.1816	2722.15	2958.26	6.839
80	0.1857	2739.43	2980.78	6.881
90	0.1896	2756.55	3003.08	6.921
100	0.1936	2773.52	3025.19	6.961
110	0.1975	2790.40	3047.16	6.999
120	0.2014	2807.19	3069.00	7.037
130	0.2053	2823.91	3090.75	7.074
140	0.2091	2840.60	3112.42	7.110
150	0.2129	2857.24	3134.03	7.145
160	0.2167	2873.87	3155.60	7.180
170	0.2205	2890.49	3177.14	7.214
180	0.2243	2907.11	3198.65	7.248
190	0.2280	2923.73	3220.15	7.281
200	0.2318	2940.36	3241.65	7.314
210	0.2355	2957.01	3263.15	7.346
220	0.2392	2973.68	3284.66	7.377
230	0.2429	2990.38	3306.18	7.409
240	0.2466	3007.11	3327.72	7.439
250	0.2503	3023.87	3349.29	7.470
260	0.2540	3040.67	3370.88	7.500
270	0.2577	3057.51	3392.51	7.529
280	0.2614	3074.39	3414.16	7.559
290	0.2650	3091.32	3435.86	7.588
300	0.2687	3108.29	3457.59	7.616
310	0.2724	3125.31	3479.37	7.644
320	0.2760	3142.38	3501.19	7.672
330	0.2797	3159.51	3523.05	7.700
340	0.2833	3176.69	3544.97	7.728
350	0.2869	3193.92	3566.93	7.755
360	0.2906	3211.21	3588.94	7.782
370	0.2942	3228.55	3611.01	7.808
380	0.2978	3245.95	3633.13	7.834
390	0.3015	3263.42	3655.30	7.861
400	0.3051	3280.94	3677.54	7.886
410	0.3087	3298.52	3699.82	7.912
420	0.3123	3316.16	3722.17	7.937
430	0.3159	3333.87	3744.57	7.963
440	0.3195	3351.64	3767.04	7.988

t (°C)	$v\left(\frac{m^3}{kg}\right)$	$u\left(\frac{kJ}{kg}\right)$	$h\left(\frac{kJ}{kg}\right)$	$s\left(\frac{kJ}{kg\,K}\right)$
Sat Vapour (g)	*0.1512*	*2589.92*	*2786.47*	*6.494*
450	0.3231	3369.47	3789.56	8.012
460	0.3268	3387.36	3812.14	8.037
470	0.3304	3405.32	3834.79	8.061
480	0.3340	3423.35	3857.50	8.085
490	0.3376	3441.44	3880.27	8.109
500	0.3412	3459.59	3903.10	8.133
510	0.3448	3477.81	3926.00	8.157
520	0.3484	3496.10	3948.96	8.180
530	0.3520	3514.45	3971.98	8.203
540	0.3555	3532.86	3995.07	8.227
550	0.3591	3551.35	4018.23	8.250
560	0.3627	3569.90	4041.45	8.272
570	0.3663	3588.52	4064.73	8.295
580	0.3699	3607.20	4088.08	8.317
590	0.3735	3625.95	4111.49	8.340
600	0.3771	3644.77	4134.97	8.362
610	0.3807	3663.66	4158.52	8.384
620	0.3842	3682.61	4182.13	8.406
630	0.3878	3701.63	4205.80	8.427
640	0.3914	3720.71	4229.55	8.449
650	0.3950	3739.86	4253.35	8.470
660	0.3986	3759.08	4277.23	8.492
670	0.4022	3778.37	4301.16	8.513
680	0.4057	3797.72	4325.17	8.534
690	0.4093	3817.14	4349.24	8.555
700	0.4129	3836.62	4373.37	8.576
710	0.4165	3856.17	4397.57	8.596
720	0.4200	3875.78	4421.83	8.617
730	0.4236	3895.46	4446.16	8.637
740	0.4272	3915.21	4470.55	8.658
750	0.4308	3935.02	4495.01	8.678
760	0.4343	3954.90	4519.53	8.698
770	0.4379	3974.84	4544.11	8.718
780	0.4415	3994.84	4568.76	8.738
790	0.4450	4014.91	4593.47	8.758
800	0.4486	4035.04	4618.24	8.777
810	0.4522	4055.24	4643.08	8.797
820	0.4558	4075.49	4667.98	8.816
830	0.4593	4095.81	4692.94	8.836
840	0.4629	4116.20	4717.96	8.855
850	0.4665	4136.64	4743.05	8.874
860	0.4700	4157.15	4768.19	8.893
870	0.4736	4177.72	4793.40	8.912

t (°C)	$v\left(\frac{m^3}{kg}\right)$	$u\left(\frac{kJ}{kg}\right)$	$h\left(\frac{kJ}{kg}\right)$	$s\left(\frac{kJ}{kg\,K}\right)$
Sat Vapour (g)	*0.1512*	*2589.92*	*2786.47*	*6.494*
880	0.4772	4198.35	4818.67	8.931
890	0.4807	4219.04	4843.99	8.950
900	0.4843	4239.79	4869.38	8.969
910	0.4879	4260.60	4894.83	8.987
920	0.4914	4281.47	4920.33	9.006
930	0.4950	4302.40	4945.90	9.024
940	0.4986	4323.39	4971.52	9.042
950	0.5021	4344.43	4997.20	9.060
960	0.5057	4365.54	5022.94	9.079
970	0.5093	4386.70	5048.73	9.097
980	0.5128	4407.92	5074.59	9.115
990	0.5164	4429.20	5100.49	9.132
1000	0.5199	4450.53	5126.46	9.150
1010	0.5235	4471.92	5152.48	9.168
1020	0.5271	4493.37	5178.56	9.186
1030	0.5306	4514.87	5204.69	9.203
1040	0.5342	4536.42	5230.87	9.221
1050	0.5378	4558.03	5257.12	9.238
1060	0.5413	4579.70	5283.41	9.255
1070	0.5449	4601.41	5309.76	9.273
1080	0.5484	4623.18	5336.16	9.290
1090	0.5520	4645.01	5362.61	9.307
1100	0.5556	4666.89	5389.12	9.324
1110	0.5591	4688.81	5415.67	9.341
1120	0.5627	4710.79	5442.28	9.357
1130	0.5662	4732.83	5468.94	9.374
1140	0.5698	4754.91	5495.65	9.391
1150	0.5734	4777.04	5522.41	9.408
1160	0.5769	4799.22	5549.22	9.424
1170	0.5805	4821.46	5576.08	9.441
1180	0.5840	4843.74	5602.99	9.457
1190	0.5876	4866.07	5629.95	9.473
1200	0.5912	4888.45	5656.95	9.490
1210	0.5947	4910.88	5684.01	9.506
1220	0.5983	4933.35	5711.11	9.522
1230	0.6018	4955.87	5738.26	9.538
1240	0.6054	4978.44	5765.45	9.554
1250	0.6089	5001.06	5792.69	9.570
1260	0.6125	5023.72	5819.98	9.586
1270	0.6161	5046.43	5847.31	9.602
1280	0.6196	5069.18	5874.69	9.617
1290	0.6232	5091.98	5902.11	9.633
1300	0.6267	5114.82	5929.58	9.649

Table 3(4) Superheated Steam Properties (Pressure Base) Pressure = 14.0 bar and saturation temperature, Tsat = 195.039°C

t (°C)	*v* $\left(\frac{m^3}{kg}\right)$	*u* $\left(\frac{kJ}{kg}\right)$	*h* $\left(\frac{kJ}{kg}\right)$	*s* $\left(\frac{kJ}{kg\,K}\right)$	*t* (°C)	*v* $\left(\frac{m^3}{kg}\right)$	*u* $\left(\frac{kJ}{kg}\right)$	*h* $\left(\frac{kJ}{kg}\right)$	*s* $\left(\frac{kJ}{kg\,K}\right)$	*t* (°C)	*v* $\left(\frac{m^3}{kg}\right)$	*u* $\left(\frac{kJ}{kg}\right)$	*h* $\left(\frac{kJ}{kg}\right)$	*s* $\left(\frac{kJ}{kg\,K}\right)$
Sat Vapour (g)	0.1408	2591.77	2788.86	6.468	Sat Vapour (g)	0.1408	2591.77	2788.86	6.468	Sat Vapour (g)	0.1408	2591.77	2788.86	6.468
10	0.1452	2613.39	2816.74	6.526	450	0.3011	3375.08	3796.62	7.986	880	0.4442	4205.22	4827.11	8.903
20	0.1495	2633.60	2842.92	6.581	460	0.3045	3393.01	3819.25	8.011	890	0.4475	4225.94	4852.46	8.922
30	0.1537	2652.94	2868.05	6.632	470	0.3078	3411.01	3841.93	8.035	900	0.4508	4246.71	4877.88	8.941
40	0.1577	2671.68	2892.44	6.680	480	0.3112	3429.06	3864.68	8.059	910	0.4541	4267.55	4903.35	8.959
50	0.1616	2689.96	2916.24	6.726	490	0.3145	3447.19	3887.49	8.083	920	0.4575	4288.44	4928.88	8.978
60	0.1655	2707.88	2939.59	6.771	500	0.3178	3465.37	3910.36	8.107	930	0.4608	4309.39	4954.47	8.996
70	0.1693	2725.52	2962.57	6.814	510	0.3212	3483.63	3933.30	8.130	940	0.4641	4330.41	4980.11	9.014
80	0.1731	2742.93	2985.26	6.856	520	0.3245	3501.95	3956.29	8.154	950	0.4674	4351.48	5005.82	9.032
90	0.1768	2760.16	3007.70	6.897	530	0.3279	3520.33	3979.35	8.177	960	0.4707	4372.60	5031.58	9.050
100	0.1805	2777.24	3029.95	6.936	540	0.3312	3538.78	4002.48	8.200	970	0.4740	4393.79	5057.40	9.068
110	0.1842	2794.22	3052.04	6.975	550	0.3346	3557.30	4025.67	8.223	980	0.4773	4415.03	5083.28	9.086
120	0.1878	2811.10	3074.01	7.012	560	0.3379	3575.88	4048.92	8.245	990	0.4806	4436.33	5109.21	9.104
130	0.1914	2827.91	3095.87	7.049	570	0.3412	3594.53	4072.24	8.268	1000	0.4839	4457.69	5135.20	9.122
140	0.1950	2844.68	3117.64	7.085	580	0.3446	3613.25	4095.63	8.290	1010	0.4872	4479.10	5161.24	9.140
150	0.1985	2861.41	3139.35	7.121	590	0.3479	3632.03	4119.08	8.313	1020	0.4906	4500.56	5187.34	9.157
160	0.2021	2878.11	3161.01	7.155	600	0.3512	3650.88	4142.59	8.335	1030	0.4939	4522.09	5213.50	9.175
170	0.2056	2894.80	3182.64	7.189	610	0.3546	3669.79	4166.17	8.357	1040	0.4972	4543.66	5239.70	9.192
180	0.2091	2911.48	3204.24	7.223	620	0.3579	3688.77	4189.81	8.379	1050	0.5005	4565.29	5265.97	9.210
190	0.2126	2928.16	3225.82	7.256	630	0.3612	3707.82	4213.52	8.400	1060	0.5038	4586.98	5292.28	9.227
200	0.2161	2944.86	3247.39	7.289	640	0.3645	3726.94	4237.29	8.422	1070	0.5071	4608.72	5318.65	9.244
210	0.2196	2961.57	3268.97	7.321	650	0.3679	3746.12	4261.13	8.443	1080	0.5104	4630.51	5345.07	9.261
220	0.2230	2978.30	3290.55	7.352	660	0.3712	3765.37	4285.04	8.465	1090	0.5137	4652.35	5371.55	9.278
230	0.2265	2995.05	3312.14	7.383	670	0.3745	3784.68	4309.01	8.486	1100	0.5170	4674.25	5398.07	9.295
240	0.2299	3011.83	3333.74	7.414	680	0.3778	3804.06	4333.04	8.507	1110	0.5203	4696.20	5424.65	9.312
250	0.2334	3028.65	3355.37	7.444	690	0.3812	3823.51	4357.14	8.528	1120	0.5236	4718.20	5451.28	9.329
260	0.2368	3045.50	3377.03	7.474	700	0.3845	3843.02	4381.31	8.548	1130	0.5269	4740.25	5477.96	9.346
270	0.2402	3062.39	3398.71	7.504	710	0.3878	3862.60	4405.54	8.569	1140	0.5302	4762.35	5504.69	9.362
280	0.2437	3079.32	3420.43	7.533	720	0.3911	3882.24	4429.83	8.590	1150	0.5336	4784.50	5531.47	9.379
290	0.2471	3096.29	3442.18	7.562	730	0.3945	3901.95	4454.19	8.610	1160	0.5369	4806.70	5558.30	9.395
300	0.2505	3113.31	3463.96	7.591	740	0.3978	3921.72	4478.61	8.630	1170	0.5402	4828.96	5585.18	9.412
310	0.2539	3130.37	3485.79	7.619	750	0.4011	3941.56	4503.09	8.651	1180	0.5435	4851.26	5612.11	9.428
320	0.2573	3147.49	3507.67	7.647	760	0.4044	3961.46	4527.64	8.671	1190	0.5468	4873.61	5639.09	9.445
330	0.2607	3164.66	3529.58	7.674	770	0.4077	3981.43	4552.26	8.691	1200	0.5501	4896.00	5666.11	9.461
340	0.2640	3181.88	3551.54	7.702	780	0.4111	4001.46	4576.93	8.710	1210	0.5534	4918.45	5693.19	9.477
350	0.2674	3199.15	3573.56	7.729	790	0.4144	4021.56	4601.67	8.730	1220	0.5567	4940.94	5720.31	9.493
360	0.2708	3216.48	3595.62	7.756	800	0.4177	4041.71	4626.47	8.750	1230	0.5600	4963.48	5747.47	9.509
370	0.2742	3233.86	3617.73	7.782	810	0.4210	4061.93	4651.34	8.769	1240	0.5633	4986.07	5774.69	9.525
380	0.2776	3251.31	3639.89	7.808	820	0.4243	4082.22	4676.26	8.789	1250	0.5666	5008.70	5801.95	9.541
390	0.2809	3268.81	3662.11	7.835	830	0.4276	4102.57	4701.25	8.808	1260	0.5699	5031.38	5829.25	9.557
400	0.2843	3286.37	3684.39	7.860	840	0.4309	4122.97	4726.30	8.827	1270	0.5732	5054.11	5856.60	9.573
410	0.2877	3303.99	3706.72	7.886	850	0.4343	4143.44	4751.41	8.846	1280	0.5765	5076.88	5884.00	9.588
420	0.2910	3321.67	3729.11	7.911	860	0.4376	4163.98	4776.58	8.865	1290	0.5798	5099.69	5911.44	9.604
430	0.2944	3339.41	3751.55	7.936	870	0.4409	4184.57	4801.82	8.884	1300	0.5831	5122.55	5938.92	9.620
440	0.2977	3357.21	3774.06	7.961										

Table 3(5) Superheated Steam Properties (Pressure Base) Pressure = 15.0 bar and saturation temperature, Tsat = 198.287°C

t (°C)	v $\left(\frac{m^3}{kg}\right)$	u $\left(\frac{kJ}{kg}\right)$	h $\left(\frac{kJ}{kg}\right)$	s $\left(\frac{kJ}{kg\,K}\right)$
Sat Vapour (g)	*0.1317*	*2593.40*	*2790.97*	*6.443*
10	0.1359	2615.36	2819.26	6.502
20	0.1400	2635.83	2845.78	6.557
30	0.1439	2655.40	2871.20	6.608
40	0.1477	2674.33	2895.83	6.657
50	0.1514	2692.78	2919.86	6.703
60	0.1550	2710.86	2943.40	6.748
70	0.1586	2728.63	2966.56	6.791
80	0.1622	2746.17	2989.40	6.833
90	0.1657	2763.51	3012.00	6.874
100	0.1691	2780.70	3034.38	6.913
110	0.1725	2797.78	3056.60	6.952
120	0.1759	2814.75	3078.67	6.989
130	0.1793	2831.65	3100.64	7.026
140	0.1827	2848.50	3122.52	7.062
150	0.1860	2865.30	3144.33	7.098
160	0.1893	2882.08	3166.08	7.132
170	0.1926	2898.83	3187.79	7.166
180	0.1959	2915.58	3209.47	7.200
190	0.1992	2932.33	3231.13	7.233
200	0.2025	2949.08	3252.78	7.265
210	0.2057	2965.85	3274.42	7.297
220	0.2090	2982.63	3296.07	7.329
230	0.2122	2999.44	3317.73	7.360
240	0.2154	3016.27	3339.40	7.391
250	0.2186	3033.14	3361.09	7.421
260	0.2218	3050.04	3382.80	7.451
270	0.2250	3066.97	3404.54	7.480
280	0.2282	3083.95	3426.31	7.510
290	0.2314	3100.97	3448.12	7.538
300	0.2346	3118.03	3469.96	7.567
310	0.2378	3135.14	3491.84	7.595
320	0.2410	3152.30	3513.76	7.623
330	0.2441	3169.51	3535.73	7.651
340	0.2473	3186.77	3557.74	7.678
350	0.2505	3204.08	3579.79	7.705
360	0.2536	3221.45	3601.90	7.732
370	0.2568	3238.87	3624.06	7.758
380	0.2599	3256.35	3646.27	7.784
390	0.2631	3273.89	3668.53	7.810
400	0.2662	3291.48	3690.85	7.836
410	0.2694	3309.14	3713.22	7.862
420	0.2725	3326.86	3735.65	7.887
430	0.2757	3344.63	3758.14	7.912
440	0.2788	3362.47	3780.68	7.937

t (°C)	v $\left(\frac{m^3}{kg}\right)$	u $\left(\frac{kJ}{kg}\right)$	h $\left(\frac{kJ}{kg}\right)$	s $\left(\frac{kJ}{kg\,K}\right)$
Sat Vapour (g)	*0.1317*	*2593.40*	*2790.97*	*6.443*
450	0.2819	3380.37	3803.28	7.962
460	0.2851	3398.34	3825.95	7.986
470	0.2882	3416.37	3848.67	8.010
480	0.2913	3434.46	3871.46	8.034
490	0.2945	3452.61	3894.30	8.058
500	0.2976	3470.83	3917.21	8.082
510	0.3007	3489.12	3940.18	8.106
520	0.3038	3507.47	3963.21	8.129
530	0.3070	3525.89	3986.31	8.152
540	0.3101	3544.37	4009.47	8.175
550	0.3132	3562.91	4032.69	8.198
560	0.3163	3581.53	4055.98	8.221
570	0.3194	3600.21	4079.34	8.243
580	0.3225	3618.95	4102.75	8.265
590	0.3256	3637.77	4126.23	8.288
600	0.3288	3656.64	4149.78	8.310
610	0.3319	3675.59	4173.39	8.332
620	0.3350	3694.60	4197.07	8.354
630	0.3381	3713.68	4220.81	8.375
640	0.3412	3732.82	4244.61	8.397
650	0.3443	3752.03	4268.48	8.418
660	0.3474	3771.31	4292.42	8.439
670	0.3505	3790.65	4316.42	8.460
680	0.3536	3810.06	4340.48	8.481
690	0.3567	3829.53	4364.61	8.502
700	0.3598	3849.07	4388.81	8.523
710	0.3629	3868.67	4413.06	8.544
720	0.3660	3888.34	4437.39	8.564
730	0.3691	3908.08	4461.77	8.585
740	0.3722	3927.88	4486.22	8.605
750	0.3753	3947.74	4510.74	8.625
760	0.3784	3967.67	4535.31	8.645
770	0.3815	3987.66	4559.95	8.665
780	0.3846	4007.72	4584.66	8.685
790	0.3877	4027.84	4609.42	8.704
800	0.3908	4048.02	4634.25	8.724
810	0.3939	4068.27	4659.14	8.744
820	0.3970	4088.57	4684.09	8.763
830	0.4001	4108.95	4709.11	8.782
840	0.4032	4129.38	4734.18	8.801
850	0.4063	4149.87	4759.32	8.821
860	0.4094	4170.43	4784.52	8.840
870	0.4125	4191.04	4809.77	8.858

t (°C)	v $\left(\frac{m^3}{kg}\right)$	u $\left(\frac{kJ}{kg}\right)$	h $\left(\frac{kJ}{kg}\right)$	s $\left(\frac{kJ}{kg\,K}\right)$
Sat Vapour (g)	*0.1317*	*2593.40*	*2790.97*	*6.443*
880	0.4156	4211.72	4835.09	8.877
890	0.4187	4232.46	4860.47	8.896
900	0.4218	4253.26	4885.91	8.915
910	0.4249	4274.11	4911.40	8.933
920	0.4280	4295.03	4936.96	8.952
930	0.4310	4316.01	4962.57	8.970
940	0.4341	4337.04	4988.24	8.988
950	0.4372	4358.13	5013.97	9.006
960	0.4403	4379.28	5039.75	9.024
970	0.4434	4400.49	5065.60	9.042
980	0.4465	4421.75	5091.49	9.060
990	0.4496	4443.07	5117.45	9.078
1000	0.4527	4464.45	5143.46	9.096
1010	0.4558	4485.88	5169.53	9.113
1020	0.4589	4507.37	5195.65	9.131
1030	0.4619	4528.91	5221.82	9.149
1040	0.4650	4550.51	5248.05	9.166
1050	0.4681	4572.16	5274.33	9.183
1060	0.4712	4593.86	5300.67	9.201
1070	0.4743	4615.62	5327.06	9.218
1080	0.4774	4637.43	5353.50	9.235
1090	0.4805	4659.29	5380.00	9.252
1100	0.4836	4681.21	5406.55	9.269
1110	0.4866	4703.18	5433.14	9.286
1120	0.4897	4725.20	5459.79	9.302
1130	0.4928	4747.27	5486.49	9.319
1140	0.4959	4769.39	5513.24	9.336
1150	0.4990	4791.56	5540.04	9.352
1160	0.5021	4813.78	5566.89	9.369
1170	0.5052	4836.05	5593.79	9.385
1180	0.5082	4858.37	5620.74	9.402
1190	0.5113	4880.73	5647.73	9.418
1200	0.5144	4903.15	5674.77	9.434
1210	0.5175	4925.61	5701.86	9.450
1220	0.5206	4948.12	5729.00	9.466
1230	0.5237	4970.68	5756.19	9.483
1240	0.5268	4993.28	5783.42	9.498
1250	0.5298	5015.93	5810.70	9.514
1260	0.5329	5038.63	5838.02	9.530
1270	0.5360	5061.37	5865.39	9.546
1280	0.5391	5084.16	5892.80	9.562
1290	0.5422	5106.99	5920.26	9.577
1300	0.5453	5129.86	5947.76	9.593

Table 3(6) Superheated Steam Properties (Pressure Base) Pressure = 16.0 bar and saturation temperature, Tsat = 201.37°C

t (°C)	$v\left(\frac{m^3}{kg}\right)$	$u\left(\frac{kJ}{kg}\right)$	$h\left(\frac{kJ}{kg}\right)$	$s\left(\frac{kJ}{kg\,K}\right)$	*t (°C)*	$v\left(\frac{m^3}{kg}\right)$	$u\left(\frac{kJ}{kg}\right)$	$h\left(\frac{kJ}{kg}\right)$	$s\left(\frac{kJ}{kg\,K}\right)$	*t (°C)*	$v\left(\frac{m^3}{kg}\right)$	$u\left(\frac{kJ}{kg}\right)$	$h\left(\frac{kJ}{kg}\right)$	$s\left(\frac{kJ}{kg\,K}\right)$
Sat Vapour (g)	*0.1237*	*2594.84*	*2792.83*	*6.420*	*Sat Vapour (g)*	*0.1237*	*2594.84*	*2792.83*	*6.420*	*Sat Vapour (g)*	*0.1237*	*2594.84*	*2792.83*	*6.420*
10	0.1278	2617.13	2821.54	6.480	450	0.2651	3385.39	3809.59	7.939	880	0.3905	4217.89	4842.67	8.853
20	0.1316	2637.86	2848.39	6.535	460	0.2681	3403.38	3832.29	7.963	890	0.3934	4238.65	4868.07	8.872
30	0.1353	2657.65	2874.09	6.586	470	0.2710	3421.44	3855.05	7.987	900	0.3963	4259.47	4893.53	8.890
40	0.1389	2676.78	2898.96	6.635	480	0.2739	3439.57	3877.87	8.011	910	0.3992	4280.35	4919.05	8.909
50	0.1424	2695.40	2923.20	6.682	490	0.2769	3457.75	3900.75	8.035	920	0.4021	4301.28	4944.62	8.927
60	0.1458	2713.62	2946.93	6.726	500	0.2798	3476.01	3923.70	8.059	930	0.4050	4322.28	4970.26	8.946
70	0.1492	2731.53	2970.26	6.770	510	0.2827	3494.32	3946.70	8.082	940	0.4079	4343.33	4995.95	8.964
80	0.1525	2749.19	2993.26	6.812	520	0.2857	3512.70	3969.77	8.106	950	0.4108	4364.45	5021.70	8.982
90	0.1558	2766.65	3016.00	6.852	530	0.2886	3531.15	3992.90	8.129	960	0.4137	4385.62	5047.51	9.000
100	0.1591	2783.94	3038.52	6.892	540	0.2915	3549.66	4016.09	8.152	970	0.4166	4406.84	5073.37	9.018
110	0.1623	2801.11	3060.85	6.930	550	0.2944	3568.24	4039.35	8.175	980	0.4195	4428.13	5099.29	9.036
120	0.1655	2818.17	3083.04	6.968	560	0.2974	3586.88	4062.67	8.197	990	0.4224	4449.47	5125.27	9.054
130	0.1687	2835.16	3105.12	7.005	570	0.3003	3605.59	4086.06	8.220	1000	0.4253	4470.86	5151.30	9.071
140	0.1719	2852.08	3127.09	7.041	580	0.3032	3624.36	4109.51	8.242	1010	0.4282	4492.32	5177.39	9.089
150	0.1750	2868.96	3149.00	7.076	590	0.3061	3643.20	4133.02	8.264	1020	0.4311	4513.82	5203.53	9.107
160	0.1781	2885.80	3170.84	7.111	600	0.3091	3662.11	4156.60	8.286	1030	0.4340	4535.39	5229.72	9.124
170	0.1813	2902.63	3192.63	7.145	610	0.3120	3681.08	4180.24	8.308	1040	0.4369	4557.00	5255.97	9.141
180	0.1843	2919.44	3214.39	7.178	620	0.3149	3700.12	4203.94	8.330	1050	0.4398	4578.67	5282.28	9.159
190	0.1874	2936.25	3236.12	7.211	630	0.3178	3719.23	4227.71	8.352	1060	0.4427	4600.39	5308.63	9.176
200	0.1905	2953.06	3257.84	7.244	640	0.3207	3738.40	4251.55	8.373	1070	0.4455	4622.17	5335.04	9.193
210	0.1935	2969.88	3279.56	7.276	650	0.3236	3757.63	4275.45	8.395	1080	0.4484	4644.00	5361.51	9.210
220	0.1966	2986.72	3301.27	7.307	660	0.3265	3776.94	4299.41	8.416	1090	0.4513	4665.88	5388.02	9.227
230	0.1996	3003.58	3322.99	7.338	670	0.3295	3796.31	4323.44	8.437	1100	0.4542	4687.82	5414.59	9.244
240	0.2027	3020.46	3344.73	7.369	680	0.3324	3815.74	4347.54	8.458	1110	0.4571	4709.80	5441.20	9.261
250	0.2057	3037.38	3366.48	7.399	690	0.3353	3835.24	4371.69	8.479	1120	0.4600	4731.84	5467.87	9.278
260	0.2087	3054.32	3388.25	7.429	700	0.3382	3854.80	4395.92	8.499	1130	0.4629	4753.93	5494.59	9.294
270	0.2117	3071.30	3410.04	7.458	710	0.3411	3874.43	4420.20	8.520	1140	0.4658	4776.06	5521.36	9.311
280	0.2147	3088.32	3431.87	7.488	720	0.3440	3894.13	4444.55	8.540	1150	0.4687	4798.25	5548.17	9.328
290	0.2177	3105.39	3453.73	7.516	730	0.3469	3913.89	4468.97	8.561	1160	0.4716	4820.49	5575.04	9.344
300	0.2207	3122.49	3475.62	7.545	740	0.3498	3933.71	4493.44	8.581	1170	0.4745	4842.78	5601.96	9.361
310	0.2237	3139.64	3497.55	7.573	750	0.3527	3953.60	4517.98	8.601	1180	0.4774	4865.11	5628.92	9.377
320	0.2267	3156.84	3519.52	7.601	760	0.3556	3973.56	4542.59	8.621	1190	0.4803	4887.50	5655.94	9.393
330	0.2297	3174.09	3541.53	7.628	770	0.3586	3993.57	4567.25	8.641	1200	0.4832	4909.93	5683.00	9.409
340	0.2326	3191.39	3563.59	7.656	780	0.3615	4013.65	4591.98	8.661	1210	0.4861	4932.41	5710.10	9.425
350	0.2356	3208.74	3585.69	7.683	790	0.3644	4033.80	4616.77	8.681	1220	0.4890	4954.93	5737.26	9.442
360	0.2386	3226.15	3607.84	7.709	800	0.3673	4054.00	4641.63	8.700	1230	0.4918	4977.51	5764.46	9.458
370	0.2415	3243.61	3630.04	7.736	810	0.3702	4074.27	4666.54	8.720	1240	0.4947	5000.13	5791.71	9.474
380	0.2445	3261.12	3652.30	7.762	820	0.3731	4094.60	4691.52	8.739	1250	0.4976	5022.79	5819.00	9.489
390	0.2474	3278.70	3674.60	7.788	830	0.3760	4115.00	4716.56	8.758	1260	0.5005	5045.51	5846.34	9.505
400	0.2504	3296.33	3696.96	7.814	840	0.3789	4135.45	4741.66	8.777	1270	0.5034	5068.26	5873.73	9.521
410	0.2533	3314.02	3719.37	7.839	850	0.3818	4155.97	4766.82	8.796	1280	0.5063	5091.06	5901.15	9.537
420	0.2563	3331.77	3741.84	7.864	860	0.3847	4176.55	4792.04	8.815	1290	0.5092	5113.91	5928.63	9.552
430	0.2592	3349.58	3764.36	7.889	870	0.3876	4197.19	4817.32	8.834	1300	0.5121	5136.80	5956.14	9.568
440	0.2622	3367.45	3786.95	7.914										

Table 3(7) Superheated Steam Properties (Pressure Base) Pressure =17.0 bar and saturation temperature, Tsat = 204.306°C

t (°C)	v $\left(\frac{m^3}{kg}\right)$	u $\left(\frac{kJ}{kg}\right)$	h $\left(\frac{kJ}{kg}\right)$	s $\left(\frac{kJ}{kg\,K}\right)$	t (°C)	v $\left(\frac{m^3}{kg}\right)$	u $\left(\frac{kJ}{kg}\right)$	h $\left(\frac{kJ}{kg}\right)$	s $\left(\frac{kJ}{kg\,K}\right)$	t (°C)	v $\left(\frac{m^3}{kg}\right)$	u $\left(\frac{kJ}{kg}\right)$	h $\left(\frac{kJ}{kg}\right)$	s $\left(\frac{kJ}{kg\,K}\right)$
Sat Vapour (g)	0.1167	2596.12	2794.47	6.398	Sat Vapour (g)	0.1167	2596.12	2794.47	6.398	Sat Vapour (g)	0.1167	2596.12	2794.47	6.398
10	0.1205	2618.73	2823.58	6.459	450	0.2503	3390.15	3815.58	7.917	880	0.3683	4223.76	4849.88	8.830
20	0.1241	2639.72	2850.76	6.514	460	0.2530	3408.18	3838.31	7.942	890	0.3710	4244.54	4875.30	8.849
30	0.1277	2659.73	2876.74	6.565	470	0.2558	3426.27	3861.11	7.966	900	0.3738	4265.38	4900.78	8.868
40	0.1311	2679.04	2901.85	6.614	480	0.2586	3444.42	3883.97	7.990	910	0.3765	4286.28	4926.32	8.886
50	0.1344	2697.83	2926.29	6.661	490	0.2613	3462.64	3906.88	8.014	920	0.3792	4307.23	4951.92	8.905
60	0.1377	2716.20	2950.22	6.706	500	0.2641	3480.92	3929.86	8.037	930	0.3820	4328.25	4977.58	8.923
70	0.1409	2734.24	2973.72	6.749	510	0.2668	3499.27	3952.90	8.061	940	0.3847	4349.33	5003.29	8.941
80	0.1440	2752.02	2996.87	6.791	520	0.2696	3517.68	3976.00	8.084	950	0.3874	4370.46	5029.06	8.959
90	0.1472	2769.58	3019.75	6.832	530	0.2724	3536.15	3999.16	8.107	960	0.3901	4391.65	5054.89	8.977
100	0.1502	2786.98	3042.39	6.872	540	0.2751	3554.69	4022.39	8.130	970	0.3929	4412.90	5080.78	8.995
110	0.1533	2804.24	3064.85	6.910	550	0.2779	3573.30	4045.68	8.153	980	0.3956	4434.20	5106.72	9.013
120	0.1563	2821.39	3087.15	6.948	560	0.2806	3591.97	4069.03	8.176	990	0.3983	4455.56	5132.71	9.031
130	0.1593	2838.46	3109.33	6.985	570	0.2834	3610.70	4092.45	8.198	1000	0.4011	4476.97	5158.76	9.048
140	0.1623	2855.46	3131.40	7.021	580	0.2861	3629.51	4115.93	8.220	1010	0.4038	4498.44	5184.87	9.066
150	0.1653	2872.41	3153.39	7.056	590	0.2889	3648.37	4139.47	8.243	1020	0.4065	4519.97	5211.03	9.084
160	0.1682	2889.32	3175.32	7.091	600	0.2916	3667.31	4163.08	8.265	1030	0.4092	4541.55	5237.25	9.101
170	0.1712	2906.21	3197.19	7.125	610	0.2944	3686.31	4186.75	8.286	1040	0.4120	4563.18	5263.52	9.118
180	0.1741	2923.08	3219.03	7.158	620	0.2971	3705.37	4210.48	8.308	1050	0.4147	4584.87	5289.84	9.136
190	0.1770	2939.95	3240.84	7.191	630	0.2999	3724.50	4234.28	8.330	1060	0.4174	4606.61	5316.22	9.153
200	0.1799	2956.82	3262.63	7.224	640	0.3026	3743.70	4258.15	8.351	1070	0.4201	4628.41	5342.64	9.170
210	0.1828	2973.69	3284.41	7.255	650	0.3054	3762.96	4282.08	8.373	1080	0.4229	4650.26	5369.13	9.187
220	0.1857	2990.59	3306.19	7.287	660	0.3081	3782.29	4306.07	8.394	1090	0.4256	4672.16	5395.66	9.204
230	0.1885	3007.49	3327.97	7.318	670	0.3108	3801.69	4330.13	8.415	1100	0.4283	4694.11	5422.24	9.221
240	0.1914	3024.43	3349.77	7.348	680	0.3136	3821.15	4354.25	8.436	1110	0.4310	4716.11	5448.88	9.238
250	0.1942	3041.39	3371.58	7.379	690	0.3163	3840.67	4378.43	8.457	1120	0.4338	4738.16	5475.56	9.255
260	0.1971	3058.38	3393.40	7.408	700	0.3191	3860.26	4402.68	8.477	1130	0.4365	4760.27	5502.30	9.271
270	0.1999	3075.41	3415.25	7.438	710	0.3218	3879.92	4426.99	8.498	1140	0.4392	4782.42	5529.08	9.288
280	0.2027	3092.47	3437.13	7.467	720	0.3246	3899.64	4451.37	8.518	1150	0.4419	4804.63	5555.92	9.304
290	0.2056	3109.58	3459.04	7.496	730	0.3273	3919.42	4475.81	8.539	1160	0.4447	4826.88	5582.80	9.321
300	0.2084	3126.72	3480.98	7.524	740	0.3300	3939.27	4500.31	8.559	1170	0.4474	4849.18	5609.74	9.337
310	0.2112	3143.91	3502.96	7.552	750	0.3328	3959.18	4524.88	8.579	1180	0.4501	4871.54	5636.72	9.354
320	0.2140	3161.15	3524.98	7.580	760	0.3355	3979.16	4549.51	8.599	1190	0.4528	4893.94	5663.75	9.370
330	0.2168	3178.44	3547.04	7.607	770	0.3382	3999.20	4574.20	8.619	1200	0.4556	4916.38	5690.83	9.386
340	0.2196	3195.77	3569.14	7.635	780	0.3410	4019.30	4598.96	8.638	1210	0.4583	4938.88	5717.95	9.402
350	0.2224	3213.16	3591.28	7.662	790	0.3437	4039.47	4623.77	8.658	1220	0.4610	4961.42	5745.12	9.418
360	0.2252	3230.60	3613.48	7.688	800	0.3464	4059.70	4648.65	8.678	1230	0.4637	4984.01	5772.34	9.434
370	0.2280	3248.10	3635.72	7.715	810	0.3492	4079.99	4673.59	8.697	1240	0.4664	5006.64	5799.60	9.450
380	0.2308	3265.65	3658.02	7.741	820	0.3519	4100.34	4698.59	8.716	1250	0.4692	5029.33	5826.91	9.466
390	0.2336	3283.26	3680.36	7.767	830	0.3546	4120.76	4723.65	8.736	1260	0.4719	5052.05	5854.27	9.482
400	0.2364	3300.93	3702.76	7.792	840	0.3574	4141.24	4748.78	8.755	1270	0.4746	5074.82	5881.67	9.497
410	0.2392	3318.65	3725.21	7.818	850	0.3601	4161.78	4773.96	8.774	1280	0.4773	5097.64	5909.11	9.513
420	0.2419	3336.43	3747.72	7.843	860	0.3628	4182.38	4799.21	8.793	1290	0.4801	5120.50	5936.60	9.529
430	0.2447	3354.28	3770.28	7.868	870	0.3656	4203.04	4824.51	8.812	1300	0.4828	5143.41	5964.13	9.544
440	0.2475	3372.18	3792.90	7.893										

Table 3(8) Superheated Steam Properties (Pressure Base) Pressure = 18.0 bar and saturation temperature, Tsat = 207.111°C

t (°C)	v $\left(\frac{m^3}{kg}\right)$	u $\left(\frac{kJ}{kg}\right)$	h $\left(\frac{kJ}{kg}\right)$	s $\left(\frac{kJ}{kg\,K}\right)$
Sat Vapour (g)	*0.1104*	***2597.25***	*2795.92*	***6.378***
10	0.1140	2620.18	2825.43	6.438
20	0.1175	2641.42	2852.93	6.494
30	0.1209	2661.64	2879.17	6.546
40	0.1241	2681.14	2904.52	6.595
50	0.1273	2700.09	2929.17	6.642
60	0.1304	2718.60	2953.28	6.687
70	0.1334	2736.78	2976.95	6.730
80	0.1364	2754.67	3000.25	6.772
90	0.1394	2772.35	3023.26	6.813
100	0.1423	2789.84	3046.03	6.853
110	0.1452	2807.19	3068.61	6.891
120	0.1481	2824.43	3091.02	6.929
130	0.1510	2841.58	3113.30	6.966
140	0.1538	2858.65	3135.47	7.002
150	0.1566	2875.67	3157.55	7.037
160	0.1594	2892.65	3179.56	7.072
170	0.1622	2909.60	3201.51	7.106
180	0.1649	2926.53	3223.42	7.139
190	0.1677	2943.46	3245.31	7.172
200	0.1704	2960.38	3267.17	7.204
210	0.1732	2977.31	3289.01	7.236
220	0.1759	2994.25	3310.86	7.268
230	0.1786	3011.21	3332.70	7.299
240	0.1813	3028.19	3354.55	7.329
250	0.1840	3045.20	3376.42	7.359
260	0.1867	3062.24	3398.30	7.389
270	0.1894	3079.31	3420.20	7.419
280	0.1921	3096.41	3442.13	7.448
290	0.1947	3113.56	3464.09	7.476
300	0.1974	3130.74	3486.08	7.505
310	0.2001	3147.97	3508.11	7.533
320	0.2027	3165.25	3530.17	7.560
330	0.2054	3182.57	3552.27	7.588
340	0.2080	3199.95	3574.42	7.615
350	0.2107	3217.37	3596.61	7.642
360	0.2133	3234.85	3618.84	7.669
370	0.2160	3252.38	3641.13	7.695
380	0.2186	3269.97	3663.46	7.721
390	0.2212	3287.61	3685.85	7.747
400	0.2239	3305.31	3708.28	7.773
410	0.2265	3323.06	3730.77	7.798
420	0.2291	3340.88	3753.32	7.823
430	0.2318	3358.75	3775.92	7.848
440	0.2344	3376.69	3798.57	7.873

t (°C)	v $\left(\frac{m^3}{kg}\right)$	u $\left(\frac{kJ}{kg}\right)$	h $\left(\frac{kJ}{kg}\right)$	s $\left(\frac{kJ}{kg\,K}\right)$
Sat Vapour (g)	*0.1104*	***2597.25***	*2795.92*	***6.378***
450	0.2370	3394.69	3821.28	7.897
460	0.2396	3412.75	3844.06	7.922
470	0.2422	3430.87	3866.89	7.946
480	0.2448	3449.05	3889.78	7.970
490	0.2475	3467.30	3912.73	7.994
500	0.2501	3485.61	3935.74	8.017
510	0.2527	3503.98	3958.81	8.041
520	0.2553	3522.42	3981.94	8.064
530	0.2579	3540.92	4005.14	8.087
540	0.2605	3559.49	4028.39	8.110
550	0.2631	3578.12	4051.71	8.133
560	0.2657	3596.82	4075.10	8.155
570	0.2683	3615.59	4098.54	8.178
580	0.2709	3634.41	4122.05	8.200
590	0.2735	3653.31	4145.62	8.222
600	0.2761	3672.27	4169.26	8.244
610	0.2787	3691.29	4192.96	8.266
620	0.2813	3710.38	4216.72	8.288
630	0.2839	3729.54	4240.55	8.309
640	0.2865	3748.76	4264.44	8.331
650	0.2891	3768.05	4288.40	8.352
660	0.2917	3787.40	4312.42	8.373
670	0.2943	3806.82	4336.50	8.394
680	0.2969	3826.31	4360.65	8.415
690	0.2994	3845.86	4384.86	8.436
700	0.3020	3865.47	4409.14	8.456
710	0.3046	3885.15	4433.48	8.477
720	0.3072	3904.89	4457.88	8.497
730	0.3098	3924.70	4482.34	8.518
740	0.3124	3944.57	4506.87	8.538
750	0.3150	3964.50	4531.46	8.558
760	0.3176	3984.50	4556.12	8.578
770	0.3201	4004.57	4580.83	8.598
780	0.3227	4024.69	4605.61	8.617
790	0.3253	4044.88	4630.45	8.637
800	0.3279	4065.13	4655.35	8.657
810	0.3305	4085.45	4680.32	8.676
820	0.3331	4105.82	4705.34	8.695
830	0.3357	4126.26	4730.43	8.715
840	0.3382	4146.76	4755.57	8.734
850	0.3408	4167.32	4780.78	8.753
860	0.3434	4187.94	4806.05	8.772
870	0.3460	4208.62	4831.37	8.790

t (°C)	v $\left(\frac{m^3}{kg}\right)$	u $\left(\frac{kJ}{kg}\right)$	h $\left(\frac{kJ}{kg}\right)$	s $\left(\frac{kJ}{kg\,K}\right)$
Sat Vapour (g)	*0.1104*	***2597.25***	*2795.92*	***6.378***
880	0.3486	4229.36	4856.76	8.809
890	0.3511	4250.16	4882.20	8.828
900	0.3537	4271.02	4907.71	8.846
910	0.3563	4291.94	4933.27	8.865
920	0.3589	4312.92	4958.89	8.883
930	0.3615	4333.95	4984.57	8.901
940	0.3640	4355.05	5010.30	8.920
950	0.3666	4376.20	5036.09	8.938
960	0.3692	4397.41	5061.94	8.956
970	0.3718	4418.67	5087.85	8.974
980	0.3743	4440.00	5113.81	8.991
990	0.3769	4461.37	5139.82	9.009
1000	0.3795	4482.81	5165.89	9.027
1010	0.3821	4504.30	5192.02	9.044
1020	0.3846	4525.84	5218.20	9.062
1030	0.3872	4547.44	5244.44	9.079
1040	0.3898	4569.09	5270.72	9.097
1050	0.3924	4590.80	5297.06	9.114
1060	0.3949	4612.55	5323.46	9.131
1070	0.3975	4634.37	5349.91	9.148
1080	0.4001	4656.23	5376.40	9.165
1090	0.4027	4678.15	5402.95	9.182
1100	0.4052	4700.11	5429.56	9.199
1110	0.4078	4722.13	5456.21	9.216
1120	0.4104	4744.20	5482.91	9.233
1130	0.4130	4766.32	5509.66	9.249
1140	0.4155	4788.50	5536.47	9.266
1150	0.4181	4810.72	5563.32	9.283
1160	0.4207	4832.99	5590.22	9.299
1170	0.4233	4855.30	5617.17	9.315
1180	0.4258	4877.67	5644.17	9.332
1190	0.4284	4900.09	5671.21	9.348
1200	0.4310	4922.55	5698.31	9.364
1210	0.4335	4945.06	5725.45	9.380
1220	0.4361	4967.62	5752.63	9.396
1230	0.4387	4990.22	5779.87	9.412
1240	0.4413	5012.87	5807.15	9.428
1250	0.4438	5035.56	5834.47	9.444
1260	0.4464	5058.31	5861.84	9.460
1270	0.4490	5081.09	5889.25	9.475
1280	0.4516	5103.92	5916.71	9.491
1290	0.4541	5126.80	5944.22	9.507
1300	0.4567	5149.72	5971.76	9.522

Table 3(9) Superheated Steam Properties (Pressure Base) Pressure = 19.0 bar and saturation temperature, Tsat = 209.798°C

t (°C)	$v\left(\frac{m^3}{kg}\right)$	$u\left(\frac{kJ}{kg}\right)$	$h\left(\frac{kJ}{kg}\right)$	$s\left(\frac{kJ}{kg\,K}\right)$	t (°C)	$v\left(\frac{m^3}{kg}\right)$	$u\left(\frac{kJ}{kg}\right)$	$h\left(\frac{kJ}{kg}\right)$	$s\left(\frac{kJ}{kg\,K}\right)$	t (°C)	$v\left(\frac{m^3}{kg}\right)$	$u\left(\frac{kJ}{kg}\right)$	$h\left(\frac{kJ}{kg}\right)$	$s\left(\frac{kJ}{kg\,K}\right)$
Sat Vapour (g)	0.1047	2598.25	2797.19	6.358	Sat Vapour (g)	0.1047	2598.25	2797.19	6.358	Sat Vapour (g)	0.1047	2598.25	2797.19	6.358
10	0.1082	2621.49	2827.09	6.419	450	0.2251	3399.02	3826.74	7.878	880	0.3309	4234.72	4863.35	8.789
20	0.1115	2642.98	2854.91	6.475	460	0.2276	3417.11	3849.54	7.903	890	0.3333	4255.54	4888.81	8.808
30	0.1147	2663.42	2881.42	6.527	470	0.2301	3435.26	3872.40	7.927	900	0.3357	4276.42	4914.34	8.826
40	0.1178	2683.10	2907.00	6.577	480	0.2326	3453.47	3895.33	7.951	910	0.3382	4297.36	4939.92	8.845
50	0.1209	2702.20	2931.86	6.624	490	0.2350	3471.75	3918.31	7.975	920	0.3406	4318.36	4965.56	8.863
60	0.1238	2720.86	2956.15	6.669	500	0.2375	3490.09	3941.35	7.998	930	0.3431	4339.41	4991.26	8.881
70	0.1267	2739.16	2979.97	6.712	510	0.2400	3508.49	3964.45	8.021	940	0.3455	4360.53	5017.01	8.899
80	0.1296	2757.17	3003.43	6.754	520	0.2425	3526.95	3987.62	8.045	950	0.3480	4381.70	5042.82	8.917
90	0.1324	2774.95	3026.57	6.795	530	0.2449	3545.48	4010.84	8.068	960	0.3504	4402.92	5068.69	8.935
100	0.1352	2792.54	3049.47	6.835	540	0.2474	3564.08	4034.13	8.091	970	0.3528	4424.21	5094.62	8.953
110	0.1380	2809.98	3072.16	6.873	550	0.2499	3582.74	4057.48	8.113	980	0.3553	4445.55	5120.60	8.971
120	0.1407	2827.30	3094.67	6.911	560	0.2523	3601.46	4080.89	8.136	990	0.3577	4466.94	5146.63	8.989
130	0.1434	2844.53	3117.05	6.948	570	0.2548	3620.25	4104.37	8.158	1000	0.3602	4488.39	5172.72	9.007
140	0.1461	2861.67	3139.31	6.984	580	0.2573	3639.11	4127.91	8.181	1010	0.3626	4509.90	5198.87	9.024
150	0.1488	2878.76	3161.48	7.019	590	0.2597	3658.03	4151.51	8.203	1020	0.3651	4531.46	5225.06	9.042
160	0.1515	2895.81	3183.58	7.054	600	0.2622	3677.01	4175.17	8.225	1030	0.3675	4553.07	5251.32	9.059
170	0.1541	2912.82	3205.61	7.088	610	0.2647	3696.06	4198.90	8.246	1040	0.3699	4574.74	5277.62	9.076
180	0.1567	2929.81	3227.60	7.121	620	0.2671	3715.18	4222.69	8.268	1050	0.3724	4596.47	5303.98	9.094
190	0.1593	2946.80	3249.55	7.154	630	0.2696	3734.36	4246.55	8.290	1060	0.3748	4618.24	5330.39	9.111
200	0.1619	2963.77	3271.48	7.186	640	0.2720	3753.61	4270.47	8.311	1070	0.3773	4640.07	5356.86	9.128
210	0.1645	2980.75	3293.39	7.218	650	0.2745	3772.92	4294.45	8.332	1080	0.3797	4661.95	5383.37	9.145
220	0.1671	2997.75	3315.29	7.250	660	0.2769	3792.29	4318.50	8.353	1090	0.3821	4683.88	5409.94	9.162
230	0.1697	3014.75	3337.20	7.281	670	0.2794	3811.74	4342.61	8.374	1100	0.3846	4705.87	5436.56	9.179
240	0.1723	3031.78	3359.10	7.311	680	0.2819	3831.24	4366.78	8.395	1110	0.3870	4727.90	5463.23	9.195
250	0.1748	3048.83	3381.02	7.341	690	0.2843	3850.82	4391.02	8.416	1120	0.3895	4749.99	5489.95	9.212
260	0.1774	3065.91	3402.96	7.371	700	0.2868	3870.45	4415.32	8.437	1130	0.3919	4772.12	5516.72	9.229
270	0.1799	3083.02	3424.91	7.400	710	0.2892	3890.15	4439.68	8.457	1140	0.3943	4794.31	5543.54	9.245
280	0.1825	3100.17	3446.89	7.429	720	0.2917	3909.92	4464.11	8.478	1150	0.3968	4816.54	5570.40	9.262
290	0.1850	3117.35	3468.90	7.458	730	0.2941	3929.75	4488.60	8.498	1160	0.3992	4838.83	5597.32	9.278
300	0.1876	3134.58	3490.94	7.486	740	0.2966	3949.64	4513.15	8.518	1170	0.4016	4861.16	5624.29	9.295
310	0.1901	3151.85	3513.01	7.514	750	0.2990	3969.60	4537.76	8.538	1180	0.4041	4883.54	5651.30	9.311
320	0.1926	3169.16	3535.12	7.542	760	0.3015	3989.62	4562.44	8.558	1190	0.4065	4905.97	5678.36	9.327
330	0.1951	3186.52	3557.27	7.569	770	0.3039	4009.70	4587.18	8.578	1200	0.4090	4928.45	5705.47	9.343
340	0.1976	3203.93	3579.45	7.596	780	0.3064	4029.85	4611.98	8.598	1210	0.4114	4950.98	5732.62	9.359
350	0.2002	3221.39	3601.69	7.623	790	0.3088	4050.06	4636.85	8.617	1220	0.4138	4973.55	5759.83	9.376
360	0.2027	3238.90	3623.96	7.650	800	0.3113	4070.33	4661.77	8.637	1230	0.4163	4996.17	5787.07	9.391
370	0.2052	3256.47	3646.29	7.676	810	0.3137	4090.67	4686.76	8.656	1240	0.4187	5018.83	5814.37	9.407
380	0.2077	3274.08	3668.66	7.702	820	0.3162	4111.06	4711.80	8.675	1250	0.4211	5041.54	5841.71	9.423
390	0.2102	3291.76	3691.08	7.728	830	0.3186	4131.52	4736.91	8.695	1260	0.4236	5064.29	5869.09	9.439
400	0.2127	3309.49	3713.55	7.754	840	0.3211	4152.04	4762.08	8.714	1270	0.4260	5087.09	5896.52	9.455
410	0.2152	3327.28	3736.08	7.779	850	0.3235	4172.62	4787.31	8.733	1280	0.4285	5109.94	5923.99	9.470
420	0.2177	3345.12	3758.66	7.804	860	0.3260	4193.26	4812.60	8.752	1290	0.4309	5132.83	5951.51	9.486
430	0.2201	3363.03	3781.30	7.829	870	0.3284	4213.96	4837.94	8.770	1300	0.4333	5155.76	5979.07	9.501
440	0.2226	3381.00	3803.99	7.854										

Table 3(10) Superheated Steam Properties (Pressure Base) Pressure = 20.0 bar and saturation temperature, Tsat = 212.377°C

t (°C)	v $\left(\frac{m^3}{kg}\right)$	u $\left(\frac{kJ}{kg}\right)$	h $\left(\frac{kJ}{kg}\right)$	s $\left(\frac{kJ}{kg\,K}\right)$	t (°C)	v $\left(\frac{m^3}{kg}\right)$	u $\left(\frac{kJ}{kg}\right)$	h $\left(\frac{kJ}{kg}\right)$	s $\left(\frac{kJ}{kg\,K}\right)$	t (°C)	v $\left(\frac{m^3}{kg}\right)$	u $\left(\frac{kJ}{kg}\right)$	h $\left(\frac{kJ}{kg}\right)$	s $\left(\frac{kJ}{kg\,K}\right)$
Sat Vapour (g)	0.0996	2599.13	2798.31	6.339	Sat Vapour (g)	0.0996	2599.13	2798.31	6.339	Sat Vapour (g)	0.0996	2599.13	2798.31	6.339
10	0.1030	2622.68	2828.60	6.401	450	0.2144	3403.17	3831.96	7.861	880	0.3149	4239.87	4869.67	8.770
20	0.1062	2644.42	2856.72	6.457	460	0.2168	3421.29	3854.79	7.885	890	0.3172	4260.71	4895.16	8.789
30	0.1092	2665.06	2883.50	6.509	470	0.2191	3439.47	3877.69	7.909	900	0.3195	4281.61	4920.70	8.807
40	0.1122	2684.92	2909.31	6.559	480	0.2215	3457.71	3900.64	7.933	910	0.3219	4302.56	4946.30	8.826
50	0.1151	2704.18	2934.37	6.606	490	0.2238	3476.01	3923.66	7.957	920	0.3242	4323.58	4971.96	8.844
60	0.1179	2722.98	2958.83	6.652	500	0.2262	3494.38	3946.73	7.980	930	0.3265	4344.65	4997.68	8.862
70	0.1207	2741.40	2982.82	6.695	510	0.2285	3512.81	3969.86	8.003	940	0.3288	4365.78	5023.45	8.880
80	0.1234	2759.53	3006.42	6.737	520	0.2309	3531.30	3993.06	8.027	950	0.3312	4386.97	5049.28	8.898
90	0.1261	2777.41	3029.70	6.778	530	0.2332	3549.86	4016.31	8.050	960	0.3335	4408.21	5075.17	8.916
100	0.1288	2795.10	3052.72	6.818	540	0.2356	3568.48	4039.63	8.072	970	0.3358	4429.51	5101.11	8.934
110	0.1314	2812.63	3075.52	6.856	550	0.2379	3587.16	4063.01	8.095	980	0.3381	4450.87	5127.11	8.952
120	0.1341	2830.03	3098.14	6.894	560	0.2403	3605.91	4086.45	8.118	990	0.3404	4472.28	5153.16	8.970
130	0.1366	2847.33	3120.61	6.931	570	0.2426	3624.73	4109.95	8.140	1000	0.3428	4493.75	5179.27	8.987
140	0.1392	2864.55	3142.97	6.967	580	0.2450	3643.60	4133.52	8.162	1010	0.3451	4515.27	5205.43	9.005
150	0.1418	2881.70	3165.22	7.002	590	0.2473	3662.55	4157.15	8.184	1020	0.3474	4536.85	5231.65	9.022
160	0.1443	2898.81	3187.39	7.037	600	0.2496	3681.56	4180.84	8.206	1030	0.3497	4558.48	5257.92	9.040
170	0.1468	2915.89	3209.50	7.071	610	0.2520	3700.63	4204.60	8.228	1040	0.3520	4580.17	5284.24	9.057
180	0.1493	2932.94	3231.56	7.104	620	0.2543	3719.77	4228.41	8.250	1050	0.3544	4601.91	5310.62	9.074
190	0.1518	2949.97	3253.59	7.137	630	0.2567	3738.98	4252.30	8.271	1060	0.3567	4623.70	5337.05	9.091
200	0.1543	2967.00	3275.58	7.169	640	0.2590	3758.25	4276.24	8.293	1070	0.3590	4645.54	5363.53	9.108
210	0.1568	2984.03	3297.56	7.201	650	0.2613	3777.58	4300.25	8.314	1080	0.3613	4667.44	5390.06	9.125
220	0.1592	3001.07	3319.52	7.232	660	0.2637	3796.98	4324.32	8.335	1090	0.3636	4689.39	5416.65	9.142
230	0.1617	3018.13	3341.48	7.263	670	0.2660	3816.45	4348.46	8.356	1100	0.3659	4711.39	5443.28	9.159
240	0.1641	3035.20	3363.45	7.294	680	0.2683	3835.98	4372.65	8.377	1110	0.3683	4733.44	5469.97	9.176
250	0.1666	3052.30	3385.42	7.324	690	0.2707	3855.57	4396.92	8.397	1120	0.3706	4755.54	5496.70	9.193
260	0.1690	3069.42	3407.41	7.354	700	0.2730	3875.23	4421.24	8.418	1130	0.3729	4777.69	5523.49	9.209
270	0.1714	3086.57	3429.41	7.383	710	0.2753	3894.95	4445.63	8.439	1140	0.3752	4799.89	5550.32	9.226
280	0.1738	3103.76	3451.44	7.412	720	0.2777	3914.74	4470.08	8.459	1150	0.3775	4822.14	5577.20	9.242
290	0.1763	3120.98	3473.50	7.441	730	0.2800	3934.59	4494.59	8.479	1160	0.3798	4844.44	5604.14	9.259
300	0.1787	3138.25	3495.58	7.469	740	0.2823	3954.51	4519.17	8.499	1170	0.3822	4866.79	5631.12	9.275
310	0.1811	3155.55	3517.70	7.497	750	0.2847	3974.49	4543.81	8.519	1180	0.3845	4889.18	5658.15	9.291
320	0.1835	3172.90	3539.85	7.524	760	0.2870	3994.53	4568.51	8.539	1190	0.3868	4911.63	5685.22	9.308
330	0.1859	3190.30	3562.04	7.552	770	0.2893	4014.63	4593.27	8.559	1200	0.3891	4934.12	5712.35	9.324
340	0.1883	3207.74	3584.27	7.579	780	0.2916	4034.80	4618.09	8.579	1210	0.3914	4956.66	5739.52	9.340
350	0.1907	3225.23	3606.54	7.606	790	0.2940	4055.03	4642.98	8.598	1220	0.3937	4979.24	5766.73	9.356
360	0.1930	3242.78	3628.86	7.632	800	0.2963	4075.32	4667.93	8.618	1230	0.3961	5001.87	5793.99	9.372
370	0.1954	3260.37	3651.22	7.659	810	0.2986	4095.68	4692.93	8.637	1240	0.3984	5024.55	5821.30	9.388
380	0.1978	3278.02	3673.63	7.685	820	0.3010	4116.09	4718.00	8.656	1250	0.4007	5047.27	5848.66	9.403
390	0.2002	3295.73	3696.09	7.710	830	0.3033	4136.57	4743.13	8.676	1260	0.4030	5070.04	5876.05	9.419
400	0.2026	3313.49	3718.60	7.736	840	0.3056	4157.11	4768.32	8.695	1270	0.4053	5092.85	5903.50	9.435
410	0.2049	3331.31	3741.16	7.761	850	0.3079	4177.71	4793.57	8.714	1280	0.4076	5115.71	5930.98	9.451
420	0.2073	3349.19	3763.78	7.786	860	0.3103	4198.37	4818.88	8.733	1290	0.4100	5138.61	5958.51	9.466
430	0.2097	3367.12	3786.45	7.811	870	0.3126	4219.09	4844.25	8.751	1300	0.4123	5161.56	5986.09	9.482
440	0.2120	3385.12	3809.17	7.836										

Table 3(11) Superheated Steam Properties (Pressure Base) Pressure = 21.0 bar and saturation temperature, Tsat = 214.857°C

t (°C)	v $\left(\frac{m^3}{kg}\right)$	u $\left(\frac{kJ}{kg}\right)$	h $\left(\frac{kJ}{kg}\right)$	s $\left(\frac{kJ}{kg\,K}\right)$	t (°C)	v $\left(\frac{m^3}{kg}\right)$	u $\left(\frac{kJ}{kg}\right)$	h $\left(\frac{kJ}{kg}\right)$	s $\left(\frac{kJ}{kg\,K}\right)$	t (°C)	v $\left(\frac{m^3}{kg}\right)$	u $\left(\frac{kJ}{kg}\right)$	h $\left(\frac{kJ}{kg}\right)$	s $\left(\frac{kJ}{kg\,K}\right)$
Sat Vapour (g)	0.0949	2599.90	2799.28	6.321	Sat Vapour (g)	0.0949	2599.90	2799.28	6.321	Sat Vapour (g)	0.0949	2599.90	2799.28	6.321
10	0.0982	2623.75	2829.95	6.383	450	0.2047	3407.16	3836.96	7.844	880	0.3004	4244.82	4875.75	8.752
20	0.1013	2645.74	2858.39	6.440	460	0.2069	3425.30	3859.83	7.868	890	0.3027	4265.67	4901.26	8.771
30	0.1042	2666.58	2885.43	6.493	470	0.2092	3443.51	3882.76	7.892	900	0.3049	4286.59	4926.82	8.789
40	0.1071	2686.62	2911.46	6.542	480	0.2114	3461.78	3905.75	7.916	910	0.3071	4307.56	4952.44	8.807
50	0.1098	2706.03	2936.72	6.590	490	0.2137	3480.11	3928.79	7.939	920	0.3093	4328.60	4978.12	8.826
60	0.1126	2724.97	2961.36	6.635	500	0.2159	3498.50	3951.89	7.963	930	0.3115	4349.69	5003.86	8.844
70	0.1152	2743.52	2985.50	6.679	510	0.2181	3516.95	3975.06	7.986	940	0.3137	4370.83	5029.65	8.862
80	0.1179	2761.75	3009.24	6.721	520	0.2204	3535.47	3998.28	8.009	950	0.3159	4392.04	5055.50	8.880
90	0.1204	2779.74	3032.66	6.762	530	0.2226	3554.05	4021.57	8.032	960	0.3181	4413.30	5081.40	8.898
100	0.1230	2797.52	3055.79	6.801	540	0.2249	3572.70	4044.91	8.055	970	0.3204	4434.62	5107.36	8.916
110	0.1255	2815.14	3078.70	6.840	550	0.2271	3591.41	4068.32	8.078	980	0.3226	4455.99	5133.38	8.934
120	0.1280	2832.62	3101.43	6.878	560	0.2293	3610.18	4091.79	8.100	990	0.3248	4477.42	5159.45	8.951
130	0.1305	2849.99	3124.00	6.915	570	0.2316	3629.02	4115.32	8.123	1000	0.3270	4498.90	5185.57	8.969
140	0.1329	2867.28	3146.44	6.951	580	0.2338	3647.93	4138.91	8.145	1010	0.3292	4520.44	5211.75	8.987
150	0.1354	2884.51	3168.78	6.986	590	0.2360	3666.89	4162.57	8.167	1020	0.3314	4542.04	5237.99	9.004
160	0.1378	2901.68	3191.03	7.021	600	0.2383	3685.93	4186.29	8.189	1030	0.3336	4563.68	5264.27	9.021
170	0.1402	2918.81	3213.22	7.055	610	0.2405	3705.02	4210.07	8.211	1040	0.3358	4585.38	5290.61	9.039
180	0.1426	2935.92	3235.35	7.088	620	0.2427	3724.19	4233.91	8.232	1050	0.3380	4607.14	5317.01	9.056
190	0.1450	2953.01	3257.44	7.121	630	0.2450	3743.42	4257.82	8.254	1060	0.3402	4628.95	5343.45	9.073
200	0.1473	2970.09	3279.50	7.153	640	0.2472	3762.71	4281.79	8.275	1070	0.3424	4650.80	5369.95	9.090
210	0.1497	2987.17	3301.54	7.185	650	0.2494	3782.07	4305.82	8.296	1080	0.3447	4672.72	5396.50	9.107
220	0.1520	3004.26	3323.56	7.216	660	0.2516	3801.49	4329.92	8.317	1090	0.3469	4694.68	5423.09	9.124
230	0.1544	3021.36	3345.58	7.247	670	0.2539	3820.98	4354.08	8.338	1100	0.3491	4716.69	5449.75	9.141
240	0.1567	3038.47	3367.60	7.277	680	0.2561	3840.53	4378.30	8.359	1110	0.3513	4738.76	5476.45	9.158
250	0.1591	3055.61	3389.63	7.308	690	0.2583	3860.14	4402.59	8.380	1120	0.3535	4760.87	5503.20	9.174
260	0.1614	3072.77	3411.66	7.337	700	0.2605	3879.82	4426.93	8.400	1130	0.3557	4783.04	5530.00	9.191
270	0.1637	3089.97	3433.72	7.366	710	0.2628	3899.57	4451.35	8.421	1140	0.3579	4805.25	5556.85	9.207
280	0.1660	3107.19	3455.80	7.395	720	0.2650	3919.38	4475.82	8.441	1150	0.3601	4827.52	5583.75	9.224
290	0.1683	3124.46	3477.90	7.424	730	0.2672	3939.25	4500.36	8.461	1160	0.3623	4849.83	5610.69	9.240
300	0.1706	3141.76	3500.03	7.452	740	0.2694	3959.18	4524.95	8.482	1170	0.3645	4872.19	5637.69	9.257
310	0.1729	3159.10	3522.19	7.480	750	0.2716	3979.18	4549.61	8.502	1180	0.3667	4894.60	5664.73	9.273
320	0.1752	3176.48	3544.38	7.508	760	0.2739	3999.25	4574.34	8.521	1190	0.3689	4917.06	5691.82	9.289
330	0.1775	3193.91	3566.61	7.535	770	0.2761	4019.37	4599.12	8.541	1200	0.3711	4939.57	5718.96	9.305
340	0.1798	3211.39	3588.88	7.562	780	0.2783	4039.56	4623.97	8.561	1210	0.3733	4962.12	5746.14	9.321
350	0.1820	3228.92	3611.19	7.589	790	0.2805	4059.81	4648.87	8.580	1220	0.3756	4984.72	5773.37	9.337
360	0.1843	3246.49	3633.55	7.616	800	0.2827	4080.12	4673.84	8.600	1230	0.3778	5007.36	5800.65	9.353
370	0.1866	3264.12	3655.95	7.642	810	0.2849	4100.49	4698.87	8.619	1240	0.3800	5030.05	5827.97	9.369
380	0.1889	3281.80	3678.40	7.668	820	0.2872	4120.93	4723.96	8.638	1250	0.3822	5052.79	5855.34	9.385
390	0.1911	3299.54	3700.89	7.694	830	0.2894	4141.43	4749.11	8.658	1260	0.3844	5075.57	5882.75	9.401
400	0.1934	3317.33	3723.44	7.719	840	0.2916	4161.98	4774.32	8.677	1270	0.3866	5098.39	5910.21	9.416
410	0.1956	3335.18	3746.04	7.744	850	0.2938	4182.60	4799.59	8.696	1280	0.3888	5121.26	5937.71	9.432
420	0.1979	3353.09	3768.69	7.770	860	0.2960	4203.28	4824.92	8.715	1290	0.3910	5144.18	5965.25	9.447
430	0.2002	3371.05	3791.39	7.794	870	0.2982	4224.02	4850.31	8.733	1300	0.3932	5167.13	5992.84	9.463
440	0.2024	3389.08	3814.15	7.819										

Table 3(12) Superheated Steam Properties (Pressure Base) Pressure = 22.0 bar and saturation temperature, Tsat = 217.248°C

t (°C)	v $\left(\frac{m^3}{kg}\right)$	u $\left(\frac{kJ}{kg}\right)$	h $\left(\frac{kJ}{kg}\right)$	s $\left(\frac{kJ}{kg\,K}\right)$	t (°C)	v $\left(\frac{m^3}{kg}\right)$	u $\left(\frac{kJ}{kg}\right)$	h $\left(\frac{kJ}{kg}\right)$	s $\left(\frac{kJ}{kg\,K}\right)$	t (°C)	v $\left(\frac{m^3}{kg}\right)$	u $\left(\frac{kJ}{kg}\right)$	h $\left(\frac{kJ}{kg}\right)$	s $\left(\frac{kJ}{kg\,K}\right)$
Sat Vapour (g)	0.0907	2600.57	2800.12	6.304	Sat Vapour (g)	0.0907	2600.57	2800.12	6.304	Sat Vapour (g)	0.0907	2600.57	2800.12	6.304
10	0.0938	2624.72	2831.17	6.366	450	0.1958	3410.99	3841.78	7.827	880	0.2873	4249.58	4881.61	8.735
20	0.0968	2646.95	2859.91	6.423	460	0.1980	3429.16	3864.68	7.852	890	0.2894	4270.46	4907.13	8.753
30	0.0996	2668.00	2887.21	6.476	470	0.2001	3447.39	3887.64	7.876	900	0.2915	4291.39	4932.71	8.772
40	0.1024	2688.21	2913.46	6.526	480	0.2023	3465.69	3910.65	7.899	910	0.2936	4312.38	4958.35	8.790
50	0.1051	2707.78	2938.92	6.574	490	0.2044	3484.04	3933.73	7.923	920	0.2957	4333.43	4984.05	8.808
60	0.1077	2726.85	2963.73	6.619	500	0.2065	3502.46	3956.86	7.947	930	0.2978	4354.54	5009.80	8.827
70	0.1102	2745.52	2988.03	6.663	510	0.2087	3520.94	3980.05	7.970	940	0.3000	4375.70	5035.61	8.845
80	0.1128	2763.86	3011.92	6.705	520	0.2108	3539.48	4003.31	7.993	950	0.3021	4396.92	5061.48	8.863
90	0.1152	2781.95	3035.46	6.746	530	0.2130	3558.09	4026.62	8.016	960	0.3042	4418.20	5087.40	8.881
100	0.1177	2799.82	3058.71	6.786	540	0.2151	3576.76	4049.99	8.039	970	0.3063	4439.53	5113.38	8.899
110	0.1201	2817.52	3081.73	6.825	550	0.2172	3595.50	4073.43	8.061	980	0.3084	4460.92	5139.41	8.916
120	0.1225	2835.09	3104.56	6.862	560	0.2194	3614.29	4096.92	8.084	990	0.3105	4482.37	5165.50	8.934
130	0.1249	2852.53	3127.22	6.899	570	0.2215	3633.16	4120.48	8.106	1000	0.3126	4503.87	5191.64	8.952
140	0.1272	2869.89	3149.75	6.935	580	0.2236	3652.08	4144.10	8.128	1010	0.3147	4525.42	5217.84	8.969
150	0.1295	2887.18	3172.17	6.971	590	0.2258	3671.07	4167.78	8.150	1020	0.3168	4547.03	5244.09	8.987
160	0.1319	2904.41	3194.51	7.005	600	0.2279	3690.13	4191.53	8.172	1030	0.3190	4568.69	5270.39	9.004
170	0.1342	2921.60	3216.76	7.039	610	0.2300	3709.25	4215.33	8.194	1040	0.3211	4590.41	5296.75	9.021
180	0.1365	2938.77	3238.97	7.072	620	0.2322	3728.44	4239.20	8.216	1050	0.3232	4612.18	5323.16	9.038
190	0.1387	2955.91	3261.12	7.105	630	0.2343	3747.69	4263.13	8.237	1060	0.3253	4634.00	5349.62	9.056
200	0.1410	2973.04	3283.25	7.138	640	0.2364	3767.00	4287.13	8.258	1070	0.3274	4655.88	5376.13	9.073
210	0.1433	2990.17	3305.34	7.169	650	0.2385	3786.38	4311.19	8.280	1080	0.3295	4677.80	5402.69	9.090
220	0.1455	3007.31	3327.43	7.201	660	0.2407	3805.83	4335.31	8.301	1090	0.3316	4699.78	5429.31	9.106
230	0.1478	3024.45	3349.50	7.231	670	0.2428	3825.33	4359.49	8.322	1100	0.3337	4721.81	5455.98	9.123
240	0.1500	3041.61	3371.58	7.262	680	0.2449	3844.91	4383.74	8.342	1110	0.3358	4743.89	5482.69	9.140
250	0.1522	3058.79	3393.65	7.292	690	0.2470	3864.54	4408.05	8.363	1120	0.3379	4766.02	5509.46	9.157
260	0.1544	3075.99	3415.74	7.322	700	0.2492	3884.24	4432.42	8.384	1130	0.3400	4788.20	5536.27	9.173
270	0.1566	3093.23	3437.85	7.351	710	0.2513	3904.01	4456.85	8.404	1140	0.3421	4810.42	5563.14	9.190
280	0.1589	3110.49	3459.97	7.380	720	0.2534	3923.84	4481.35	8.424	1150	0.3442	4832.70	5590.05	9.206
290	0.1611	3127.79	3482.12	7.408	730	0.2555	3943.73	4505.90	8.445	1160	0.3464	4855.03	5617.01	9.223
300	0.1633	3145.13	3504.29	7.437	740	0.2577	3963.69	4530.53	8.465	1170	0.3485	4877.40	5644.02	9.239
310	0.1655	3162.50	3526.49	7.464	750	0.2598	3983.71	4555.21	8.485	1180	0.3506	4899.83	5671.08	9.255
320	0.1676	3179.92	3548.73	7.492	760	0.2619	4003.79	4579.95	8.504	1190	0.3527	4922.30	5698.18	9.271
330	0.1698	3197.39	3571.00	7.519	770	0.2640	4023.93	4604.76	8.524	1200	0.3548	4944.82	5725.33	9.287
340	0.1720	3214.90	3593.31	7.546	780	0.2661	4044.14	4629.62	8.544	1210	0.3569	4967.38	5752.53	9.304
350	0.1742	3232.45	3615.66	7.573	790	0.2682	4064.41	4654.55	8.563	1220	0.3590	4989.99	5779.78	9.320
360	0.1764	3250.06	3638.06	7.600	800	0.2704	4084.74	4679.54	8.583	1230	0.3611	5012.65	5807.06	9.335
370	0.1785	3267.72	3660.49	7.626	810	0.2725	4105.13	4704.59	8.602	1240	0.3632	5035.35	5834.40	9.351
380	0.1807	3285.44	3682.98	7.652	820	0.2746	4125.59	4729.70	8.621	1250	0.3653	5058.10	5861.78	9.367
390	0.1829	3303.20	3705.51	7.678	830	0.2767	4146.10	4754.87	8.641	1260	0.3674	5080.89	5889.20	9.383
400	0.1850	3321.02	3728.09	7.703	840	0.2788	4166.68	4780.10	8.660	1270	0.3695	5103.73	5916.67	9.398
410	0.1872	3338.90	3750.72	7.728	850	0.2809	4187.31	4805.39	8.679	1280	0.3716	5126.61	5944.19	9.414
420	0.1894	3356.84	3773.41	7.753	860	0.2831	4208.01	4830.74	8.697	1290	0.3737	5149.54	5971.74	9.430
430	0.1915	3374.83	3796.14	7.778	870	0.2852	4228.77	4856.14	8.716	1300	0.3758	5172.51	5999.34	9.445
440	0.1937	3392.88	3818.93	7.803										

Table 3(13) Superheated Steam Properties (Pressure Base) Pressure = 23.0 bar and saturation temperature, Tsat = 219.556°C

t (°C)	$v\left(\frac{m^3}{kg}\right)$	$u\left(\frac{kJ}{kg}\right)$	$h\left(\frac{kJ}{kg}\right)$	$s\left(\frac{kJ}{kg\,K}\right)$	t (°C)	$v\left(\frac{m^3}{kg}\right)$	$u\left(\frac{kJ}{kg}\right)$	$h\left(\frac{kJ}{kg}\right)$	$s\left(\frac{kJ}{kg\,K}\right)$	t (°C)	$v\left(\frac{m^3}{kg}\right)$	$u\left(\frac{kJ}{kg}\right)$	$h\left(\frac{kJ}{kg}\right)$	$s\left(\frac{kJ}{kg\,K}\right)$
Sat Vapour (g)	0.0868	2601.16	2800.83	6.287	Sat Vapour (g)	0.0868	2601.16	2800.83	6.287	Sat Vapour (g)	0.0868	2601.16	2800.83	6.287
10	0.0899	2625.60	2832.26	6.350	450	0.1877	3414.68	3846.42	7.812	880	0.2753	4254.18	4887.26	8.718
20	0.0927	2648.07	2861.31	6.408	460	0.1898	3432.88	3869.35	7.836	890	0.2773	4275.07	4912.80	8.737
30	0.0955	2669.32	2888.86	6.461	470	0.1918	3451.14	3892.34	7.860	900	0.2793	4296.02	4938.40	8.755
40	0.0981	2689.70	2915.34	6.511	480	0.1939	3469.46	3915.38	7.884	910	0.2813	4317.03	4964.06	8.774
50	0.1007	2709.42	2940.98	6.559	490	0.1959	3487.84	3938.48	7.908	920	0.2833	4338.09	4989.77	8.792
60	0.1032	2728.62	2965.97	6.604	500	0.1980	3506.28	3961.65	7.931	930	0.2854	4359.22	5015.55	8.810
70	0.1057	2747.41	2990.43	6.648	510	0.2000	3524.78	3984.87	7.954	940	0.2874	4380.40	5041.37	8.828
80	0.1081	2765.87	3014.45	6.690	520	0.2021	3543.35	4008.15	7.977	950	0.2894	4401.63	5067.26	8.846
90	0.1105	2784.05	3038.11	6.731	530	0.2041	3561.98	4031.49	8.000	960	0.2914	4422.93	5093.20	8.864
100	0.1128	2802.02	3061.49	6.771	540	0.2062	3580.68	4054.89	8.023	970	0.2934	4444.28	5119.19	8.882
110	0.1151	2819.80	3084.61	6.810	550	0.2082	3599.43	4078.35	8.046	980	0.2955	4465.68	5145.24	8.900
120	0.1174	2837.44	3107.54	6.848	560	0.2103	3618.26	4101.87	8.068	990	0.2975	4487.14	5171.34	8.917
130	0.1197	2854.96	3130.30	6.884	570	0.2123	3637.14	4125.46	8.090	1000	0.2995	4508.66	5197.50	8.935
140	0.1220	2872.39	3152.91	6.921	580	0.2144	3656.09	4149.10	8.113	1010	0.3015	4530.23	5223.71	8.953
150	0.1242	2889.74	3175.42	6.956	590	0.2164	3675.11	4172.81	8.135	1020	0.3035	4551.85	5249.98	8.970
160	0.1264	2907.03	3197.83	6.990	600	0.2184	3694.18	4196.58	8.156	1030	0.3056	4573.53	5276.30	8.987
170	0.1286	2924.28	3220.16	7.024	610	0.2205	3713.33	4220.41	8.178	1040	0.3076	4595.26	5302.67	9.005
180	0.1308	2941.50	3242.43	7.058	620	0.2225	3732.53	4244.30	8.200	1050	0.3096	4617.04	5329.10	9.022
190	0.1330	2958.69	3264.65	7.091	630	0.2245	3751.81	4268.26	8.221	1060	0.3116	4638.88	5355.57	9.039
200	0.1352	2975.87	3286.84	7.123	640	0.2266	3771.14	4292.28	8.242	1070	0.3136	4660.77	5382.10	9.056
210	0.1374	2993.05	3308.99	7.155	650	0.2286	3790.54	4316.36	8.264	1080	0.3156	4682.71	5408.68	9.073
220	0.1395	3010.23	3331.13	7.186	660	0.2307	3810.01	4340.50	8.285	1090	0.3177	4704.70	5435.31	9.090
230	0.1417	3027.42	3353.26	7.217	670	0.2327	3829.54	4364.71	8.306	1100	0.3197	4726.74	5461.99	9.107
240	0.1438	3044.62	3375.39	7.247	680	0.2347	3849.13	4388.98	8.326	1110	0.3217	4748.84	5488.72	9.123
250	0.1459	3061.84	3397.52	7.277	690	0.2367	3868.79	4413.31	8.347	1120	0.3237	4770.98	5515.50	9.140
260	0.1481	3079.08	3419.66	7.307	700	0.2388	3888.51	4437.70	8.368	1130	0.3257	4793.17	5542.33	9.156
270	0.1502	3096.36	3441.81	7.336	710	0.2408	3908.29	4462.16	8.388	1140	0.3277	4815.41	5569.21	9.173
280	0.1523	3113.66	3463.98	7.365	720	0.2428	3928.14	4486.68	8.408	1150	0.3298	4837.70	5596.13	9.189
290	0.1544	3130.99	3486.17	7.393	730	0.2449	3948.06	4511.26	8.428	1160	0.3318	4860.04	5623.11	9.206
300	0.1565	3148.37	3508.39	7.422	740	0.2469	3968.03	4535.90	8.449	1170	0.3338	4882.43	5650.13	9.222
310	0.1586	3165.78	3530.63	7.449	750	0.2489	3988.07	4560.60	8.468	1180	0.3358	4904.87	5677.20	9.238
320	0.1607	3183.23	3552.91	7.477	760	0.2510	4008.17	4585.37	8.488	1190	0.3378	4927.35	5704.32	9.255
330	0.1628	3200.73	3575.22	7.504	770	0.2530	4028.33	4610.19	8.508	1200	0.3398	4949.88	5731.49	9.271
340	0.1649	3218.27	3597.57	7.531	780	0.2550	4048.56	4635.08	8.528	1210	0.3418	4972.46	5758.70	9.287
350	0.1670	3235.86	3619.96	7.558	790	0.2570	4068.85	4660.03	8.547	1220	0.3439	4995.08	5785.95	9.303
360	0.1691	3253.50	3642.39	7.584	800	0.2591	4089.20	4685.04	8.567	1230	0.3459	5017.75	5813.26	9.318
370	0.1712	3271.19	3664.86	7.611	810	0.2611	4109.61	4710.11	8.586	1240	0.3479	5040.47	5840.60	9.334
380	0.1732	3288.93	3687.38	7.637	820	0.2631	4130.08	4735.24	8.605	1250	0.3499	5063.23	5868.00	9.350
390	0.1753	3306.73	3709.95	7.662	830	0.2651	4150.61	4760.43	8.624	1260	0.3519	5086.03	5895.43	9.366
400	0.1774	3324.58	3732.57	7.688	840	0.2672	4171.21	4785.68	8.643	1270	0.3539	5108.88	5922.92	9.381
410	0.1795	3342.48	3755.23	7.713	850	0.2692	4191.86	4810.98	8.662	1280	0.3559	5131.78	5950.44	9.397
420	0.1815	3360.45	3777.95	7.738	860	0.2712	4212.57	4836.35	8.681	1290	0.3580	5154.71	5978.01	9.412
430	0.1836	3378.47	3800.72	7.763	870	0.2732	4233.35	4861.78	8.700	1300	0.3600	5177.69	6005.62	9.428
440	0.1856	3396.54	3823.54	7.787										

Table 3(14) Superheated Steam Properties (Pressure Base) Pressure = 24.0 bar and saturation temperature, Tsat = 221.788°C

t (°C)	v $\left(\frac{m^3}{kg}\right)$	u $\left(\frac{kJ}{kg}\right)$	h $\left(\frac{kJ}{kg}\right)$	s $\left(\frac{kJ}{kg\,K}\right)$	t (°C)	v $\left(\frac{m^3}{kg}\right)$	u $\left(\frac{kJ}{kg}\right)$	h $\left(\frac{kJ}{kg}\right)$	s $\left(\frac{kJ}{kg\,K}\right)$	t (°C)	v $\left(\frac{m^3}{kg}\right)$	u $\left(\frac{kJ}{kg}\right)$	h $\left(\frac{kJ}{kg}\right)$	s $\left(\frac{kJ}{kg\,K}\right)$
Sat Vapour (g)	0.0832	2601.65	2801.44	6.271	Sat Vapour (g)	0.0832	2601.65	2801.44	6.271	Sat Vapour (g)	0.0832	2601.65	2801.44	6.271
10	0.0862	2626.39	2833.24	6.335	450	0.1803	3418.24	3850.89	7.797	880	0.2642	4258.62	4892.72	8.703
20	0.0890	2649.10	2862.59	6.392	460	0.1822	3436.47	3873.85	7.821	890	0.2661	4279.53	4918.28	8.721
30	0.0916	2670.54	2890.40	6.446	470	0.1842	3454.75	3896.87	7.845	900	0.2681	4300.50	4943.90	8.740
40	0.0942	2691.09	2917.09	6.496	480	0.1862	3473.09	3919.94	7.869	910	0.2700	4321.52	4969.57	8.758
50	0.0967	2710.97	2942.93	6.544	490	0.1882	3491.50	3943.07	7.893	920	0.2720	4342.60	4995.31	8.776
60	0.0991	2730.30	2968.09	6.590	500	0.1901	3509.97	3966.26	7.916	930	0.2739	4363.74	5021.09	8.794
70	0.1015	2749.21	2992.69	6.634	510	0.1921	3528.49	3989.51	7.939	940	0.2758	4384.94	5046.94	8.813
80	0.1038	2767.77	3016.85	6.676	520	0.1941	3547.09	4012.82	7.962	950	0.2778	4406.19	5072.84	8.831
90	0.1061	2786.05	3040.64	6.717	530	0.1960	3565.74	4036.19	7.985	960	0.2797	4427.50	5098.79	8.848
100	0.1083	2804.11	3064.13	6.757	540	0.1980	3584.46	4059.61	8.008	970	0.2816	4448.86	5124.81	8.866
110	0.1106	2821.97	3087.36	6.796	550	0.1999	3603.24	4083.10	8.031	980	0.2836	4470.28	5150.87	8.884
120	0.1128	2839.69	3110.39	6.834	560	0.2019	3622.08	4106.65	8.053	990	0.2855	4491.76	5176.99	8.902
130	0.1150	2857.28	3133.23	6.870	570	0.2039	3640.99	4130.26	8.075	1000	0.2874	4513.29	5203.17	8.919
140	0.1172	2874.77	3155.94	6.906	580	0.2058	3659.96	4153.93	8.098	1010	0.2894	4534.87	5229.39	8.937
150	0.1193	2892.19	3178.52	6.942	590	0.2078	3679.00	4177.66	8.119	1020	0.2913	4556.51	5255.68	8.954
160	0.1214	2909.54	3201.01	6.976	600	0.2097	3698.10	4201.46	8.141	1030	0.2933	4578.20	5282.01	8.971
170	0.1236	2926.85	3223.41	7.010	610	0.2117	3717.26	4225.31	8.163	1040	0.2952	4599.95	5308.40	8.989
180	0.1257	2944.12	3245.75	7.044	620	0.2136	3736.49	4249.23	8.185	1050	0.2971	4621.75	5334.84	9.006
190	0.1278	2961.36	3268.03	7.076	630	0.2156	3755.78	4273.21	8.206	1060	0.2991	4643.60	5361.33	9.023
200	0.1299	2978.59	3290.28	7.109	640	0.2175	3775.14	4297.25	8.227	1070	0.3010	4665.50	5387.87	9.040
210	0.1320	2995.82	3312.50	7.140	650	0.2195	3794.56	4321.36	8.248	1080	0.3029	4687.45	5414.46	9.057
220	0.1340	3013.04	3334.70	7.172	660	0.2214	3814.05	4345.52	8.269	1090	0.3049	4709.46	5441.11	9.074
230	0.1361	3030.27	3356.88	7.202	670	0.2234	3833.60	4369.75	8.290	1100	0.3068	4731.51	5467.80	9.091
240	0.1381	3047.52	3379.06	7.233	680	0.2253	3853.21	4394.04	8.311	1110	0.3087	4753.62	5494.55	9.107
250	0.1402	3064.78	3401.24	7.263	690	0.2273	3872.89	4418.40	8.332	1120	0.3107	4775.78	5521.34	9.124
260	0.1422	3082.06	3423.42	7.292	700	0.2292	3892.63	4442.81	8.352	1130	0.3126	4797.98	5548.18	9.140
270	0.1443	3099.37	3445.62	7.322	710	0.2312	3912.43	4467.29	8.373	1140	0.3145	4820.24	5575.08	9.157
280	0.1463	3116.71	3467.84	7.350	720	0.2331	3932.30	4491.83	8.393	1150	0.3164	4842.54	5602.02	9.173
290	0.1483	3134.08	3490.07	7.379	730	0.2351	3952.23	4516.43	8.413	1160	0.3184	4864.89	5629.01	9.190
300	0.1504	3151.48	3512.33	7.407	740	0.2370	3972.23	4541.09	8.433	1170	0.3203	4887.29	5656.04	9.206
310	0.1524	3168.93	3534.62	7.435	750	0.2390	3992.28	4565.82	8.453	1180	0.3222	4909.74	5683.13	9.222
320	0.1544	3186.42	3556.94	7.463	760	0.2409	4012.40	4590.60	8.473	1190	0.3242	4932.24	5710.26	9.238
330	0.1564	3203.94	3579.29	7.490	770	0.2429	4032.59	4615.45	8.493	1200	0.3261	4954.78	5737.44	9.254
340	0.1584	3221.52	3601.67	7.517	780	0.2448	4052.83	4640.36	8.512	1210	0.3280	4977.37	5764.66	9.270
350	0.1604	3239.14	3624.10	7.544	790	0.2467	4073.14	4665.32	8.532	1220	0.3300	5000.01	5791.93	9.286
360	0.1624	3256.81	3646.57	7.570	800	0.2487	4093.50	4690.35	8.551	1230	0.3319	5022.69	5819.24	9.302
370	0.1644	3274.53	3669.08	7.596	810	0.2506	4113.93	4715.44	8.570	1240	0.3338	5045.41	5846.60	9.318
380	0.1664	3292.30	3691.63	7.622	820	0.2526	4134.42	4740.59	8.590	1250	0.3358	5068.19	5874.01	9.334
390	0.1684	3310.13	3714.23	7.648	830	0.2545	4154.97	4765.80	8.609	1260	0.3377	5091.00	5901.46	9.350
400	0.1704	3328.00	3736.88	7.673	840	0.2565	4175.58	4791.06	8.628	1270	0.3396	5113.86	5928.95	9.365
410	0.1724	3345.94	3759.58	7.698	850	0.2584	4196.25	4816.39	8.647	1280	0.3416	5136.77	5956.49	9.381
420	0.1743	3363.93	3782.33	7.723	860	0.2603	4216.98	4841.78	8.665	1290	0.3435	5159.72	5984.07	9.396
430	0.1763	3381.97	3805.13	7.748	870	0.2623	4237.77	4867.22	8.684	1300	0.3454	5182.71	6011.69	9.412
440	0.1783	3400.08	3827.98	7.773										

Table 3(15) Superheated Steam Properties (Pressure Base) Pressure = 25.0 bar and saturation temperature, Tsat = 223.95°C

t (°C)	v $\left(\frac{m^3}{kg}\right)$	u $\left(\frac{kJ}{kg}\right)$	h $\left(\frac{kJ}{kg}\right)$	s $\left(\frac{kJ}{kg\,K}\right)$
Sat Vapour (g)	0.0800	2602.07	2801.95	6.256
10	0.0828	2627.11	2834.12	6.320
20	0.0855	2650.04	2863.77	6.378
30	0.0881	2671.69	2891.82	6.432
40	0.0905	2692.41	2918.73	6.482
50	0.0929	2712.42	2944.76	6.530
60	0.0953	2731.89	2970.08	6.576
70	0.0976	2750.91	2994.84	6.620
80	0.0998	2769.58	3019.14	6.663
90	0.1020	2787.96	3043.05	6.704
100	0.1042	2806.10	3066.65	6.743
110	0.1064	2824.05	3089.99	6.782
120	0.1085	2841.84	3113.11	6.820
130	0.1106	2859.50	3136.05	6.857
140	0.1127	2877.06	3158.83	6.893
150	0.1148	2894.54	3181.50	6.928
160	0.1168	2911.95	3204.06	6.963
170	0.1189	2929.31	3226.53	6.997
180	0.1209	2946.63	3248.94	7.030
190	0.1229	2963.93	3271.29	7.063
200	0.1250	2981.21	3293.59	7.095
210	0.1270	2998.48	3315.87	7.127
220	0.1290	3015.75	3338.12	7.158
230	0.1309	3033.02	3360.36	7.189
240	0.1329	3050.30	3382.59	7.219
250	0.1349	3067.60	3404.82	7.249
260	0.1369	3084.93	3427.05	7.279
270	0.1388	3102.27	3449.30	7.308
280	0.1408	3119.65	3471.55	7.337
290	0.1427	3137.05	3493.83	7.365
300	0.1447	3154.49	3516.13	7.393
310	0.1466	3171.97	3538.46	7.421
320	0.1485	3189.49	3560.82	7.449
330	0.1505	3207.05	3583.21	7.476
340	0.1524	3224.66	3605.63	7.503
350	0.1543	3242.31	3628.10	7.530
360	0.1562	3260.01	3650.60	7.556
370	0.1582	3277.75	3673.14	7.582
380	0.1601	3295.55	3695.73	7.608
390	0.1620	3313.41	3718.37	7.634
400	0.1639	3331.31	3741.05	7.659
410	0.1658	3349.27	3763.78	7.684
420	0.1677	3367.29	3786.56	7.709
430	0.1696	3385.36	3809.39	7.734
440	0.1715	3403.49	3832.27	7.759

t (°C)	v $\left(\frac{m^3}{kg}\right)$	u $\left(\frac{kJ}{kg}\right)$	h $\left(\frac{kJ}{kg}\right)$	s $\left(\frac{kJ}{kg\,K}\right)$
Sat Vapour (g)	0.0800	2602.07	2801.95	6.256
450	0.1734	3421.68	3855.21	7.783
460	0.1753	3439.93	3878.20	7.807
470	0.1772	3458.24	3901.25	7.831
480	0.1791	3476.61	3924.35	7.855
490	0.1810	3495.04	3947.51	7.878
500	0.1829	3513.53	3970.73	7.902
510	0.1848	3532.08	3994.00	7.925
520	0.1867	3550.69	4017.34	7.948
530	0.1885	3569.37	4040.73	7.971
540	0.1904	3588.11	4064.18	7.994
550	0.1923	3606.91	4087.70	8.016
560	0.1942	3625.78	4111.27	8.039
570	0.1961	3644.71	4134.90	8.061
580	0.1980	3663.70	4158.60	8.083
590	0.1998	3682.76	4182.36	8.105
600	0.2017	3701.88	4206.17	8.127
610	0.2036	3721.07	4230.05	8.149
620	0.2055	3740.32	4253.99	8.170
630	0.2073	3759.63	4278.00	8.191
640	0.2092	3779.01	4302.06	8.213
650	0.2111	3798.45	4326.19	8.234
660	0.2130	3817.96	4350.38	8.255
670	0.2148	3837.52	4374.63	8.276
680	0.2167	3857.16	4398.94	8.296
690	0.2186	3876.85	4423.32	8.317
700	0.2205	3896.61	4447.75	8.338
710	0.2223	3916.44	4472.25	8.358
720	0.2242	3936.32	4496.81	8.378
730	0.2261	3956.27	4521.43	8.398
740	0.2279	3976.29	4546.12	8.418
750	0.2298	3996.36	4570.86	8.438
760	0.2317	4016.50	4595.67	8.458
770	0.2335	4036.70	4620.53	8.478
780	0.2354	4056.96	4645.46	8.497
790	0.2373	4077.28	4670.45	8.517
800	0.2391	4097.67	4695.49	8.536
810	0.2410	4118.11	4720.60	8.556
820	0.2429	4138.62	4745.77	8.575
830	0.2447	4159.19	4770.99	8.594
840	0.2466	4179.82	4796.28	8.613
850	0.2484	4200.50	4821.62	8.632
860	0.2503	4221.25	4847.03	8.650
870	0.2522	4242.06	4872.49	8.669

t (°C)	v $\left(\frac{m^3}{kg}\right)$	u $\left(\frac{kJ}{kg}\right)$	h $\left(\frac{kJ}{kg}\right)$	s $\left(\frac{kJ}{kg\,K}\right)$
Sat Vapour (g)	0.0800	2602.07	2801.95	6.256
880	0.2540	4262.92	4898.01	8.688
890	0.2559	4283.84	4923.59	8.706
900	0.2578	4304.83	4949.22	8.725
910	0.2596	4325.87	4974.91	8.743
920	0.2615	4346.96	5000.66	8.761
930	0.2633	4368.12	5026.47	8.779
940	0.2652	4389.33	5052.33	8.797
950	0.2671	4410.60	5078.24	8.815
960	0.2689	4431.92	5104.21	8.833
970	0.2708	4453.30	5130.24	8.851
980	0.2726	4474.74	5156.32	8.869
990	0.2745	4496.23	5182.46	8.886
1000	0.2764	4517.77	5208.65	8.904
1010	0.2782	4539.37	5234.89	8.921
1020	0.2801	4561.02	5261.19	8.939
1030	0.2819	4582.73	5287.54	8.956
1040	0.2838	4604.49	5313.94	8.973
1050	0.2856	4626.30	5340.39	8.991
1060	0.2875	4648.16	5366.90	9.008
1070	0.2894	4670.08	5393.46	9.025
1080	0.2912	4692.04	5420.06	9.042
1090	0.2931	4714.06	5446.72	9.058
1100	0.2949	4736.13	5473.43	9.075
1110	0.2968	4758.25	5500.19	9.092
1120	0.2986	4780.42	5527.00	9.109
1130	0.3005	4802.64	5553.85	9.125
1140	0.3023	4824.91	5580.76	9.142
1150	0.3042	4847.22	5607.71	9.158
1160	0.3061	4869.59	5634.72	9.174
1170	0.3079	4892.00	5661.77	9.191
1180	0.3098	4914.46	5688.86	9.207
1190	0.3116	4936.97	5716.01	9.223
1200	0.3135	4959.52	5743.20	9.239
1210	0.3153	4982.12	5770.43	9.255
1220	0.3172	5004.77	5797.71	9.271
1230	0.3190	5027.46	5825.04	9.287
1240	0.3209	5050.20	5852.41	9.303
1250	0.3227	5072.99	5879.83	9.318
1260	0.3246	5095.81	5907.29	9.334
1270	0.3264	5118.68	5934.80	9.350
1280	0.3283	5141.60	5962.35	9.365
1290	0.3302	5164.56	5989.94	9.381
1300	0.3320	5187.56	6017.57	9.396

Table 3(16) Superheated Steam Properties (Pressure Base) Pressure = 26 bar and saturation temperature, Tsat = 226.045°C

t (°C)	v $\left(\frac{m^3}{kg}\right)$	u $\left(\frac{kJ}{kg}\right)$	h $\left(\frac{kJ}{kg}\right)$	s $\left(\frac{kJ}{kg\,K}\right)$	t (°C)	v $\left(\frac{m^3}{kg}\right)$	u $\left(\frac{kJ}{kg}\right)$	h $\left(\frac{kJ}{kg}\right)$	s $\left(\frac{kJ}{kg\,K}\right)$	t (°C)	v $\left(\frac{m^3}{kg}\right)$	u $\left(\frac{kJ}{kg}\right)$	h $\left(\frac{kJ}{kg}\right)$	s $\left(\frac{kJ}{kg\,K}\right)$
Sat Vapour (g)	0.0769	2602.42	2802.36	6.241	Sat Vapour (g)	0.0769	2602.42	2802.36	6.241	Sat Vapour (g)	0.0769	2602.42	2802.36	6.241
10	0.0797	2627.74	2834.90	6.305	450	0.1671	3425.01	3859.39	7.769	880	0.2446	4267.09	4903.13	8.673
20	0.0823	2650.91	2864.84	6.364	460	0.1689	3443.28	3882.41	7.794	890	0.2464	4288.03	4928.73	8.692
30	0.0848	2672.75	2893.15	6.418	470	0.1707	3461.62	3905.48	7.818	900	0.2482	4309.02	4954.38	8.710
40	0.0872	2693.64	2920.27	6.468	480	0.1725	3480.01	3928.61	7.841	910	0.2500	4330.08	4980.09	8.729
50	0.0895	2713.80	2946.48	6.517	490	0.1744	3498.46	3951.80	7.865	920	0.2518	4351.19	5005.85	8.747
60	0.0918	2733.40	2971.97	6.563	500	0.1762	3516.97	3975.05	7.888	930	0.2536	4372.36	5031.67	8.765
70	0.0940	2752.54	2996.88	6.607	510	0.1780	3535.55	3998.35	7.911	940	0.2554	4393.59	5057.55	8.783
80	0.0962	2771.31	3021.31	6.649	520	0.1798	3554.19	4021.71	7.934	950	0.2572	4414.87	5083.48	8.801
90	0.0983	2789.78	3045.35	6.690	530	0.1816	3572.89	4045.13	7.957	960	0.2589	4436.21	5109.47	8.819
100	0.1004	2808.01	3069.06	6.730	540	0.1834	3591.65	4068.61	7.980	970	0.2607	4457.60	5135.51	8.837
110	0.1025	2826.04	3092.50	6.769	550	0.1853	3610.47	4092.14	8.003	980	0.2625	4479.05	5161.61	8.854
120	0.1045	2843.90	3115.71	6.807	560	0.1871	3629.36	4115.74	8.025	990	0.2643	4500.55	5187.76	8.872
130	0.1066	2861.64	3138.74	6.844	570	0.1889	3648.31	4139.40	8.047	1000	0.2661	4522.11	5213.96	8.889
140	0.1086	2879.26	3161.61	6.880	580	0.1907	3667.33	4163.12	8.069	1010	0.2679	4543.73	5240.22	8.907
150	0.1106	2896.80	3184.35	6.915	590	0.1925	3686.41	4186.90	8.091	1020	0.2697	4565.39	5266.53	8.924
160	0.1126	2914.27	3206.99	6.950	600	0.1943	3705.55	4210.74	8.113	1030	0.2715	4587.11	5292.90	8.942
170	0.1146	2931.68	3229.53	6.984	610	0.1961	3724.75	4234.64	8.135	1040	0.2732	4608.88	5319.31	8.959
180	0.1165	2949.05	3252.00	7.017	620	0.1979	3744.02	4258.61	8.156	1050	0.2750	4630.71	5345.78	8.976
190	0.1185	2966.40	3274.41	7.050	630	0.1997	3763.36	4282.63	8.178	1060	0.2768	4652.58	5372.30	8.993
200	0.1204	2983.73	3296.78	7.082	640	0.2015	3782.75	4306.72	8.199	1070	0.2786	4674.51	5398.87	9.010
210	0.1223	3001.04	3319.11	7.114	650	0.2033	3802.21	4330.87	8.220	1080	0.2804	4696.49	5425.49	9.027
220	0.1243	3018.35	3341.42	7.145	660	0.2051	3821.74	4355.08	8.241	1090	0.2822	4718.52	5452.16	9.044
230	0.1262	3035.67	3363.71	7.176	670	0.2069	3841.33	4379.35	8.262	1100	0.2840	4740.61	5478.89	9.061
240	0.1281	3052.99	3385.99	7.206	680	0.2087	3860.98	4403.69	8.282	1110	0.2857	4762.74	5505.66	9.077
250	0.1300	3070.33	3408.27	7.236	690	0.2105	3880.70	4428.08	8.303	1120	0.2875	4784.92	5532.48	9.094
260	0.1319	3087.69	3430.55	7.266	700	0.2123	3900.47	4452.54	8.323	1130	0.2893	4807.15	5559.35	9.110
270	0.1338	3105.07	3452.84	7.295	710	0.2141	3920.32	4477.06	8.344	1140	0.2911	4829.43	5586.27	9.127
280	0.1356	3122.48	3475.14	7.324	720	0.2159	3940.22	4501.64	8.364	1150	0.2929	4851.76	5613.24	9.143
290	0.1375	3139.92	3497.46	7.352	730	0.2177	3960.19	4526.28	8.384	1160	0.2947	4874.14	5640.25	9.160
300	0.1394	3157.40	3519.80	7.380	740	0.2195	3980.22	4550.98	8.404	1170	0.2964	4896.56	5667.31	9.176
310	0.1413	3174.91	3542.17	7.408	750	0.2213	4000.31	4575.75	8.424	1180	0.2982	4919.03	5694.42	9.192
320	0.1431	3192.46	3564.57	7.436	760	0.2231	4020.47	4600.57	8.444	1190	0.3000	4941.55	5721.58	9.208
330	0.1450	3210.05	3587.00	7.463	770	0.2249	4040.68	4625.46	8.464	1200	0.3018	4964.12	5748.78	9.224
340	0.1468	3227.69	3609.46	7.490	780	0.2267	4060.96	4650.40	8.483	1210	0.3036	4986.73	5776.03	9.240
350	0.1487	3245.37	3631.96	7.516	790	0.2285	4081.30	4675.41	8.503	1220	0.3054	5009.39	5803.32	9.256
360	0.1505	3263.09	3654.49	7.543	800	0.2303	4101.71	4700.47	8.522	1230	0.3071	5032.09	5830.66	9.272
370	0.1524	3280.87	3677.07	7.569	810	0.2321	4122.17	4725.60	8.541	1240	0.3089	5054.84	5858.05	9.288
380	0.1542	3298.70	3699.69	7.595	820	0.2339	4142.69	4750.79	8.560	1250	0.3107	5077.64	5885.48	9.303
390	0.1561	3316.58	3722.36	7.620	830	0.2357	4163.28	4776.03	8.580	1260	0.3125	5100.48	5912.95	9.319
400	0.1579	3334.51	3745.07	7.646	840	0.2375	4183.92	4801.33	8.598	1270	0.3143	5123.36	5940.46	9.335
410	0.1597	3352.50	3767.84	7.671	850	0.2393	4204.62	4826.70	8.617	1280	0.3161	5146.28	5968.02	9.350
420	0.1616	3370.54	3790.65	7.696	860	0.2411	4225.38	4852.12	8.636	1290	0.3178	5169.25	5995.63	9.366
430	0.1634	3388.64	3813.51	7.721	870	0.2428	4246.21	4877.60	8.655	1300	0.3196	5192.27	6023.27	9.381
440	0.1652	3406.80	3836.42	7.745										

Table 3(17) Superheated Steam Properties (Pressure Base) Pressure = 27 bar and saturation temperature, Tsat = 228.079°C

t (°C)	v $\left(\frac{m^3}{kg}\right)$	u $\left(\frac{kJ}{kg}\right)$	h $\left(\frac{kJ}{kg}\right)$	s $\left(\frac{kJ}{kg\,K}\right)$
Sat Vapour (g)	0.0741	2602.70	2802.68	6.226
10	0.0768	2628.31	2835.58	6.291
20	0.0793	2651.71	2865.82	6.350
30	0.0817	2673.74	2894.38	6.404
40	0.0840	2694.80	2921.71	6.455
50	0.0863	2715.11	2948.11	6.504
60	0.0885	2734.83	2973.77	6.550
70	0.0906	2754.08	2998.82	6.594
80	0.0928	2772.96	3023.38	6.637
90	0.0948	2791.52	3047.54	6.678
100	0.0969	2809.84	3071.36	6.718
110	0.0989	2827.95	3094.90	6.757
120	0.1009	2845.88	3118.21	6.794
130	0.1028	2863.68	3141.33	6.831
140	0.1048	2881.37	3164.28	6.867
150	0.1067	2898.97	3187.10	6.903
160	0.1086	2916.50	3209.81	6.937
170	0.1105	2933.96	3232.42	6.971
180	0.1124	2951.39	3254.95	7.005
190	0.1143	2968.78	3277.43	7.037
200	0.1162	2986.15	3299.86	7.070
210	0.1180	3003.51	3322.24	7.101
220	0.1199	3020.87	3344.60	7.133
230	0.1217	3038.22	3366.95	7.163
240	0.1236	3055.59	3389.28	7.194
250	0.1254	3072.97	3411.60	7.224
260	0.1272	3090.36	3433.93	7.253
270	0.1291	3107.78	3456.26	7.282
280	0.1309	3125.22	3478.61	7.311
290	0.1327	3142.70	3500.97	7.339
300	0.1345	3160.21	3523.35	7.368
310	0.1363	3177.75	3545.76	7.395
320	0.1381	3195.33	3568.19	7.423
330	0.1399	3212.95	3590.66	7.450
340	0.1417	3230.62	3613.16	7.477
350	0.1435	3248.33	3635.69	7.504
360	0.1453	3266.08	3658.26	7.530
370	0.1470	3283.89	3680.88	7.556
380	0.1488	3301.74	3703.53	7.582
390	0.1506	3319.65	3726.23	7.607
400	0.1524	3337.61	3748.97	7.633
410	0.1541	3355.62	3771.77	7.658
420	0.1559	3373.69	3794.61	7.683
430	0.1577	3391.82	3817.50	7.708
440	0.1594	3410.00	3840.44	7.732

t (°C)	v $\left(\frac{m^3}{kg}\right)$	u $\left(\frac{kJ}{kg}\right)$	h $\left(\frac{kJ}{kg}\right)$	s $\left(\frac{kJ}{kg\,K}\right)$
Sat Vapour (g)	0.0741	2602.70	2802.68	6.226
450	0.1612	3428.24	3863.44	7.756
460	0.1629	3446.53	3886.49	7.781
470	0.1647	3464.89	3909.59	7.804
480	0.1665	3483.30	3932.75	7.828
490	0.1682	3501.78	3955.96	7.852
500	0.1700	3520.32	3979.23	7.875
510	0.1717	3538.91	4002.56	7.898
520	0.1735	3557.57	4025.94	7.921
530	0.1752	3576.29	4049.39	7.944
540	0.1770	3595.08	4072.89	7.967
550	0.1787	3613.92	4096.45	7.989
560	0.1805	3632.83	4120.08	8.012
570	0.1822	3651.80	4143.76	8.034
580	0.1839	3670.84	4167.50	8.056
590	0.1857	3689.94	4191.30	8.078
600	0.1874	3709.10	4215.17	8.100
610	0.1892	3728.33	4239.09	8.121
620	0.1909	3747.62	4263.08	8.143
630	0.1927	3766.97	4287.13	8.164
640	0.1944	3786.39	4311.24	8.185
650	0.1961	3805.87	4335.41	8.206
660	0.1979	3825.41	4359.64	8.227
670	0.1996	3845.02	4383.93	8.248
680	0.2013	3864.69	4408.29	8.269
690	0.2031	3884.42	4432.70	8.290
700	0.2048	3904.22	4457.18	8.310
710	0.2065	3924.08	4481.72	8.330
720	0.2083	3944.00	4506.32	8.351
730	0.2100	3963.99	4530.98	8.371
740	0.2117	3984.03	4555.70	8.391
750	0.2135	4004.14	4580.49	8.411
760	0.2152	4024.32	4605.33	8.430
770	0.2169	4044.55	4630.24	8.450
780	0.2186	4064.85	4655.20	8.470
790	0.2204	4085.20	4680.22	8.489
800	0.2221	4105.62	4705.31	8.508
810	0.2238	4126.10	4730.45	8.528
820	0.2256	4146.64	4755.65	8.547
830	0.2273	4167.24	4780.92	8.566
840	0.2290	4187.90	4806.24	8.585
850	0.2307	4208.62	4831.62	8.604
860	0.2325	4229.40	4857.05	8.622
870	0.2342	4250.23	4882.55	8.641

t (°C)	v $\left(\frac{m^3}{kg}\right)$	u $\left(\frac{kJ}{kg}\right)$	h $\left(\frac{kJ}{kg}\right)$	s $\left(\frac{kJ}{kg\,K}\right)$
Sat Vapour (g)	0.0741	2602.70	2802.68	6.226
880	0.2359	4271.13	4908.10	8.660
890	0.2376	4292.08	4933.71	8.678
900	0.2394	4313.10	4959.38	8.696
910	0.2411	4334.17	4985.11	8.715
920	0.2428	4355.29	5010.89	8.733
930	0.2445	4376.48	5036.72	8.751
940	0.2463	4397.72	5062.61	8.769
950	0.2480	4419.01	5088.56	8.787
960	0.2497	4440.37	5114.57	8.805
970	0.2514	4461.77	5140.62	8.823
980	0.2531	4483.24	5166.73	8.840
990	0.2549	4504.76	5192.90	8.858
1000	0.2566	4526.33	5219.12	8.875
1010	0.2583	4547.95	5245.39	8.893
1020	0.2600	4569.63	5271.72	8.910
1030	0.2618	4591.36	5298.10	8.928
1040	0.2635	4613.15	5324.53	8.945
1050	0.2652	4634.99	5351.01	8.962
1060	0.2669	4656.88	5377.54	8.979
1070	0.2686	4678.82	5404.13	8.996
1080	0.2704	4700.81	5430.76	9.013
1090	0.2721	4722.86	5457.45	9.030
1100	0.2738	4744.95	5484.18	9.046
1110	0.2755	4767.09	5510.97	9.063
1120	0.2772	4789.29	5537.80	9.080
1130	0.2789	4811.53	5564.69	9.096
1140	0.2807	4833.82	5591.62	9.113
1150	0.2824	4856.16	5618.60	9.129
1160	0.2841	4878.55	5645.62	9.145
1170	0.2858	4900.99	5672.70	9.162
1180	0.2875	4923.47	5699.82	9.178
1190	0.2893	4946.00	5726.99	9.194
1200	0.2910	4968.58	5754.20	9.210
1210	0.2927	4991.20	5781.46	9.226
1220	0.2944	5013.87	5808.77	9.242
1230	0.2961	5036.59	5836.12	9.258
1240	0.2978	5059.35	5863.51	9.273
1250	0.2996	5082.15	5890.95	9.289
1260	0.3013	5105.00	5918.44	9.305
1270	0.3030	5127.89	5945.97	9.320
1280	0.3047	5150.83	5973.54	9.336
1290	0.3064	5173.81	6001.15	9.351
1300	0.3081	5196.83	6028.81	9.367

Table 3(18) Superheated Steam Properties (Pressure Base) Pressure = 28 bar and saturation temperature, Tsat = 230.056°C

t (°C)	$v\left(\frac{m^3}{kg}\right)$	$u\left(\frac{kJ}{kg}\right)$	$h\left(\frac{kJ}{kg}\right)$	$s\left(\frac{kJ}{kg\,K}\right)$
Sat Vapour (g)	0.0714	2602.91	2802.92	6.212
10	0.0741	2628.81	2836.19	6.278
20	0.0765	2652.45	2866.72	6.337
30	0.0789	2674.67	2895.52	6.391
40	0.0811	2695.89	2923.06	6.443
50	0.0833	2716.34	2949.65	6.491
60	0.0855	2736.19	2975.46	6.537
70	0.0875	2755.55	3000.66	6.582
80	0.0896	2774.53	3025.36	6.624
90	0.0916	2793.19	3049.63	6.666
100	0.0936	2811.59	3073.57	6.706
110	0.0955	2829.78	3097.21	6.745
120	0.0974	2847.79	3120.61	6.782
130	0.0993	2865.65	3143.81	6.819
140	0.1012	2883.41	3166.85	6.855
150	0.1031	2901.06	3189.74	6.891
160	0.1050	2918.64	3212.52	6.925
170	0.1068	2936.16	3235.20	6.959
180	0.1086	2953.64	3257.80	6.993
190	0.1104	2971.08	3280.34	7.025
200	0.1123	2988.50	3302.82	7.058
210	0.1141	3005.90	3325.26	7.089
220	0.1159	3023.30	3347.68	7.120
230	0.1176	3040.70	3370.07	7.151
240	0.1194	3058.10	3392.45	7.182
250	0.1212	3075.51	3414.82	7.211
260	0.1229	3092.95	3437.19	7.241
270	0.1247	3110.40	3459.57	7.270
280	0.1265	3127.88	3481.96	7.299
290	0.1282	3145.38	3504.36	7.327
300	0.1300	3162.92	3526.79	7.355
310	0.1317	3180.50	3549.23	7.383
320	0.1334	3198.11	3571.70	7.411
330	0.1352	3215.76	3594.20	7.438
340	0.1369	3233.46	3616.74	7.465
350	0.1386	3251.20	3639.31	7.491
360	0.1403	3268.98	3661.91	7.518
370	0.1421	3286.81	3684.56	7.544
380	0.1438	3304.70	3707.25	7.570
390	0.1455	3322.63	3729.98	7.595
400	0.1472	3340.61	3752.75	7.620
410	0.1489	3358.65	3775.58	7.646
420	0.1506	3376.75	3798.45	7.670
430	0.1523	3394.89	3821.37	7.695
440	0.1540	3413.10	3844.34	7.720

t (°C)	$v\left(\frac{m^3}{kg}\right)$	$u\left(\frac{kJ}{kg}\right)$	$h\left(\frac{kJ}{kg}\right)$	$s\left(\frac{kJ}{kg\,K}\right)$
Sat Vapour (g)	0.0714	2602.91	2802.92	6.212
450	0.1557	3431.36	3867.36	7.744
460	0.1574	3449.68	3890.44	7.768
470	0.1591	3468.06	3913.57	7.792
480	0.1608	3486.50	3936.75	7.816
490	0.1625	3505.00	3959.99	7.839
500	0.1642	3523.56	3983.29	7.862
510	0.1659	3542.18	4006.64	7.886
520	0.1676	3560.86	4030.05	7.909
530	0.1693	3579.60	4053.52	7.931
540	0.1709	3598.40	4077.05	7.954
550	0.1726	3617.27	4100.64	7.977
560	0.1743	3636.20	4124.28	7.999
570	0.1760	3655.19	4147.99	8.021
580	0.1777	3674.25	4171.75	8.043
590	0.1794	3693.37	4195.58	8.065
600	0.1810	3712.55	4219.47	8.087
610	0.1827	3731.79	4243.41	8.108
620	0.1844	3751.10	4267.42	8.130
630	0.1861	3770.48	4291.49	8.151
640	0.1878	3789.91	4315.62	8.172
650	0.1894	3809.41	4339.81	8.194
660	0.1911	3828.97	4364.07	8.214
670	0.1928	3848.60	4388.38	8.235
680	0.1945	3868.29	4412.76	8.256
690	0.1961	3888.04	4437.19	8.277
700	0.1978	3907.85	4461.69	8.297
710	0.1995	3927.73	4486.25	8.317
720	0.2011	3947.67	4510.87	8.338
730	0.2028	3967.67	4535.55	8.358
740	0.2045	3987.74	4560.29	8.378
750	0.2062	4007.87	4585.09	8.397
760	0.2078	4028.05	4609.95	8.417
770	0.2095	4048.30	4634.88	8.437
780	0.2112	4068.62	4659.86	8.456
790	0.2128	4088.99	4684.90	8.476
800	0.2145	4109.42	4710.00	8.495
810	0.2162	4129.92	4735.16	8.514
820	0.2178	4150.47	4760.38	8.534
830	0.2195	4171.09	4785.66	8.553
840	0.2212	4191.76	4811.00	8.572
850	0.2228	4212.50	4836.40	8.590
860	0.2245	4233.29	4861.85	8.609
870	0.2261	4254.14	4887.36	8.628

t (°C)	$v\left(\frac{m^3}{kg}\right)$	$u\left(\frac{kJ}{kg}\right)$	$h\left(\frac{kJ}{kg}\right)$	$s\left(\frac{kJ}{kg\,K}\right)$
Sat Vapour (g)	0.0714	2602.91	2802.92	6.212
880	0.2278	4275.06	4912.93	8.646
890	0.2295	4296.02	4938.56	8.665
900	0.2311	4317.05	4964.24	8.683
910	0.2328	4338.14	4989.98	8.701
920	0.2345	4359.28	5015.78	8.720
930	0.2361	4380.48	5041.63	8.738
940	0.2378	4401.73	5067.54	8.756
950	0.2395	4423.04	5093.50	8.774
960	0.2411	4444.41	5119.52	8.791
970	0.2428	4465.83	5145.59	8.809
980	0.2444	4487.30	5171.72	8.827
990	0.2461	4508.84	5197.90	8.845
1000	0.2478	4530.42	5224.13	8.862
1010	0.2494	4552.06	5250.42	8.879
1020	0.2511	4573.75	5276.76	8.897
1030	0.2527	4595.50	5303.15	8.914
1040	0.2544	4617.30	5329.59	8.931
1050	0.2561	4639.15	5356.09	8.948
1060	0.2577	4661.05	5382.64	8.965
1070	0.2594	4683.00	5409.23	8.982
1080	0.2610	4705.01	5435.88	8.999
1090	0.2627	4727.06	5462.58	9.016
1100	0.2643	4749.17	5489.33	9.033
1110	0.2660	4771.32	5516.13	9.049
1120	0.2677	4793.53	5542.97	9.066
1130	0.2693	4815.78	5569.87	9.083
1140	0.2710	4838.09	5596.81	9.099
1150	0.2726	4860.44	5623.80	9.115
1160	0.2743	4882.84	5650.84	9.132
1170	0.2759	4905.29	5677.93	9.148
1180	0.2776	4927.78	5705.06	9.164
1190	0.2793	4950.32	5732.24	9.180
1200	0.2809	4972.91	5759.47	9.196
1210	0.2826	4995.55	5786.74	9.212
1220	0.2842	5018.23	5814.06	9.228
1230	0.2859	5040.95	5841.42	9.244
1240	0.2875	5063.72	5868.83	9.260
1250	0.2892	5086.54	5896.28	9.275
1260	0.2908	5109.40	5923.77	9.291
1270	0.2925	5132.30	5951.31	9.307
1280	0.2942	5155.25	5978.89	9.322
1290	0.2958	5178.24	6006.52	9.338
1300	0.2975	5201.27	6034.18	9.353

Table 3(19) Superheated Steam Properties (Pressure Base) Pressure = 29 bar and saturation temperature, Tsat = 231.979°C

t (°C)	v $\left(\frac{m^3}{kg}\right)$	u $\left(\frac{kJ}{kg}\right)$	h $\left(\frac{kJ}{kg}\right)$	s $\left(\frac{kJ}{kg\,K}\right)$	t (°C)	v $\left(\frac{m^3}{kg}\right)$	u $\left(\frac{kJ}{kg}\right)$	h $\left(\frac{kJ}{kg}\right)$	s $\left(\frac{kJ}{kg\,K}\right)$	t (°C)	v $\left(\frac{m^3}{kg}\right)$	u $\left(\frac{kJ}{kg}\right)$	h $\left(\frac{kJ}{kg}\right)$	s $\left(\frac{kJ}{kg\,K}\right)$
Sat Vapour (g)	0.0690	2603.07	2803.08	6.199	Sat Vapour (g)	0.0690	2603.07	2803.08	6.199	Sat Vapour (g)	0.0690	2603.07	2803.08	6.199
10	0.0715	2629.26	2836.71	6.265	450	0.1506	3434.40	3871.17	7.732	880	0.2203	4278.87	4917.63	8.634
20	0.0739	2653.12	2867.54	6.324	460	0.1523	3452.74	3894.27	7.756	890	0.2219	4299.86	4943.27	8.652
30	0.0762	2675.53	2896.58	6.379	470	0.1539	3471.14	3917.43	7.780	900	0.2235	4320.90	4968.97	8.670
40	0.0784	2696.91	2924.33	6.430	480	0.1555	3489.60	3940.64	7.804	910	0.2251	4342.00	4994.72	8.689
50	0.0805	2717.51	2951.10	6.479	490	0.1572	3508.12	3963.91	7.827	920	0.2267	4363.15	5020.54	8.707
60	0.0826	2737.48	2977.08	6.525	500	0.1588	3526.70	3987.23	7.850	930	0.2283	4384.36	5046.40	8.725
70	0.0846	2756.96	3002.42	6.570	510	0.1604	3545.34	4010.61	7.873	940	0.2299	4405.63	5072.33	8.743
80	0.0866	2776.04	3027.24	6.613	520	0.1621	3564.05	4034.04	7.896	950	0.2315	4426.96	5098.30	8.761
90	0.0886	2794.79	3051.64	6.654	530	0.1637	3582.81	4057.54	7.919	960	0.2331	4448.34	5124.33	8.779
100	0.0905	2813.27	3075.68	6.694	540	0.1653	3601.64	4081.09	7.942	970	0.2347	4469.77	5150.42	8.796
110	0.0924	2831.53	3099.42	6.733	550	0.1670	3620.52	4104.70	7.964	980	0.2363	4491.26	5176.56	8.814
120	0.0942	2849.62	3122.92	6.771	560	0.1686	3639.47	4128.37	7.987	990	0.2379	4512.80	5202.76	8.832
130	0.0961	2867.55	3146.21	6.808	570	0.1702	3658.49	4152.10	8.009	1000	0.2395	4534.40	5229.00	8.849
140	0.0979	2885.36	3169.32	6.844	580	0.1718	3677.56	4175.89	8.031	1010	0.2411	4556.05	5255.31	8.866
150	0.0997	2903.08	3192.29	6.879	590	0.1735	3696.70	4199.73	8.053	1020	0.2427	4577.76	5281.66	8.884
160	0.1015	2920.72	3215.14	6.914	600	0.1751	3715.90	4223.64	8.075	1030	0.2443	4599.52	5308.06	8.901
170	0.1033	2938.29	3237.88	6.948	610	0.1767	3735.16	4247.61	8.096	1040	0.2459	4621.33	5334.52	8.918
180	0.1051	2955.81	3260.55	6.981	620	0.1783	3754.49	4271.64	8.118	1050	0.2475	4643.19	5361.03	8.935
190	0.1068	2973.30	3283.14	7.014	630	0.1799	3773.88	4295.73	8.139	1060	0.2491	4665.10	5387.59	8.952
200	0.1086	2990.77	3305.69	7.046	640	0.1816	3793.34	4319.88	8.160	1070	0.2507	4687.07	5414.20	8.969
210	0.1103	3008.21	3328.18	7.078	650	0.1832	3812.85	4344.10	8.181	1080	0.2523	4709.09	5440.86	8.986
220	0.1121	3025.65	3350.65	7.109	660	0.1848	3832.43	4368.37	8.202	1090	0.2539	4731.15	5467.57	9.003
230	0.1138	3043.09	3373.09	7.140	670	0.1864	3852.08	4392.70	8.223	1100	0.2555	4753.27	5494.33	9.020
240	0.1155	3060.53	3395.52	7.170	680	0.1880	3871.78	4417.10	8.243	1110	0.2571	4775.44	5521.14	9.036
250	0.1172	3077.98	3417.94	7.200	690	0.1897	3891.55	4441.55	8.264	1120	0.2587	4797.66	5548.00	9.053
260	0.1189	3095.45	3440.36	7.229	700	0.1913	3911.38	4466.07	8.284	1130	0.2603	4819.92	5574.91	9.069
270	0.1206	3112.94	3462.78	7.258	710	0.1929	3931.28	4490.65	8.305	1140	0.2619	4842.24	5601.87	9.086
280	0.1223	3130.45	3485.21	7.287	720	0.1945	3951.24	4515.29	8.325	1150	0.2635	4864.60	5628.87	9.102
290	0.1240	3147.99	3507.65	7.316	730	0.1961	3971.26	4539.99	8.345	1160	0.2651	4887.01	5655.92	9.119
300	0.1257	3165.56	3530.11	7.344	740	0.1977	3991.34	4564.75	8.365	1170	0.2667	4909.47	5683.02	9.135
310	0.1274	3183.16	3552.60	7.371	750	0.1993	4011.48	4589.57	8.385	1180	0.2683	4931.98	5710.17	9.151
320	0.1291	3200.81	3575.10	7.399	760	0.2010	4031.69	4614.45	8.405	1190	0.2699	4954.53	5737.36	9.167
330	0.1307	3218.49	3597.64	7.426	770	0.2026	4051.95	4639.39	8.424	1200	0.2715	4977.13	5764.59	9.183
340	0.1324	3236.21	3620.21	7.453	780	0.2042	4072.28	4664.39	8.444	1210	0.2731	4999.77	5791.88	9.199
350	0.1341	3253.98	3642.81	7.479	790	0.2058	4092.67	4689.45	8.463	1220	0.2747	5022.46	5819.21	9.215
360	0.1357	3271.79	3665.45	7.506	800	0.2074	4113.12	4714.57	8.483	1230	0.2763	5045.20	5846.58	9.231
370	0.1374	3289.65	3688.13	7.532	810	0.2090	4133.63	4739.74	8.502	1240	0.2779	5067.98	5874.00	9.246
380	0.1391	3307.56	3710.85	7.558	820	0.2106	4154.20	4764.98	8.521	1250	0.2795	5090.80	5901.46	9.262
390	0.1407	3325.52	3733.61	7.583	830	0.2122	4174.83	4790.28	8.540	1260	0.2811	5113.67	5928.96	9.278
400	0.1424	3343.53	3756.42	7.608	840	0.2138	4195.52	4815.63	8.559	1270	0.2827	5136.59	5956.51	9.293
410	0.1440	3361.59	3779.27	7.634	850	0.2154	4216.27	4841.04	8.578	1280	0.2843	5159.54	5984.10	9.309
420	0.1457	3379.71	3802.17	7.658	860	0.2170	4237.08	4866.51	8.596	1290	0.2859	5182.54	6011.74	9.324
430	0.1473	3397.88	3825.12	7.683	870	0.2187	4257.95	4892.04	8.615	1300	0.2875	5205.59	6039.42	9.340
440	0.1490	3416.11	3848.12	7.708										

Table 3(20) Superheated Steam Properties (Pressure Base) Pressure = 30.0 bar and saturation temperature, Tsat = 233.853°C

t (°C)	v (/kg)	u (kJ/kg)	h (kJ/kg)	s (kJ/kg K)	t (°C)	v (/kg)	u (kJ/kg)	h (kJ/kg)	s (kJ/kg K)	t (°C)	v (/kg)	u (kJ/kg)	h (kJ/kg)	s (kJ/kg K)
Sat Vapour (g)	0.0667	2603.17	2803.17	6.186	Sat Vapour (g)	0.0667	2603.17	2803.17	6.186	Sat Vapour (g)	0.0667	2603.17	2803.17	6.186
10	0.0692	2629.65	2837.16	6.252	450	0.1458	3437.34	3874.87	7.720	880	0.2132	4282.59	4922.20	8.621
20	0.0715	2653.73	2868.28	6.312	460	0.1474	3455.71	3898.00	7.744	890	0.2148	4303.59	4947.86	8.640
30	0.0737	2676.34	2897.56	6.367	470	0.1490	3474.13	3921.19	7.768	900	0.2163	4324.64	4973.57	8.658
40	0.0759	2697.88	2925.52	6.418	480	0.1506	3492.62	3944.42	7.792	910	0.2179	4345.75	4999.34	8.676
50	0.0780	2718.61	2952.47	6.467	490	0.1522	3511.16	3967.71	7.815	920	0.2194	4366.92	5025.17	8.694
60	0.0800	2738.71	2978.61	6.514	500	0.1538	3529.76	3991.06	7.839	930	0.2210	4388.15	5051.05	8.712
70	0.0819	2758.30	3004.09	6.558	510	0.1553	3548.42	4014.46	7.862	940	0.2225	4409.43	5076.99	8.730
80	0.0839	2777.48	3029.05	6.601	520	0.1569	3567.15	4037.92	7.885	950	0.2241	4430.77	5102.98	8.748
90	0.0857	2796.32	3053.56	6.642	530	0.1585	3585.93	4061.44	7.907	960	0.2256	4452.16	5129.02	8.766
100	0.0876	2814.89	3077.71	6.683	540	0.1601	3604.78	4085.02	7.930	970	0.2272	4473.81	5155.12	8.784
110	0.0894	2833.22	3101.55	6.722	550	0.1617	3623.68	4108.65	7.953	980	0.2287	4495.11	5181.28	8.802
120	0.0913	2851.38	3125.13	6.759	560	0.1632	3642.65	4132.34	7.975	990	0.2303	4516.67	5207.49	8.819
130	0.0930	2869.38	3148.51	6.796	570	0.1648	3661.69	4156.09	7.997	1000	0.2318	4538.28	5233.75	8.837
140	0.0948	2887.25	3171.70	6.833	580	0.1664	3680.78	4179.90	8.019	1010	0.2334	4559.94	5260.06	8.854
150	0.0966	2905.03	3194.74	6.868	590	0.1679	3699.94	4203.77	8.041	1020	0.2349	4581.66	5286.43	8.871
160	0.0983	2922.72	3217.66	6.903	600	0.1695	3719.16	4227.70	8.063	1030	0.2365	4603.43	5312.85	8.889
170	0.1000	2940.34	3240.48	6.936	610	0.1711	3738.44	4251.69	8.084	1040	0.2380	4625.25	5339.32	8.906
180	0.1018	2957.92	3263.20	6.970	620	0.1727	3757.79	4275.74	8.106	1050	0.2396	4647.13	5365.84	8.923
190	0.1035	2975.45	3285.86	7.003	630	0.1742	3777.20	4299.86	8.127	1060	0.2411	4669.05	5392.41	8.940
200	0.1052	2992.96	3308.46	7.035	640	0.1758	3796.67	4324.03	8.148	1070	0.2427	4691.03	5419.04	8.957
210	0.1069	3010.45	3331.01	7.066	650	0.1774	3816.20	4348.26	8.169	1080	0.2442	4713.06	5445.71	8.974
220	0.1085	3027.93	3353.53	7.098	660	0.1789	3835.80	4372.55	8.190	1090	0.2458	4735.14	5472.43	8.990
230	0.1102	3045.40	3376.02	7.128	670	0.1805	3855.46	4396.91	8.211	1100	0.2473	4757.27	5499.21	9.007
240	0.1119	3062.88	3398.49	7.159	680	0.1820	3875.19	4421.32	8.231	1110	0.2489	4779.45	5526.03	9.024
250	0.1135	3080.37	3420.96	7.188	690	0.1836	3894.97	4445.80	8.252	1120	0.2504	4801.67	5552.90	9.040
260	0.1152	3097.87	3443.42	7.218	700	0.1852	3914.82	4470.33	8.272	1130	0.2520	4823.95	5579.82	9.057
270	0.1168	3115.39	3465.88	7.247	710	0.1867	3934.73	4494.93	8.293	1140	0.2535	4846.28	5606.79	9.073
280	0.1185	3132.94	3488.35	7.276	720	0.1883	3954.71	4519.59	8.313	1150	0.2551	4868.65	5633.80	9.090
290	0.1201	3150.51	3510.84	7.304	730	0.1899	3974.74	4544.30	8.333	1160	0.2566	4891.07	5660.87	9.106
300	0.1217	3168.11	3533.34	7.332	740	0.1914	3994.84	4569.08	8.353	1170	0.2581	4913.54	5687.98	9.122
310	0.1234	3185.75	3555.86	7.360	750	0.1930	4015.00	4593.92	8.373	1180	0.2597	4936.06	5715.13	9.138
320	0.1250	3203.42	3578.40	7.387	760	0.1945	4035.22	4618.82	8.392	1190	0.2612	4958.62	5742.34	9.154
330	0.1266	3221.13	3600.98	7.415	770	0.1961	4055.51	4643.78	8.412	1200	0.2628	4981.23	5769.58	9.170
340	0.1282	3238.88	3623.58	7.441	780	0.1976	4075.85	4668.79	8.432	1210	0.2643	5003.88	5796.88	9.186
350	0.1298	3256.68	3646.22	7.468	790	0.1992	4096.25	4693.87	8.451	1220	0.2659	5026.59	5824.22	9.202
360	0.1315	3274.52	3668.89	7.494	800	0.2008	4116.72	4719.01	8.470	1230	0.2674	5049.33	5851.60	9.218
370	0.1331	3292.40	3691.60	7.520	810	0.2023	4137.24	4744.20	8.490	1240	0.2690	5072.12	5879.03	9.234
380	0.1347	3310.34	3714.35	7.546	820	0.2039	4157.83	4769.46	8.509	1250	0.2705	5094.96	5906.50	9.249
390	0.1363	3328.32	3737.14	7.572	830	0.2054	4178.48	4794.77	8.528	1260	0.2721	5117.84	5934.02	9.265
400	0.1379	3346.36	3759.98	7.597	840	0.2070	4199.18	4820.14	8.547	1270	0.2736	5140.76	5961.58	9.281
410	0.1395	3364.45	3782.86	7.622	850	0.2085	4219.94	4845.57	8.565	1280	0.2752	5163.73	5989.18	9.296
420	0.1411	3382.59	3805.79	7.647	860	0.2101	4240.77	4871.05	8.584	1290	0.2767	5186.74	6016.82	9.311
430	0.1427	3400.78	3828.77	7.672	870	0.2116	4261.65	4896.60	8.603	1300	0.2782	5209.79	6044.51	9.327
440	0.1443	3419.04	3851.80	7.696										

Table 3(21) Superheated Steam Properties (Pressure Base) Pressure = 31.0 bar and saturation temperature, Tsat = 235.678°C

t (°C)	v $\left(\frac{m^3}{kg}\right)$	u $\left(\frac{kJ}{kg}\right)$	h $\left(\frac{kJ}{kg}\right)$	s $\left(\frac{kJ}{kg\,K}\right)$
Sat Vapour (g)	0.0645	2603.22	2803.20	6.173
10	0.0670	2629.98	2837.54	6.240
20	0.0692	2654.29	2868.95	6.300
30	0.0714	2677.08	2898.47	6.355
40	0.0735	2698.79	2926.64	6.407
50	0.0755	2719.66	2953.77	6.456
60	0.0775	2739.88	2980.06	6.502
70	0.0794	2759.58	3005.69	6.547
80	0.0813	2778.86	3030.77	6.590
90	0.0831	2797.79	3055.40	6.631
100	0.0849	2816.44	3079.65	6.672
110	0.0867	2834.85	3103.59	6.711
120	0.0884	2853.08	3127.27	6.749
130	0.0902	2871.14	3150.73	6.786
140	0.0919	2889.08	3174.00	6.822
150	0.0936	2906.91	3197.11	6.857
160	0.0953	2924.65	3220.10	6.892
170	0.0970	2942.33	3242.98	6.926
180	0.0987	2959.95	3265.77	6.959
190	0.1003	2977.53	3288.48	6.992
200	0.1020	2995.08	3311.14	7.024
210	0.1036	3012.61	3333.74	7.056
220	0.1052	3030.13	3356.31	7.087
230	0.1068	3047.65	3378.85	7.117
240	0.1085	3065.16	3401.38	7.148
250	0.1101	3082.69	3423.89	7.178
260	0.1117	3100.23	3446.39	7.207
270	0.1133	3117.78	3468.90	7.236
280	0.1149	3135.36	3491.41	7.265
290	0.1164	3152.96	3513.93	7.293
300	0.1180	3170.59	3536.47	7.321
310	0.1196	3188.26	3559.03	7.349
320	0.1212	3205.96	3581.61	7.376
330	0.1227	3223.70	3604.22	7.404
340	0.1243	3241.48	3626.85	7.430
350	0.1259	3259.30	3649.52	7.457
360	0.1274	3277.17	3672.23	7.483
370	0.1290	3295.08	3694.97	7.509
380	0.1306	3313.04	3717.75	7.535
390	0.1321	3331.05	3740.57	7.561
400	0.1337	3349.11	3763.44	7.586
410	0.1352	3367.22	3786.35	7.611
420	0.1367	3385.39	3809.31	7.636
430	0.1383	3403.61	3832.31	7.660
440	0.1398	3421.88	3855.37	7.685

t (°C)	v $\left(\frac{m^3}{kg}\right)$	u $\left(\frac{kJ}{kg}\right)$	h $\left(\frac{kJ}{kg}\right)$	s $\left(\frac{kJ}{kg\,K}\right)$
Sat Vapour (g)	0.0645	2603.22	2803.20	6.173
450	0.1414	3440.21	3878.47	7.709
460	0.1429	3458.60	3901.63	7.733
470	0.1444	3477.05	3924.84	7.757
480	0.1460	3495.55	3948.10	7.781
490	0.1475	3514.11	3971.42	7.804
500	0.1490	3532.74	3994.79	7.827
510	0.1506	3551.42	4018.22	7.850
520	0.1521	3570.16	4041.70	7.873
530	0.1536	3588.97	4065.24	7.896
540	0.1552	3607.83	4088.84	7.919
550	0.1567	3626.76	4112.49	7.941
560	0.1582	3645.75	4136.21	7.963
570	0.1597	3664.80	4159.98	7.986
580	0.1613	3683.92	4183.81	8.008
590	0.1628	3703.09	4207.71	8.029
600	0.1643	3722.33	4231.66	8.051
610	0.1658	3741.63	4255.67	8.073
620	0.1673	3761.00	4279.74	8.094
630	0.1689	3780.42	4303.87	8.115
640	0.1704	3799.91	4328.06	8.137
650	0.1719	3819.46	4352.32	8.158
660	0.1734	3839.08	4376.63	8.178
670	0.1749	3858.76	4401.00	8.199
680	0.1764	3878.50	4425.44	8.220
690	0.1779	3898.30	4449.93	8.240
700	0.1795	3918.17	4474.48	8.261
710	0.1810	3938.10	4499.10	8.281
720	0.1825	3958.09	4523.77	8.301
730	0.1840	3978.14	4548.51	8.321
740	0.1855	3998.25	4573.31	8.341
750	0.1870	4018.43	4598.16	8.361
760	0.1885	4038.67	4623.08	8.381
770	0.1900	4058.96	4648.05	8.400
780	0.1915	4079.32	4673.09	8.420
790	0.1930	4099.74	4698.18	8.439
800	0.1946	4120.22	4723.33	8.459
810	0.1961	4140.76	4748.54	8.478
820	0.1976	4161.36	4773.81	8.497
830	0.1991	4182.02	4799.14	8.516
840	0.2006	4202.74	4824.53	8.535
850	0.2021	4223.52	4849.97	8.553
860	0.2036	4244.36	4875.48	8.572
870	0.2051	4265.25	4901.03	8.591

t (°C)	v $\left(\frac{m^3}{kg}\right)$	u $\left(\frac{kJ}{kg}\right)$	h $\left(\frac{kJ}{kg}\right)$	s $\left(\frac{kJ}{kg\,K}\right)$
Sat Vapour (g)	0.0645	2603.22	2803.20	6.173
880	0.2066	4286.21	4926.65	8.609
890	0.2081	4307.22	4952.32	8.628
900	0.2096	4328.29	4978.05	8.646
910	0.2111	4349.41	5003.84	8.664
920	0.2126	4370.59	5029.68	8.682
930	0.2141	4391.83	5055.57	8.700
940	0.2156	4413.13	5081.53	8.718
950	0.2171	4434.48	5107.53	8.736
960	0.2186	4455.88	5133.59	8.754
970	0.2201	4477.34	5159.71	8.772
980	0.2216	4498.86	5185.88	8.789
990	0.2231	4520.43	5212.10	8.807
1000	0.2246	4542.05	5238.37	8.825
1010	0.2261	4563.73	5264.70	8.842
1020	0.2276	4585.46	5291.08	8.859
1030	0.2291	4607.24	5317.51	8.876
1040	0.2306	4629.07	5343.99	8.894
1050	0.2321	4650.96	5370.53	8.911
1060	0.2336	4672.90	5397.11	8.928
1070	0.2351	4694.89	5423.75	8.945
1080	0.2366	4716.93	5450.44	8.961
1090	0.2381	4739.02	5477.17	8.978
1100	0.2396	4761.16	5503.96	8.995
1110	0.2411	4783.35	5530.79	9.012
1120	0.2426	4805.59	5557.67	9.028
1130	0.2441	4827.88	5584.61	9.045
1140	0.2456	4850.21	5611.59	9.061
1150	0.2471	4872.60	5638.61	9.077
1160	0.2486	4895.03	5665.69	9.094
1170	0.2501	4917.51	5692.81	9.110
1180	0.2516	4940.03	5719.97	9.126
1190	0.2531	4962.61	5747.19	9.142
1200	0.2546	4985.23	5774.45	9.158
1210	0.2561	5007.89	5801.75	9.174
1220	0.2576	5030.60	5829.10	9.190
1230	0.2591	5053.36	5856.50	9.206
1240	0.2606	5076.16	5883.93	9.221
1250	0.2621	5099.00	5911.42	9.237
1260	0.2636	5121.89	5938.94	9.253
1270	0.2651	5144.83	5966.51	9.268
1280	0.2666	5167.80	5994.13	9.284
1290	0.2681	5190.82	6021.78	9.299
1300	0.2695	5213.88	6049.48	9.314

Table 3(22) Superheated Steam Properties (Pressure Base) Pressure = 32.0 bar and saturation temperature, Tsat = 237.459°C

t (°C)	$v\left(\frac{m^3}{kg}\right)$	$u\left(\frac{kJ}{kg}\right)$	$h\left(\frac{kJ}{kg}\right)$	$s\left(\frac{kJ}{kg\,K}\right)$
Sat Vapour (g)	*0.0625*	*2603.22*	*2803.15*	*6.160*
10	0.0649	2630.26	2837.86	6.228
20	0.0671	2654.80	2869.55	6.288
30	0.0692	2677.78	2899.31	6.343
40	0.0713	2699.64	2927.69	6.395
50	0.0732	2720.66	2954.99	6.445
60	0.0751	2741.00	2981.45	6.491
70	0.0770	2760.81	3007.22	6.536
80	0.0788	2780.18	3032.42	6.579
90	0.0806	2799.20	3057.17	6.621
100	0.0824	2817.93	3081.52	6.661
110	0.0841	2836.42	3105.56	6.700
120	0.0858	2854.71	3129.33	6.738
130	0.0875	2872.84	3152.87	6.775
140	0.0892	2890.84	3176.22	6.811
150	0.0908	2908.73	3199.41	6.847
160	0.0925	2926.52	3222.46	6.881
170	0.0941	2944.25	3245.40	6.915
180	0.0957	2961.92	3268.25	6.948
190	0.0973	2979.55	3291.03	6.981
200	0.0989	2997.14	3313.73	7.013
210	0.1005	3014.71	3336.39	7.045
220	0.1021	3032.27	3359.01	7.076
230	0.1037	3049.82	3381.60	7.107
240	0.1052	3067.38	3404.17	7.137
250	0.1068	3084.94	3426.73	7.167
260	0.1084	3102.51	3449.27	7.197
270	0.1099	3120.10	3471.82	7.226
280	0.1115	3137.71	3494.37	7.254
290	0.1130	3155.34	3516.93	7.283
300	0.1145	3173.01	3539.51	7.311
310	0.1161	3190.70	3562.11	7.338
320	0.1176	3208.43	3584.72	7.366
330	0.1191	3226.20	3607.36	7.393
340	0.1206	3244.00	3630.04	7.420
350	0.1222	3261.85	3652.74	7.446
360	0.1237	3279.74	3675.47	7.472
370	0.1252	3297.68	3698.25	7.498
380	0.1267	3315.67	3721.06	7.524
390	0.1282	3333.70	3743.91	7.550
400	0.1297	3351.79	3766.81	7.575
410	0.1312	3369.92	3789.75	7.600
420	0.1327	3388.11	3812.73	7.625
430	0.1342	3406.35	3835.77	7.650
440	0.1357	3424.65	3858.85	7.674

t (°C)	$v\left(\frac{m^3}{kg}\right)$	$u\left(\frac{kJ}{kg}\right)$	$h\left(\frac{kJ}{kg}\right)$	$s\left(\frac{kJ}{kg\,K}\right)$
Sat Vapour (g)	*0.0625*	*2603.22*	*2803.15*	*6.160*
450	0.1372	3443.00	3881.98	7.698
460	0.1387	3461.41	3905.16	7.722
470	0.1402	3479.88	3928.39	7.746
480	0.1416	3498.41	3951.68	7.770
490	0.1431	3516.99	3975.02	7.793
500	0.1446	3535.64	3998.42	7.816
510	0.1461	3554.34	4021.87	7.839
520	0.1476	3573.10	4045.38	7.862
530	0.1491	3591.93	4068.94	7.885
540	0.1505	3610.81	4092.56	7.908
550	0.1520	3629.76	4116.24	7.930
560	0.1535	3648.77	4139.98	7.952
570	0.1550	3667.84	4163.77	7.975
580	0.1565	3686.97	4187.62	7.996
590	0.1579	3706.16	4211.54	8.018
600	0.1594	3725.42	4235.51	8.040
610	0.1609	3744.74	4259.54	8.062
620	0.1623	3764.12	4283.63	8.083
630	0.1638	3783.57	4307.78	8.104
640	0.1653	3803.07	4332.00	8.125
650	0.1668	3822.64	4356.27	8.146
660	0.1682	3842.28	4380.60	8.167
670	0.1697	3861.97	4404.99	8.188
680	0.1712	3881.73	4429.44	8.209
690	0.1726	3901.55	4453.96	8.229
700	0.1741	3921.43	4478.53	8.249
710	0.1756	3941.37	4503.16	8.270
720	0.1770	3961.38	4527.86	8.290
730	0.1785	3981.45	4552.61	8.310
740	0.1800	4001.58	4577.42	8.330
750	0.1814	4021.77	4602.30	8.350
760	0.1829	4042.02	4627.23	8.369
770	0.1843	4062.33	4652.22	8.389
780	0.1858	4082.71	4677.27	8.408
790	0.1873	4103.14	4702.38	8.428
800	0.1887	4123.64	4727.55	8.447
810	0.1902	4144.19	4752.78	8.466
820	0.1916	4164.81	4778.06	8.485
830	0.1931	4185.48	4803.41	8.504
840	0.1946	4206.22	4828.81	8.523
850	0.1960	4227.01	4854.27	8.542
860	0.1975	4247.86	4879.79	8.561
870	0.1989	4268.77	4905.36	8.579

t (°C)	$v\left(\frac{m^3}{kg}\right)$	$u\left(\frac{kJ}{kg}\right)$	$h\left(\frac{kJ}{kg}\right)$	$s\left(\frac{kJ}{kg\,K}\right)$
Sat Vapour (g)	*0.0625*	*2603.22*	*2803.15*	*6.160*
880	0.2004	4289.73	4930.99	8.598
890	0.2019	4310.76	4956.68	8.616
900	0.2033	4331.84	4982.42	8.634
910	0.2048	4352.98	5008.22	8.653
920	0.2062	4374.18	5034.08	8.671
930	0.2077	4395.43	5059.99	8.689
940	0.2091	4416.73	5085.95	8.707
950	0.2106	4438.10	5111.97	8.725
960	0.2120	4459.52	5138.05	8.743
970	0.2135	4480.99	5164.18	8.760
980	0.2150	4502.52	5190.36	8.778
990	0.2164	4524.10	5216.59	8.795
1000	0.2179	4545.73	5242.88	8.813
1010	0.2193	4567.42	5269.22	8.830
1020	0.2208	4589.16	5295.61	8.848
1030	0.2222	4610.96	5322.06	8.865
1040	0.2237	4632.80	5348.55	8.882
1050	0.2251	4654.70	5375.10	8.899
1060	0.2266	4676.65	5401.70	8.916
1070	0.2280	4698.65	5428.35	8.933
1080	0.2295	4720.70	5455.04	8.950
1090	0.2309	4742.80	5481.79	8.966
1100	0.2324	4764.95	5508.59	8.983
1110	0.2338	4787.15	5535.44	9.000
1120	0.2353	4809.40	5562.33	9.016
1130	0.2367	4831.70	5589.27	9.033
1140	0.2382	4854.05	5616.26	9.049
1150	0.2396	4876.44	5643.30	9.066
1160	0.2411	4898.89	5670.39	9.082
1170	0.2425	4921.38	5697.52	9.098
1180	0.2440	4943.91	5724.70	9.114
1190	0.2454	4966.50	5751.92	9.130
1200	0.2469	4989.13	5779.19	9.146
1210	0.2483	5011.80	5806.51	9.162
1220	0.2498	5034.52	5833.87	9.178
1230	0.2512	5057.29	5861.27	9.194
1240	0.2527	5080.10	5888.72	9.209
1250	0.2541	5102.95	5916.21	9.225
1260	0.2556	5125.85	5943.75	9.241
1270	0.2570	5148.79	5971.33	9.256
1280	0.2585	5171.78	5998.95	9.272
1290	0.2599	5194.80	6026.61	9.287
1300	0.2614	5217.88	6054.32	9.302

Table 3(23) Superheated Steam Properties (Pressure Base) Pressure = 33.0 bar and saturation temperature, Tsat = 239.198°C

t (°C)	v ($\frac{m^3}{kg}$)	u ($\frac{kJ}{kg}$)	h ($\frac{kJ}{kg}$)	s ($\frac{kJ}{kg\,K}$)	t (°C)	v ($\frac{m^3}{kg}$)	u ($\frac{kJ}{kg}$)	h ($\frac{kJ}{kg}$)	s ($\frac{kJ}{kg\,K}$)	t (°C)	v ($\frac{m^3}{kg}$)	u ($\frac{kJ}{kg}$)	h ($\frac{kJ}{kg}$)	s ($\frac{kJ}{kg\,K}$)
Sat Vapour (g)	0.0606	2603.18	2803.05	6.148	Sat Vapour (g)	0.0606	2603.18	2803.05	6.148	Sat Vapour (g)	0.0606	2603.18	2803.05	6.148
10	0.0629	2630.50	2838.11	6.216	450	0.1332	3445.72	3885.39	7.688	880	0.1946	4293.18	4935.23	8.587
20	0.0651	2655.26	2870.09	6.276	460	0.1347	3464.16	3908.60	7.712	890	0.1960	4314.21	4960.93	8.605
30	0.0672	2678.43	2900.09	6.332	470	0.1361	3482.64	3931.86	7.735	900	0.1974	4335.31	4986.69	8.623
40	0.0692	2700.45	2928.67	6.384	480	0.1376	3501.19	3955.17	7.759	910	0.1988	4356.46	5012.50	8.642
50	0.0711	2721.60	2956.15	6.434	490	0.1390	3519.80	3978.54	7.782	920	0.2002	4377.67	5038.37	8.660
60	0.0729	2742.07	2982.76	6.481	500	0.1405	3538.46	4001.95	7.806	930	0.2016	4398.93	5064.30	8.678
70	0.0748	2761.98	3008.67	6.526	510	0.1419	3557.18	4025.43	7.829	940	0.2030	4420.25	5090.27	8.696
80	0.0765	2781.45	3034.00	6.569	520	0.1433	3575.97	4048.96	7.852	950	0.2044	4441.63	5116.31	8.713
90	0.0783	2800.56	3058.86	6.610	530	0.1448	3594.81	4072.55	7.874	960	0.2059	4463.06	5142.40	8.731
100	0.0800	2819.37	3083.32	6.651	540	0.1462	3613.71	4096.19	7.897	970	0.2073	4484.54	5168.54	8.749
110	0.0817	2837.93	3107.45	6.690	550	0.1476	3632.68	4119.89	7.919	980	0.2087	4506.08	5194.73	8.767
120	0.0833	2856.29	3131.31	6.728	560	0.1491	3651.71	4143.65	7.942	990	0.2101	4527.68	5220.98	8.784
130	0.0850	2874.49	3154.93	6.765	570	0.1505	3670.80	4167.46	7.964	1000	0.2115	4549.32	5247.28	8.802
140	0.0866	2892.54	3178.36	6.801	580	0.1519	3689.95	4191.34	7.986	1010	0.2129	4571.02	5273.63	8.819
150	0.0882	2910.49	3201.62	6.836	590	0.1534	3709.16	4215.27	8.008	1020	0.2143	4592.78	5300.04	8.836
160	0.0898	2928.34	3224.74	6.871	600	0.1548	3728.43	4239.26	8.029	1030	0.2157	4614.58	5326.50	8.853
170	0.0914	2946.11	3247.75	6.905	610	0.1562	3747.77	4263.32	8.051	1040	0.2171	4636.44	5353.00	8.871
180	0.0930	2963.83	3270.66	6.938	620	0.1577	3767.17	4287.43	8.072	1050	0.2186	4658.35	5379.56	8.888
190	0.0945	2981.50	3293.49	6.971	630	0.1591	3786.63	4311.60	8.093	1060	0.2200	4680.31	5406.17	8.905
200	0.0961	2999.14	3316.25	7.003	640	0.1605	3806.16	4335.83	8.115	1070	0.2214	4702.32	5432.83	8.921
210	0.0976	3016.75	3338.97	7.035	650	0.1619	3825.74	4360.12	8.136	1080	0.2228	4724.38	5459.54	8.938
220	0.0992	3034.35	3361.63	7.066	660	0.1634	3845.39	4384.47	8.156	1090	0.2242	4746.50	5486.30	8.955
230	0.1007	3051.94	3384.27	7.097	670	0.1648	3865.10	4408.88	8.177	1100	0.2256	4768.66	5513.11	8.972
240	0.1022	3069.53	3406.89	7.127	680	0.1662	3884.88	4433.35	8.198	1110	0.2270	4790.87	5539.97	8.988
250	0.1037	3087.12	3429.48	7.157	690	0.1676	3904.71	4457.88	8.218	1120	0.2284	4813.13	5566.87	9.005
260	0.1053	3104.73	3452.07	7.186	700	0.1691	3924.61	4482.48	8.239	1130	0.2298	4835.44	5593.83	9.021
270	0.1068	3122.35	3474.66	7.215	710	0.1705	3944.57	4507.13	8.259	1140	0.2312	4857.79	5620.83	9.038
280	0.1083	3139.99	3497.25	7.244	720	0.1719	3964.59	4531.84	8.279	1150	0.2326	4880.20	5647.88	9.054
290	0.1098	3157.66	3519.85	7.272	730	0.1733	3984.68	4556.61	8.299	1160	0.2340	4902.65	5674.97	9.070
300	0.1112	3175.35	3542.47	7.300	740	0.1747	4004.82	4581.44	8.319	1170	0.2354	4925.15	5702.12	9.086
310	0.1127	3193.07	3565.10	7.328	750	0.1762	4025.03	4606.33	8.339	1180	0.2369	4947.70	5729.31	9.103
320	0.1142	3210.83	3587.75	7.355	760	0.1776	4045.29	4631.28	8.358	1190	0.2383	4970.29	5756.54	9.119
330	0.1157	3228.63	3610.43	7.383	770	0.1790	4065.62	4656.29	8.378	1200	0.2397	4992.93	5783.82	9.135
340	0.1172	3246.46	3633.13	7.409	780	0.1804	4086.01	4681.35	8.397	1210	0.2411	5015.62	5811.15	9.151
350	0.1186	3264.34	3655.87	7.436	790	0.1818	4106.46	4706.48	8.417	1220	0.2425	5038.35	5838.52	9.166
360	0.1201	3282.25	3678.63	7.462	800	0.1832	4126.97	4731.66	8.436	1230	0.2439	5061.12	5865.93	9.182
370	0.1216	3300.22	3701.44	7.488	810	0.1847	4147.54	4756.91	8.455	1240	0.2453	5083.94	5893.39	9.198
380	0.1230	3318.23	3724.28	7.514	820	0.1861	4168.17	4782.21	8.474	1250	0.2467	5106.80	5920.89	9.214
390	0.1245	3336.28	3747.16	7.539	830	0.1875	4188.86	4807.57	8.493	1260	0.2481	5129.71	5948.44	9.229
400	0.1260	3354.39	3770.09	7.565	840	0.1889	4209.60	4832.99	8.512	1270	0.2495	5152.66	5976.03	9.245
410	0.1274	3372.55	3793.05	7.590	850	0.1903	4230.41	4858.46	8.531	1280	0.2509	5175.66	6003.66	9.260
420	0.1289	3390.76	3816.07	7.614	860	0.1917	4251.27	4883.99	8.550	1290	0.2523	5198.69	6031.33	9.275
430	0.1303	3409.03	3839.13	7.639	870	0.1931	4272.20	4909.58	8.568	1300	0.2537	5221.77	6059.05	9.291
440	0.1318	3427.35	3862.23	7.663										

Table 3(24) Superheated Steam Properties (Pressure Base) Pressure = 34.0 bar and saturation temperature, Tsat = 240.896°C

t (°C)	$v\left(\frac{m^3}{kg}\right)$	$u\left(\frac{kJ}{kg}\right)$	$h\left(\frac{kJ}{kg}\right)$	$s\left(\frac{kJ}{kg\,K}\right)$	t (°C)	$v\left(\frac{m^3}{kg}\right)$	$u\left(\frac{kJ}{kg}\right)$	$h\left(\frac{kJ}{kg}\right)$	$s\left(\frac{kJ}{kg\,K}\right)$	t (°C)	$v\left(\frac{m^3}{kg}\right)$	$u\left(\frac{kJ}{kg}\right)$	$h\left(\frac{kJ}{kg}\right)$	$s\left(\frac{kJ}{kg\,K}\right)$
Sat Vapour (g)	0.0588	2603.09	2802.89	6.136	Sat Vapour (g)	0.0588	2603.09	2802.89	6.136	Sat Vapour (g)	0.0588	2603.09	2802.89	6.136
10	0.0611	2630.69	2838.31	6.204	450	0.1295	3448.38	3888.72	7.677	880	0.1891	4296.54	4939.37	8.576
20	0.0632	2655.68	2870.57	6.265	460	0.1309	3466.83	3911.95	7.701	890	0.1904	4317.59	4965.08	8.594
30	0.0652	2679.03	2900.81	6.321	470	0.1323	3485.34	3935.24	7.725	900	0.1918	4338.70	4990.85	8.612
40	0.0672	2701.21	2929.59	6.374	480	0.1337	3503.91	3958.57	7.749	910	0.1932	4359.86	5016.68	8.631
50	0.0690	2722.50	2957.25	6.423	490	0.1351	3522.53	3981.96	7.772	920	0.1946	4381.08	5042.56	8.649
60	0.0709	2743.08	2984.02	6.470	500	0.1365	3541.21	4005.40	7.795	930	0.1959	4402.36	5068.50	8.667
70	0.0726	2763.10	3010.06	6.515	510	0.1379	3559.96	4028.90	7.819	940	0.1973	4423.69	5094.49	8.685
80	0.0744	2782.67	3035.52	6.559	520	0.1393	3578.76	4052.45	7.841	950	0.1987	4445.08	5120.54	8.703
90	0.0761	2801.87	3060.49	6.600	530	0.1407	3597.62	4076.06	7.864	960	0.2000	4466.52	5146.64	8.720
100	0.0777	2820.76	3085.05	6.641	540	0.1421	3616.55	4099.73	7.887	970	0.2014	4488.02	5172.80	8.738
110	0.0794	2839.39	3109.28	6.680	550	0.1435	3635.53	4123.45	7.909	980	0.2028	4509.57	5199.00	8.756
120	0.0810	2857.82	3133.22	6.718	560	0.1449	3654.58	4147.23	7.931	990	0.2041	4531.17	5225.26	8.773
130	0.0826	2876.08	3156.93	6.755	570	0.1463	3673.68	4171.07	7.953	1000	0.2055	4552.83	5251.58	8.791
140	0.0842	2894.19	3180.43	6.791	580	0.1477	3692.85	4194.96	7.975	1010	0.2069	4574.54	5277.94	8.808
150	0.0858	2912.19	3203.76	6.826	590	0.1491	3712.08	4218.92	7.997	1020	0.2083	4596.31	5304.36	8.825
160	0.0873	2930.09	3226.95	6.861	600	0.1505	3731.37	4242.93	8.019	1030	0.2096	4618.12	5330.83	8.842
170	0.0889	2947.92	3250.02	6.895	610	0.1518	3750.73	4267.00	8.040	1040	0.2110	4639.99	5357.35	8.860
180	0.0904	2965.68	3272.99	6.928	620	0.1532	3770.14	4291.13	8.062	1050	0.2124	4661.91	5383.92	8.877
190	0.0919	2983.40	3295.88	6.961	630	0.1546	3789.62	4315.32	8.083	1060	0.2137	4683.88	5410.54	8.894
200	0.0934	3001.08	3318.70	6.993	640	0.1560	3809.16	4339.57	8.104	1070	0.2151	4705.91	5437.21	8.910
210	0.0949	3018.73	3341.46	7.025	650	0.1574	3828.77	4363.88	8.125	1080	0.2165	4727.98	5463.94	8.927
220	0.0964	3036.37	3364.18	7.056	660	0.1588	3848.43	4388.25	8.146	1090	0.2178	4750.10	5490.71	8.944
230	0.0979	3053.99	3386.86	7.087	670	0.1602	3868.16	4412.68	8.167	1100	0.2192	4772.27	5517.53	8.961
240	0.0994	3071.62	3409.52	7.117	680	0.1615	3887.95	4437.17	8.187	1110	0.2206	4794.50	5544.40	8.977
250	0.1009	3089.25	3432.17	7.147	690	0.1629	3907.80	4461.72	8.208	1120	0.2219	4816.77	5571.31	8.994
260	0.1023	3106.89	3454.80	7.176	700	0.1643	3927.72	4486.33	8.228	1130	0.2233	4839.09	5598.28	9.010
270	0.1038	3124.54	3477.43	7.205	710	0.1657	3947.69	4511.00	8.248	1140	0.2247	4861.45	5625.29	9.027
280	0.1052	3142.21	3500.06	7.234	720	0.1671	3967.73	4535.72	8.268	1150	0.2260	4883.87	5652.35	9.043
290	0.1067	3159.91	3522.70	7.262	730	0.1684	3987.83	4560.51	8.288	1160	0.2274	4906.33	5679.46	9.059
300	0.1082	3177.63	3545.35	7.290	740	0.1698	4007.99	4585.36	8.308	1170	0.2288	4928.84	5706.61	9.075
310	0.1096	3195.39	3568.01	7.318	750	0.1712	4028.21	4610.27	8.328	1180	0.2301	4951.40	5733.81	9.091
320	0.1110	3213.17	3590.70	7.346	760	0.1726	4048.49	4635.23	8.348	1190	0.2315	4974.00	5761.05	9.108
330	0.1125	3230.99	3613.41	7.373	770	0.1739	4068.83	4660.26	8.367	1200	0.2329	4996.65	5788.34	9.123
340	0.1139	3248.85	3636.15	7.399	780	0.1753	4089.24	4685.34	8.387	1210	0.2342	5019.34	5815.68	9.139
350	0.1153	3266.75	3658.91	7.426	790	0.1767	4109.70	4710.48	8.406	1220	0.2356	5042.08	5843.06	9.155
360	0.1168	3284.70	3681.71	7.452	800	0.1781	4130.22	4735.68	8.425	1230	0.2369	5064.87	5870.48	9.171
370	0.1182	3302.69	3704.55	7.478	810	0.1795	4150.81	4760.94	8.445	1240	0.2383	5087.69	5897.95	9.187
380	0.1196	3320.72	3727.42	7.504	820	0.1808	4171.45	4786.26	8.464	1250	0.2397	5110.57	5925.47	9.202
390	0.1210	3338.80	3750.33	7.529	830	0.1822	4192.15	4811.63	8.483	1260	0.2410	5133.48	5953.02	9.218
400	0.1225	3356.93	3773.28	7.555	840	0.1836	4212.91	4837.07	8.501	1270	0.2424	5156.44	5980.62	9.233
410	0.1239	3375.12	3796.28	7.580	850	0.1849	4233.73	4862.56	8.520	1280	0.2438	5179.45	6008.26	9.249
420	0.1253	3393.35	3819.32	7.604	860	0.1863	4254.61	4888.10	8.539	1290	0.2451	5202.49	6035.94	9.264
430	0.1267	3411.64	3842.40	7.629	870	0.1877	4275.54	4913.71	8.557	1300	0.2465	5225.58	6063.67	9.280
440	0.1281	3429.98	3865.54	7.653										

Table 3(25) Superheated Steam Properties (Pressure Base) Pressure = 35.0 bar and saturation temperature, Tsat = 242.557°C

t (°C)	$v\left(\frac{m^3}{kg}\right)$	$u\left(\frac{kJ}{kg}\right)$	$h\left(\frac{kJ}{kg}\right)$	$s\left(\frac{kJ}{kg\,K}\right)$	t (°C)	$v\left(\frac{m^3}{kg}\right)$	$u\left(\frac{kJ}{kg}\right)$	$h\left(\frac{kJ}{kg}\right)$	$s\left(\frac{kJ}{kg\,K}\right)$	t (°C)	$v\left(\frac{m^3}{kg}\right)$	$u\left(\frac{kJ}{kg}\right)$	$h\left(\frac{kJ}{kg}\right)$	$s\left(\frac{kJ}{kg\,K}\right)$
Sat Vapour (g)	0.0571	2602.96	2802.67	6.124	Sat Vapour (g)	0.0571	2602.96	2802.67	6.124	Sat Vapour (g)	0.0571	2602.96	2802.67	6.124
10	0.0593	2630.84	2838.45	6.193	450	0.1260	3450.96	3891.97	7.668	880	0.1839	4299.82	4943.41	8.565
20	0.0614	2656.05	2870.99	6.254	460	0.1274	3469.44	3915.22	7.692	890	0.1852	4320.88	4969.14	8.584
30	0.0634	2679.59	2901.47	6.311	470	0.1287	3487.97	3938.53	7.715	900	0.1865	4342.01	4994.92	8.602
40	0.0653	2701.92	2930.46	6.363	480	0.1301	3506.55	3961.89	7.739	910	0.1879	4363.18	5020.77	8.620
50	0.0671	2723.35	2958.29	6.413	490	0.1315	3525.20	3985.30	7.762	920	0.1892	4384.41	5046.66	8.638
60	0.0689	2744.05	2985.21	6.460	500	0.1328	3543.90	4008.77	7.786	930	0.1905	4405.70	5072.61	8.656
70	0.0706	2764.18	3011.39	6.505	510	0.1342	3562.66	4032.29	7.809	940	0.1919	4427.05	5098.62	8.674
80	0.0723	2783.84	3036.97	6.549	520	0.1355	3581.49	4055.87	7.831	950	0.1932	4448.45	5124.68	8.692
90	0.0740	2803.12	3062.05	6.590	530	0.1369	3600.37	4079.50	7.854	960	0.1945	4469.90	5150.79	8.710
100	0.0756	2822.09	3086.72	6.631	540	0.1382	3619.31	4103.18	7.877	970	0.1959	4491.41	5176.96	8.727
110	0.0772	2840.80	3111.04	6.670	550	0.1396	3638.31	4126.93	7.899	980	0.1972	4512.97	5203.18	8.745
120	0.0788	2859.30	3135.07	6.708	560	0.1410	3657.38	4150.73	7.921	990	0.1985	4534.59	5229.45	8.763
130	0.0804	2877.62	3158.86	6.745	570	0.1423	3676.50	4174.58	7.943	1000	0.1999	4556.26	5255.78	8.780
140	0.0819	2895.79	3182.44	6.781	580	0.1437	3695.69	4198.50	7.965	1010	0.2012	4577.98	5282.16	8.797
150	0.0834	2913.84	3205.84	6.817	590	0.1450	3714.93	4222.47	7.987	1020	0.2025	4599.76	5308.59	8.815
160	0.0849	2931.80	3229.10	6.852	600	0.1464	3734.24	4246.51	8.009	1030	0.2039	4621.59	5335.07	8.832
170	0.0864	2949.67	3252.23	6.886	610	0.1477	3753.62	4270.60	8.030	1040	0.2052	4643.46	5361.60	8.849
180	0.0879	2967.48	3275.26	6.919	620	0.1491	3773.05	4294.75	8.052	1050	0.2065	4665.40	5388.18	8.866
190	0.0894	2985.24	3298.20	6.952	630	0.1504	3792.54	4318.96	8.073	1060	0.2078	4687.38	5414.81	8.883
200	0.0909	3002.96	3321.07	6.984	640	0.1518	3812.10	4343.23	8.094	1070	0.2092	4709.41	5441.50	8.900
210	0.0924	3020.66	3343.89	7.015	650	0.1531	3831.72	4367.56	8.115	1080	0.2105	4731.49	5468.23	8.917
220	0.0938	3038.33	3366.65	7.047	660	0.1544	3851.40	4391.94	8.136	1090	0.2118	4753.63	5495.01	8.933
230	0.0953	3055.99	3389.39	7.077	670	0.1558	3871.15	4416.39	8.156	1100	0.2132	4775.81	5521.84	8.950
240	0.0967	3073.65	3412.09	7.108	680	0.1571	3890.95	4440.90	8.177	1110	0.2145	4798.04	5548.72	8.967
250	0.0981	3091.32	3434.77	7.137	690	0.1585	3910.82	4465.46	8.197	1120	0.2158	4820.32	5575.65	8.983
260	0.0996	3108.99	3457.45	7.167	700	0.1598	3930.75	4490.09	8.218	1130	0.2171	4842.65	5602.63	9.000
270	0.1010	3126.68	3480.12	7.196	710	0.1612	3950.74	4514.78	8.238	1140	0.2185	4865.03	5629.65	9.016
280	0.1024	3144.38	3502.79	7.225	720	0.1625	3970.79	4539.52	8.258	1150	0.2198	4887.45	5656.72	9.032
290	0.1038	3162.10	3525.46	7.253	730	0.1638	3990.90	4564.33	8.278	1160	0.2211	4909.93	5683.84	9.048
300	0.1052	3179.86	3548.15	7.281	740	0.1652	4011.08	4589.19	8.298	1170	0.2224	4932.44	5711.00	9.065
310	0.1066	3197.64	3570.85	7.308	750	0.1665	4031.31	4614.11	8.318	1180	0.2238	4955.01	5738.21	9.081
320	0.1080	3215.45	3593.57	7.336	760	0.1679	4051.61	4639.09	8.337	1190	0.2251	4977.62	5765.47	9.097
330	0.1094	3233.30	3616.31	7.363	770	0.1692	4071.97	4664.13	8.357	1200	0.2264	5000.28	5792.77	9.113
340	0.1108	3251.19	3639.08	7.390	780	0.1705	4092.39	4689.23	8.376	1210	0.2278	5022.98	5820.11	9.129
350	0.1122	3269.11	3661.88	7.416	790	0.1719	4112.86	4714.39	8.396	1220	0.2291	5045.73	5847.50	9.144
360	0.1136	3287.08	3684.71	7.442	800	0.1732	4133.40	4739.61	8.415	1230	0.2304	5068.52	5874.94	9.160
370	0.1150	3305.09	3707.58	7.468	810	0.1745	4154.00	4764.88	8.434	1240	0.2317	5091.36	5902.41	9.176
380	0.1164	3323.15	3730.48	7.494	820	0.1759	4174.65	4790.21	8.453	1250	0.2331	5114.24	5929.94	9.191
390	0.1178	3341.26	3753.42	7.519	830	0.1772	4195.37	4815.60	8.472	1260	0.2344	5137.17	5957.50	9.207
400	0.1191	3359.41	3776.40	7.545	840	0.1785	4216.14	4841.05	8.491	1270	0.2357	5160.14	5985.11	9.223
410	0.1205	3377.62	3799.42	7.570	850	0.1799	4236.98	4866.56	8.510	1280	0.2370	5183.15	6012.76	9.238
420	0.1219	3395.87	3822.49	7.595	860	0.1812	4257.87	4892.12	8.528	1290	0.2384	5206.20	6040.45	9.253
430	0.1233	3414.18	3845.60	7.619	870	0.1825	4278.82	4917.73	8.547	1300	0.2397	5229.30	6068.19	9.269
440	0.1246	3432.55	3868.76	7.643										

Table 3(26) Superheated Steam Properties (Pressure Base) Pressure = 36.0 bar and saturation temperature, Tsat = 244.182°C

t (°C)	v $\left(\frac{m^3}{kg}\right)$	u $\left(\frac{kJ}{kg}\right)$	h $\left(\frac{kJ}{kg}\right)$	s $\left(\frac{kJ}{kg\ K}\right)$	t (°C)	v $\left(\frac{m^3}{kg}\right)$	u $\left(\frac{kJ}{kg}\right)$	h $\left(\frac{kJ}{kg}\right)$	s $\left(\frac{kJ}{kg\ K}\right)$	t (°C)	v $\left(\frac{m^3}{kg}\right)$	u $\left(\frac{kJ}{kg}\right)$	h $\left(\frac{kJ}{kg}\right)$	s $\left(\frac{kJ}{kg\ K}\right)$
Sat Vapour (g)	*0.0554*	*2602.78*	*2802.40*	*6.113*	*Sat Vapour (g)*	*0.0554*	*2602.78*	*2802.40*	*6.113*	*Sat Vapour (g)*	*0.0554*	*2602.78*	*2802.40*	*6.113*
10	0.0577	2630.95	2838.53	6.182	450	0.1227	3453.49	3895.14	7.658	880	0.1790	4303.03	4947.36	8.555
20	0.0597	2656.38	2871.37	6.244	460	0.1240	3471.98	3918.42	7.682	890	0.1803	4324.11	4973.11	8.574
30	0.0617	2680.10	2902.08	6.300	470	0.1253	3490.53	3941.75	7.706	900	0.1816	4345.24	4998.91	8.592
40	0.0635	2702.59	2931.26	6.353	480	0.1267	3509.14	3965.13	7.729	910	0.1829	4366.43	5024.76	8.610
50	0.0653	2724.15	2959.27	6.403	490	0.1280	3527.81	3988.57	7.753	920	0.1842	4387.67	5050.67	8.628
60	0.0670	2744.98	2986.35	6.450	500	0.1293	3546.53	4012.06	7.776	930	0.1855	4408.98	5076.63	8.646
70	0.0687	2765.21	3012.67	6.496	510	0.1306	3565.31	4035.60	7.799	940	0.1868	4430.33	5102.65	8.664
80	0.0704	2784.97	3038.37	6.539	520	0.1320	3584.15	4059.20	7.822	950	0.1881	4451.74	5128.72	8.682
90	0.0720	2804.34	3063.56	6.581	530	0.1333	3603.05	4082.85	7.844	960	0.1893	4473.21	5154.85	8.700
100	0.0736	2823.38	3088.33	6.621	540	0.1346	3622.01	4106.56	7.867	970	0.1906	4494.73	5181.03	8.717
110	0.0752	2842.16	3112.74	6.661	550	0.1359	3641.03	4130.32	7.889	980	0.1919	4516.30	5207.26	8.735
120	0.0767	2860.73	3136.86	6.699	560	0.1372	3660.11	4154.14	7.912	990	0.1932	4537.93	5233.55	8.752
130	0.0782	2879.11	3160.72	6.736	570	0.1385	3679.25	4178.02	7.934	1000	0.1945	4559.61	5259.89	8.770
140	0.0797	2897.34	3184.38	6.772	580	0.1399	3698.46	4201.96	7.956	1010	0.1958	4581.35	5286.28	8.787
150	0.0812	2915.45	3207.85	6.808	590	0.1412	3717.72	4225.95	7.977	1020	0.1971	4603.13	5312.72	8.804
160	0.0827	2933.45	3231.17	6.842	600	0.1425	3737.05	4250.00	7.999	1030	0.1984	4624.97	5339.21	8.821
170	0.0842	2951.37	3254.37	6.876	610	0.1438	3756.44	4274.11	8.020	1040	0.1997	4646.86	5365.75	8.839
180	0.0856	2969.23	3277.46	6.910	620	0.1451	3775.89	4298.28	8.042	1050	0.2010	4668.80	5392.35	8.856
190	0.0871	2987.03	3300.45	6.942	630	0.1464	3795.40	4322.51	8.063	1060	0.2023	4690.79	5418.99	8.873
200	0.0885	3004.79	3323.38	6.975	640	0.1477	3814.97	4346.80	8.084	1070	0.2036	4712.84	5445.69	8.889
210	0.0899	3022.53	3346.24	7.006	650	0.1490	3834.61	4371.15	8.105	1080	0.2049	4734.93	5472.43	8.906
220	0.0913	3040.24	3369.06	7.037	660	0.1503	3854.30	4395.55	8.126	1090	0.2062	4757.07	5499.22	8.923
230	0.0927	3057.94	3391.84	7.068	670	0.1517	3874.06	4420.02	8.146	1100	0.2074	4779.27	5526.07	8.940
240	0.0942	3075.63	3414.59	7.098	680	0.1530	3893.88	4444.54	8.167	1110	0.2087	4801.51	5552.96	8.956
250	0.0956	3093.33	3437.31	7.128	690	0.1543	3913.77	4469.13	8.188	1120	0.2100	4823.80	5579.90	8.973
260	0.0969	3111.04	3460.03	7.157	700	0.1556	3933.71	4493.77	8.208	1130	0.2113	4846.14	5606.88	8.989
270	0.0983	3128.75	3482.74	7.187	710	0.1569	3953.72	4518.47	8.228	1140	0.2126	4868.52	5633.92	9.005
280	0.0997	3146.49	3505.45	7.215	720	0.1582	3973.78	4543.23	8.248	1150	0.2139	4890.96	5661.00	9.022
290	0.1011	3164.24	3528.16	7.243	730	0.1595	3993.91	4568.05	8.268	1160	0.2152	4913.44	5688.12	9.038
300	0.1025	3182.02	3550.88	7.271	740	0.1608	4014.10	4592.93	8.288	1170	0.2165	4935.97	5715.30	9.054
310	0.1038	3199.83	3573.62	7.299	750	0.1621	4034.35	4617.87	8.308	1180	0.2178	4958.54	5742.52	9.070
320	0.1052	3217.67	3596.37	7.326	760	0.1634	4054.66	4642.87	8.327	1190	0.2191	4981.17	5769.78	9.086
330	0.1066	3235.55	3619.15	7.353	770	0.1647	4075.03	4667.93	8.347	1200	0.2204	5003.83	5797.09	9.102
340	0.1079	3253.46	3641.95	7.380	780	0.1660	4095.47	4693.04	8.366	1210	0.2216	5026.54	5824.45	9.118
350	0.1093	3271.41	3664.78	7.407	790	0.1673	4115.96	4718.21	8.386	1220	0.2229	5049.30	5851.85	9.134
360	0.1106	3289.40	3687.64	7.433	800	0.1686	4136.51	4743.45	8.405	1230	0.2242	5072.10	5879.29	9.150
370	0.1120	3307.44	3710.53	7.459	810	0.1699	4157.12	4768.74	8.424	1240	0.2255	5094.95	5906.78	9.165
380	0.1133	3325.52	3733.46	7.485	820	0.1712	4177.79	4794.08	8.443	1250	0.2268	5117.84	5934.31	9.181
390	0.1147	3343.65	3756.43	7.510	830	0.1725	4198.52	4819.49	8.462	1260	0.2281	5140.77	5961.88	9.196
400	0.1160	3361.83	3779.44	7.535	840	0.1738	4219.30	4844.95	8.481	1270	0.2294	5163.75	5989.50	9.212
410	0.1173	3380.06	3802.49	7.560	850	0.1751	4240.15	4870.47	8.500	1280	0.2307	5186.77	6017.16	9.227
420	0.1187	3398.34	3825.58	7.585	860	0.1764	4261.05	4896.04	8.518	1290	0.2320	5209.83	6044.86	9.243
430	0.1200	3416.67	3848.72	7.610	870	0.1777	4282.01	4921.67	8.537	1300	0.2332	5232.94	6072.60	9.258
440	0.1213	3435.05	3871.90	7.634										

Table 3(27) Superheated Steam Properties (Pressure Base) Pressure = 37.0 bar and saturation temperature, Tsat = 245.772°C

t (°C)	v $\left(\frac{m^3}{kg}\right)$	u $\left(\frac{kJ}{kg}\right)$	h $\left(\frac{kJ}{kg}\right)$	s $\left(\frac{kJ}{kg\,K}\right)$	t (°C)	v $\left(\frac{m^3}{kg}\right)$	u $\left(\frac{kJ}{kg}\right)$	h $\left(\frac{kJ}{kg}\right)$	s $\left(\frac{kJ}{kg\,K}\right)$	t (°C)	v $\left(\frac{m^3}{kg}\right)$	u $\left(\frac{kJ}{kg}\right)$	h $\left(\frac{kJ}{kg}\right)$	s $\left(\frac{kJ}{kg\,K}\right)$
Sat Vapour (g)	0.0539	2602.57	2802.08	6.102	Sat Vapour (g)	0.0539	2602.57	2802.08	6.102	Sat Vapour (g)	0.0539	2602.57	2802.08	6.102
10	0.0561	2631.02	2838.57	6.171	450	0.1195	3455.96	3898.23	7.649	880	0.1743	4306.17	4951.23	8.545
20	0.0581	2656.67	2871.69	6.234	460	0.1208	3474.47	3921.54	7.673	890	0.1756	4327.26	4976.99	8.564
30	0.0600	2680.58	2902.64	6.290	470	0.1221	3493.04	3944.89	7.696	900	0.1769	4348.41	5002.80	8.582
40	0.0618	2703.23	2932.02	6.343	480	0.1234	3511.67	3968.30	7.720	910	0.1781	4369.61	5028.67	8.600
50	0.0636	2724.92	2960.20	6.393	490	0.1247	3530.35	3991.76	7.743	920	0.1794	4390.86	5054.59	8.618
60	0.0653	2745.86	2987.43	6.441	500	0.1260	3549.09	4015.27	7.766	930	0.1806	4412.18	5080.57	8.636
70	0.0669	2766.20	3013.88	6.486	510	0.1273	3567.89	4038.83	7.790	940	0.1819	4433.54	5106.60	8.654
80	0.0686	2786.05	3039.71	6.530	520	0.1286	3586.75	4062.45	7.812	950	0.1832	4454.97	5132.68	8.672
90	0.0701	2805.50	3065.01	6.572	530	0.1299	3605.67	4086.13	7.835	960	0.1844	4476.44	5158.82	8.690
100	0.0717	2824.63	3089.88	6.612	540	0.1311	3624.65	4109.86	7.858	970	0.1857	4497.98	5185.02	8.707
110	0.0732	2843.48	3114.39	6.651	550	0.1324	3643.69	4133.64	7.880	980	0.1869	4519.56	5211.26	8.725
120	0.0747	2862.11	3138.59	6.690	560	0.1337	3662.79	4157.48	7.902	990	0.1882	4541.20	5237.56	8.742
130	0.0762	2880.55	3162.53	6.727	570	0.1350	3681.95	4181.38	7.924	1000	0.1895	4562.89	5263.91	8.760
140	0.0777	2898.84	3186.26	6.763	580	0.1363	3701.17	4205.34	7.946	1010	0.1907	4584.64	5290.31	8.777
150	0.0791	2917.00	3209.80	6.798	590	0.1375	3720.45	4229.35	7.968	1020	0.1920	4606.43	5316.76	8.794
160	0.0806	2935.06	3233.19	6.833	600	0.1388	3739.79	4253.42	7.989	1030	0.1932	4628.28	5343.27	8.811
170	0.0820	2953.03	3256.45	6.867	610	0.1401	3759.20	4277.55	8.011	1040	0.1945	4650.18	5369.82	8.829
180	0.0834	2970.93	3279.59	6.901	620	0.1414	3778.66	4301.74	8.032	1050	0.1958	4672.13	5396.43	8.846
190	0.0848	2988.77	3302.65	6.933	630	0.1426	3798.19	4325.99	8.053	1060	0.1970	4694.14	5423.08	8.862
200	0.0862	3006.58	3325.62	6.965	640	0.1439	3817.78	4350.29	8.074	1070	0.1983	4716.19	5449.79	8.879
210	0.0876	3024.35	3348.54	6.997	650	0.1452	3837.43	4374.66	8.095	1080	0.1995	4738.29	5476.54	8.896
220	0.0890	3042.10	3371.40	7.028	660	0.1465	3857.14	4399.08	8.116	1090	0.2008	4760.45	5503.35	8.913
230	0.0904	3059.83	3394.22	7.059	670	0.1477	3876.92	4423.56	8.137	1100	0.2020	4782.65	5530.20	8.929
240	0.0917	3077.56	3417.02	7.089	680	0.1490	3896.75	4448.10	8.157	1110	0.2033	4804.90	5557.10	8.946
250	0.0931	3095.29	3439.79	7.119	690	0.1503	3916.65	4472.71	8.178	1120	0.2046	4827.20	5584.05	8.963
260	0.0945	3113.03	3462.54	7.148	700	0.1516	3936.61	4497.37	8.198	1130	0.2058	4849.55	5611.05	8.979
270	0.0958	3130.78	3485.29	7.177	710	0.1528	3956.63	4522.08	8.218	1140	0.2071	4871.95	5638.09	8.995
280	0.0972	3148.54	3508.04	7.206	720	0.1541	3976.71	4546.86	8.238	1150	0.2083	4894.39	5665.18	9.012
290	0.0985	3166.33	3530.79	7.234	730	0.1554	3996.85	4571.70	8.258	1160	0.2096	4916.88	5692.32	9.028
300	0.0998	3184.14	3553.54	7.262	740	0.1566	4017.06	4596.60	8.278	1170	0.2108	4939.42	5719.50	9.044
310	0.1012	3201.97	3576.31	7.290	750	0.1579	4037.32	4621.55	8.298	1180	0.2121	4962.00	5746.73	9.060
320	0.1025	3219.84	3599.10	7.317	760	0.1592	4057.65	4646.56	8.318	1190	0.2133	4984.63	5774.01	9.076
330	0.1038	3237.74	3621.91	7.344	770	0.1604	4078.03	4671.64	8.337	1200	0.2146	5007.31	5801.33	9.092
340	0.1052	3255.68	3644.74	7.371	780	0.1617	4098.48	4696.77	8.357	1210	0.2159	5030.03	5828.69	9.108
350	0.1065	3273.66	3667.60	7.398	790	0.1630	4118.98	4721.95	8.376	1220	0.2171	5052.80	5856.10	9.124
360	0.1078	3291.67	3690.49	7.424	800	0.1642	4139.55	4747.20	8.395	1230	0.2184	5075.61	5883.55	9.139
370	0.1091	3309.73	3713.42	7.450	810	0.1655	4160.17	4772.51	8.414	1240	0.2196	5098.46	5911.05	9.155
380	0.1104	3327.84	3736.37	7.475	820	0.1668	4180.85	4797.87	8.433	1250	0.2209	5121.36	5938.59	9.171
390	0.1117	3345.99	3759.37	7.501	830	0.1680	4201.59	4823.29	8.452	1260	0.2221	5144.30	5966.17	9.186
400	0.1130	3364.19	3782.40	7.526	840	0.1693	4222.39	4848.76	8.471	1270	0.2234	5167.29	5993.80	9.202
410	0.1143	3382.44	3805.48	7.551	850	0.1706	4243.25	4874.29	8.490	1280	0.2246	5190.32	6021.47	9.217
420	0.1156	3400.74	3828.60	7.576	860	0.1718	4264.17	4899.88	8.508	1290	0.2259	5213.39	6049.18	9.233
430	0.1169	3419.09	3851.76	7.600	870	0.1731	4285.14	4925.53	8.527	1300	0.2271	5236.50	6076.93	9.248
440	0.1182	3437.50	3874.97	7.625										

Table 3(28) Superheated Steam Properties (Pressure Base) Pressure = 38.0 bar and saturation temperature, Tsat = 247.33°C

t (°C)	v (m^3/kg)	u (kJ/kg)	h (kJ/kg)	s (kJ/kg K)	t (°C)	v (m^3/kg)	u (kJ/kg)	h (kJ/kg)	s (kJ/kg K)	t (°C)	v (m^3/kg)	u (kJ/kg)	h (kJ/kg)	s (kJ/kg K)
Sat Vapour (g)	0.0525	2602.33	2801.71	6.091	Sat Vapour (g)	0.0525	2602.33	2801.71	6.091	Sat Vapour (g)	0.0525	2602.33	2801.71	6.091
10	0.0546	2631.05	2838.56	6.161	450	0.1166	3458.37	3901.26	7.640	880	0.1699	4309.25	4955.02	8.536
20	0.0566	2656.93	2871.96	6.223	460	0.1178	3476.90	3924.59	7.664	890	0.1712	4330.35	4980.79	8.554
30	0.0585	2681.02	2903.15	6.281	470	0.1191	3495.49	3947.97	7.687	900	0.1724	4351.50	5006.61	8.572
40	0.0602	2703.82	2932.73	6.334	480	0.1203	3514.14	3971.40	7.711	910	0.1736	4372.72	5032.49	8.590
50	0.0620	2725.64	2961.08	6.384	490	0.1216	3532.84	3994.88	7.734	920	0.1749	4393.99	5058.43	8.609
60	0.0636	2746.70	2988.46	6.432	500	0.1228	3551.60	4018.41	7.757	930	0.1761	4415.31	5084.42	8.627
70	0.0652	2767.15	3015.05	6.477	510	0.1241	3570.42	4042.00	7.780	940	0.1773	4436.69	5110.46	8.644
80	0.0668	2787.09	3040.99	6.521	520	0.1254	3589.30	4065.64	7.803	950	0.1785	4458.12	5136.56	8.662
90	0.0684	2806.63	3066.41	6.563	530	0.1266	3608.23	4089.33	7.826	960	0.1798	4479.61	5162.71	8.680
100	0.0699	2825.83	3091.38	6.603	540	0.1279	3627.23	4113.08	7.848	970	0.1810	4501.15	5188.92	8.698
110	0.0714	2844.76	3115.97	6.642	550	0.1291	3646.29	4136.89	7.871	980	0.1822	4522.75	5215.17	8.715
120	0.0728	2863.45	3140.26	6.681	560	0.1304	3665.40	4160.75	7.893	990	0.1834	4544.40	5241.48	8.733
130	0.0743	2881.95	3164.28	6.718	570	0.1316	3684.58	4184.67	7.915	1000	0.1847	4566.10	5267.84	8.750
140	0.0757	2900.30	3188.08	6.754	580	0.1328	3703.82	4208.64	7.937	1010	0.1859	4587.86	5294.26	8.767
150	0.0772	2918.51	3211.70	6.790	590	0.1341	3723.11	4232.68	7.959	1020	0.1871	4609.66	5320.72	8.785
160	0.0786	2936.62	3235.15	6.824	600	0.1353	3742.47	4256.77	7.980	1030	0.1883	4631.52	5347.24	8.802
170	0.0800	2954.63	3258.47	6.858	610	0.1366	3761.89	4280.91	8.002	1040	0.1896	4653.43	5373.80	8.819
180	0.0813	2972.58	3281.67	6.892	620	0.1378	3781.38	4305.12	8.023	1050	0.1908	4675.40	5400.42	8.836
190	0.0827	2990.47	3304.78	6.924	630	0.1391	3800.92	4329.39	8.044	1060	0.1920	4697.41	5427.09	8.853
200	0.0841	3008.31	3327.81	6.957	640	0.1403	3820.53	4353.71	8.065	1070	0.1932	4719.47	5453.80	8.870
210	0.0854	3026.12	3350.77	6.988	650	0.1416	3840.19	4378.09	8.086	1080	0.1945	4741.59	5480.57	8.886
220	0.0868	3043.91	3373.68	7.019	660	0.1428	3859.92	4402.53	8.107	1090	0.1957	4763.75	5507.38	8.903
230	0.0881	3061.68	3396.55	7.050	670	0.1440	3879.71	4427.03	8.128	1100	0.1969	4785.96	5534.25	8.920
240	0.0895	3079.44	3419.39	7.080	680	0.1453	3899.56	4451.59	8.148	1110	0.1981	4808.22	5561.16	8.936
250	0.0908	3097.21	3442.20	7.110	690	0.1465	3919.47	4476.21	8.169	1120	0.1994	4830.53	5588.12	8.953
260	0.0921	3114.98	3464.99	7.140	700	0.1477	3939.45	4500.89	8.189	1130	0.2006	4852.89	5615.12	8.969
270	0.0934	3132.75	3487.78	7.169	710	0.1490	3959.48	4525.62	8.209	1140	0.2018	4875.30	5642.18	8.985
280	0.0947	3150.55	3510.56	7.197	720	0.1502	3979.58	4550.42	8.229	1150	0.2030	4897.75	5669.28	9.002
290	0.0960	3168.36	3533.35	7.226	730	0.1515	3999.73	4575.27	8.249	1160	0.2043	4920.25	5696.43	9.018
300	0.0974	3186.20	3556.14	7.253	740	0.1527	4019.95	4600.18	8.269	1170	0.2055	4942.80	5723.62	9.034
310	0.0987	3204.06	3578.95	7.281	750	0.1539	4040.23	4625.15	8.289	1180	0.2067	4965.39	5750.86	9.050
320	0.1000	3221.96	3601.77	7.308	760	0.1552	4060.57	4650.18	8.308	1190	0.2079	4988.03	5778.14	9.066
330	0.1012	3239.88	3624.61	7.335	770	0.1564	4080.97	4675.27	8.328	1200	0.2091	5010.71	5805.47	9.082
340	0.1025	3257.85	3647.47	7.362	780	0.1576	4101.43	4700.41	8.347	1210	0.2104	5033.44	5832.85	9.098
350	0.1038	3275.85	3670.36	7.389	790	0.1589	4121.94	4725.62	8.366	1220	0.2116	5056.22	5860.26	9.114
360	0.1051	3293.89	3693.28	7.415	800	0.1601	4142.52	4750.88	8.386	1230	0.2128	5079.04	5887.73	9.129
370	0.1064	3311.97	3716.23	7.441	810	0.1613	4163.16	4776.20	8.405	1240	0.2140	5101.90	5915.23	9.145
380	0.1077	3330.10	3739.22	7.466	820	0.1626	4183.85	4801.57	8.424	1250	0.2153	5124.81	5942.78	9.161
390	0.1089	3348.28	3762.24	7.492	830	0.1638	4204.61	4827.00	8.443	1260	0.2165	5147.76	5970.37	9.176
400	0.1102	3366.50	3785.30	7.517	840	0.1650	4225.42	4852.49	8.461	1270	0.2177	5170.75	5998.01	9.192
410	0.1115	3384.77	3808.41	7.542	850	0.1663	4246.29	4878.04	8.480	1280	0.2189	5193.79	6025.69	9.207
420	0.1128	3403.09	3831.55	7.567	860	0.1675	4267.22	4903.64	8.499	1290	0.2201	5216.87	6053.40	9.223
430	0.1140	3421.46	3854.74	7.591	870	0.1687	4288.21	4929.30	8.517	1300	0.2214	5239.99	6081.16	9.238
440	0.1153	3439.89	3877.98	7.616										

Table 3(29) Superheated Steam Properties (Pressure Base) Pressure = 39.0 bar and saturation temperature, Tsat = 248.857°C

t (°C)	v $\left(\frac{m^3}{kg}\right)$	u $\left(\frac{kJ}{kg}\right)$	h $\left(\frac{kJ}{kg}\right)$	s $\left(\frac{kJ}{kg\,K}\right)$	t (°C)	v $\left(\frac{m^3}{kg}\right)$	u $\left(\frac{kJ}{kg}\right)$	h $\left(\frac{kJ}{kg}\right)$	s $\left(\frac{kJ}{kg\,K}\right)$	t (°C)	v $\left(\frac{m^3}{kg}\right)$	u $\left(\frac{kJ}{kg}\right)$	h $\left(\frac{kJ}{kg}\right)$	s $\left(\frac{kJ}{kg\,K}\right)$
Sat Vapour (g)	*0.0511*	*2602.05*	*2801.30*	*6.080*	*Sat Vapour (g)*	*0.0511*	*2602.05*	*2801.30*	*6.080*	*Sat Vapour (g)*	*0.0511*	*2602.05*	*2801.30*	*6.080*
10	0.0532	2631.05	2838.51	6.151	450	0.1137	3460.73	3904.22	7.631	880	0.1658	4312.26	4958.73	8.526
20	0.0551	2657.15	2872.19	6.213	460	0.1149	3479.28	3927.57	7.655	890	0.1670	4333.37	4984.51	8.545
30	0.0570	2681.42	2903.61	6.271	470	0.1162	3497.89	3950.97	7.678	900	0.1682	4354.54	5010.35	8.563
40	0.0587	2704.38	2933.39	6.324	480	0.1174	3516.56	3974.43	7.702	910	0.1694	4375.76	5036.24	8.581
50	0.0604	2726.33	2961.91	6.375	490	0.1186	3535.28	3997.93	7.725	920	0.1706	4397.04	5062.19	8.599
60	0.0620	2747.51	2989.44	6.422	500	0.1199	3554.06	4021.48	7.748	930	0.1717	4418.38	5088.19	8.617
70	0.0636	2768.06	3016.17	6.468	510	0.1211	3572.89	4045.09	7.771	940	0.1729	4439.77	5114.25	8.635
80	0.0652	2788.10	3042.23	6.512	520	0.1223	3591.79	4068.75	7.794	950	0.1741	4461.21	5140.36	8.653
90	0.0667	2807.72	3067.75	6.554	530	0.1235	3610.74	4092.47	7.817	960	0.1753	4482.71	5166.52	8.671
100	0.0682	2826.99	3092.82	6.594	540	0.1247	3629.76	4116.24	7.839	970	0.1765	4504.27	5192.74	8.688
110	0.0696	2845.99	3117.51	6.634	550	0.1260	3648.83	4140.07	7.862	980	0.1777	4525.87	5219.01	8.706
120	0.0711	2864.75	3141.88	6.672	560	0.1272	3667.96	4163.95	7.884	990	0.1789	4547.53	5245.33	8.723
130	0.0725	2883.31	3165.98	6.709	570	0.1284	3687.16	4187.88	7.906	1000	0.1801	4569.25	5271.70	8.741
140	0.0739	2901.71	3189.85	6.746	580	0.1296	3706.41	4211.88	7.928	1010	0.1813	4591.01	5298.13	8.758
150	0.0753	2919.98	3213.53	6.781	590	0.1308	3725.72	4235.93	7.950	1020	0.1825	4612.83	5324.60	8.775
160	0.0766	2938.13	3237.05	6.816	600	0.1320	3745.10	4260.04	7.971	1030	0.1837	4634.70	5351.13	8.792
170	0.0780	2956.20	3260.43	6.850	610	0.1332	3764.54	4284.21	7.993	1040	0.1849	4656.62	5377.71	8.809
180	0.0794	2974.19	3283.69	6.883	620	0.1345	3784.03	4308.43	8.014	1050	0.1861	4678.59	5404.33	8.826
190	0.0807	2992.12	3306.85	6.916	630	0.1357	3803.59	4332.71	8.035	1060	0.1873	4700.62	5431.01	8.843
200	0.0820	3010.00	3329.93	6.948	640	0.1369	3823.21	4357.06	8.056	1070	0.1885	4722.69	5457.74	8.860
210	0.0834	3027.85	3352.94	6.980	650	0.1381	3842.90	4381.46	8.077	1080	0.1897	4744.81	5484.51	8.877
220	0.0847	3045.67	3375.90	7.011	660	0.1393	3862.64	4405.91	8.098	1090	0.1909	4766.98	5511.34	8.894
230	0.0860	3063.48	3398.82	7.042	670	0.1405	3882.44	4430.43	8.118	1100	0.1921	4789.21	5538.21	8.910
240	0.0873	3081.28	3421.70	7.072	680	0.1417	3902.31	4455.01	8.139	1110	0.1932	4811.48	5565.13	8.927
250	0.0886	3099.07	3444.55	7.102	690	0.1429	3922.24	4479.64	8.159	1120	0.1944	4833.80	5592.10	8.943
260	0.0899	3116.87	3467.38	7.131	700	0.1441	3942.22	4504.33	8.180	1130	0.1956	4856.16	5619.12	8.960
270	0.0912	3134.68	3490.21	7.160	710	0.1453	3962.27	4529.09	8.200	1140	0.1968	4878.58	5646.18	8.976
280	0.0924	3152.51	3513.03	7.189	720	0.1465	3982.38	4553.90	8.220	1150	0.1980	4901.04	5673.29	8.992
290	0.0937	3170.35	3535.85	7.217	730	0.1477	4002.55	4578.76	8.240	1160	0.1992	4923.55	5700.45	9.008
300	0.0950	3188.21	3558.68	7.245	740	0.1490	4022.79	4603.69	8.260	1170	0.2004	4946.10	5727.65	9.024
310	0.0963	3206.10	3581.52	7.272	750	0.1502	4043.08	4628.68	8.279	1180	0.2016	4968.71	5754.90	9.041
320	0.0975	3224.02	3604.37	7.300	760	0.1514	4063.43	4653.72	8.299	1190	0.2028	4991.35	5782.20	9.057
330	0.0988	3241.98	3627.24	7.327	770	0.1526	4083.84	4678.82	8.319	1200	0.2040	5014.05	5809.54	9.072
340	0.1000	3259.96	3650.14	7.353	780	0.1538	4104.31	4703.98	8.338	1210	0.2052	5036.79	5836.92	9.088
350	0.1013	3277.99	3673.06	7.380	790	0.1550	4124.84	4729.20	8.357	1220	0.2064	5059.57	5864.35	9.104
360	0.1026	3296.05	3696.01	7.406	800	0.1562	4145.43	4754.48	8.376	1230	0.2075	5082.40	5891.82	9.120
370	0.1038	3314.16	3718.99	7.432	810	0.1574	4166.08	4779.81	8.395	1240	0.2087	5105.27	5919.33	9.135
380	0.1050	3332.31	3742.00	7.458	820	0.1586	4186.79	4805.20	8.414	1250	0.2099	5128.18	5946.89	9.151
390	0.1063	3350.51	3765.05	7.483	830	0.1598	4207.56	4830.65	8.433	1260	0.2111	5151.14	5974.49	9.167
400	0.1075	3368.75	3788.14	7.508	840	0.1610	4228.38	4856.15	8.452	1270	0.2123	5174.14	6002.13	9.182
410	0.1088	3387.05	3811.27	7.533	850	0.1622	4249.27	4881.71	8.471	1280	0.2135	5197.19	6029.82	9.197
420	0.1100	3405.39	3834.44	7.558	860	0.1634	4270.21	4907.33	8.489	1290	0.2147	5220.27	6057.55	9.213
430	0.1112	3423.78	3857.65	7.582	870	0.1646	4291.21	4933.00	8.508	1300	0.2159	5243.40	6085.32	9.228
440	0.1125	3442.23	3880.91	7.607										

Table 3(30) Superheated Steam Properties (Pressure Base) Pressure = 40.0 bar and saturation temperature, Tsat = 250.354°C

t (°C)	v $\left(\frac{m^3}{kg}\right)$	u $\left(\frac{kJ}{kg}\right)$	h $\left(\frac{kJ}{kg}\right)$	s $\left(\frac{kJ}{kg\,K}\right)$
Sat Vapour (g)	0.0498	2601.73	2800.85	6.070
10	0.0518	2631.02	2838.41	6.141
20	0.0538	2657.34	2872.37	6.204
30	0.0556	2681.79	2904.02	6.262
40	0.0573	2704.90	2934.00	6.315
50	0.0589	2726.99	2962.70	6.366
60	0.0605	2748.28	2990.38	6.414
70	0.0621	2768.93	3017.23	6.459
80	0.0636	2789.06	3043.42	6.503
90	0.0651	2808.77	3069.05	6.545
100	0.0665	2828.12	3094.22	6.586
110	0.0680	2847.18	3119.00	6.625
120	0.0694	2866.01	3143.45	6.663
130	0.0708	2884.63	3167.63	6.701
140	0.0721	2903.09	3191.57	6.737
150	0.0735	2921.40	3215.32	6.773
160	0.0748	2939.61	3238.90	6.807
170	0.0762	2957.72	3262.34	6.841
180	0.0775	2975.75	3285.66	6.875
190	0.0788	2993.72	3308.87	6.908
200	0.0801	3011.65	3332.00	6.940
210	0.0814	3029.53	3355.06	6.971
220	0.0827	3047.39	3378.07	7.003
230	0.0839	3065.24	3401.03	7.033
240	0.0852	3083.07	3423.95	7.063
250	0.0865	3100.90	3446.84	7.093
260	0.0877	3118.73	3469.72	7.123
270	0.0890	3136.57	3492.58	7.152
280	0.0903	3154.42	3515.44	7.180
290	0.0915	3172.29	3538.29	7.209
300	0.0927	3190.18	3561.16	7.236
310	0.0940	3208.10	3584.03	7.264
320	0.0952	3226.04	3606.91	7.291
330	0.0964	3244.02	3629.82	7.318
340	0.0977	3262.03	3652.74	7.345
350	0.0989	3280.08	3675.69	7.371
360	0.1001	3298.17	3698.67	7.398
370	0.1013	3316.30	3721.68	7.424
380	0.1026	3334.48	3744.72	7.449
390	0.1038	3352.70	3767.80	7.475
400	0.1050	3370.96	3790.91	7.500
410	0.1062	3389.27	3814.07	7.525
420	0.1074	3407.64	3837.26	7.549
430	0.1086	3426.05	3860.50	7.574
440	0.1098	3444.52	3883.79	7.598

t (°C)	v $\left(\frac{m^3}{kg}\right)$	u $\left(\frac{kJ}{kg}\right)$	h $\left(\frac{kJ}{kg}\right)$	s $\left(\frac{kJ}{kg\,K}\right)$
Sat Vapour (g)	0.0498	2601.73	2800.85	6.070
450	0.1110	3463.04	3907.11	7.622
460	0.1122	3481.61	3930.49	7.646
470	0.1134	3500.24	3953.92	7.670
480	0.1146	3518.92	3977.39	7.693
490	0.1158	3537.66	4000.92	7.717
500	0.1170	3556.46	4024.49	7.740
510	0.1182	3575.31	4048.12	7.763
520	0.1194	3594.23	4071.80	7.786
530	0.1206	3613.20	4095.54	7.808
540	0.1218	3632.23	4119.33	7.831
550	0.1230	3651.32	4143.18	7.853
560	0.1242	3670.47	4167.08	7.875
570	0.1253	3689.68	4191.03	7.897
580	0.1265	3708.95	4215.05	7.919
590	0.1277	3728.28	4239.12	7.941
600	0.1289	3747.67	4263.25	7.962
610	0.1301	3767.12	4287.43	7.984
620	0.1313	3786.64	4311.67	8.005
630	0.1324	3806.21	4335.97	8.026
640	0.1336	3825.85	4360.33	8.047
650	0.1348	3845.54	4384.75	8.068
660	0.1360	3865.30	4409.23	8.089
670	0.1372	3885.12	4433.76	8.110
680	0.1383	3905.00	4458.35	8.130
690	0.1395	3924.94	4483.00	8.151
700	0.1407	3944.95	4507.71	8.171
710	0.1419	3965.01	4532.48	8.191
720	0.1430	3985.13	4557.30	8.211
730	0.1442	4005.32	4582.19	8.231
740	0.1454	4025.56	4607.13	8.251
750	0.1466	4045.87	4632.13	8.270
760	0.1477	4066.23	4657.19	8.290
770	0.1489	4086.66	4682.31	8.310
780	0.1501	4107.14	4707.48	8.329
790	0.1513	4127.69	4732.72	8.348
800	0.1524	4148.29	4758.01	8.367
810	0.1536	4168.95	4783.35	8.386
820	0.1548	4189.67	4808.76	8.405
830	0.1559	4210.45	4834.22	8.424
840	0.1571	4231.29	4859.73	8.443
850	0.1583	4252.18	4885.31	8.462
860	0.1595	4273.14	4910.94	8.480
870	0.1606	4294.14	4936.62	8.499

t (°C)	v $\left(\frac{m^3}{kg}\right)$	u $\left(\frac{kJ}{kg}\right)$	h $\left(\frac{kJ}{kg}\right)$	s $\left(\frac{kJ}{kg\,K}\right)$
Sat Vapour (g)	0.0498	2601.73	2800.85	6.070
880	0.1618	4315.21	4962.36	8.517
890	0.1630	4336.33	4988.16	8.536
900	0.1641	4357.51	5014.01	8.554
910	0.1653	4378.75	5039.92	8.572
920	0.1665	4400.04	5065.88	8.590
930	0.1676	4421.39	5091.89	8.608
940	0.1688	4442.79	5117.96	8.626
950	0.1700	4464.24	5144.08	8.644
960	0.1711	4485.75	5170.26	8.661
970	0.1723	4507.32	5196.49	8.679
980	0.1735	4528.94	5222.77	8.697
990	0.1746	4550.61	5249.10	8.714
1000	0.1758	4572.33	5275.48	8.731
1010	0.1770	4594.11	5301.92	8.749
1020	0.1781	4615.93	5328.41	8.766
1030	0.1793	4637.81	5354.94	8.783
1040	0.1804	4659.74	5381.53	8.800
1050	0.1816	4681.73	5408.17	8.817
1060	0.1828	4703.76	5434.86	8.834
1070	0.1839	4725.84	5461.60	8.851
1080	0.1851	4747.97	5488.38	8.868
1090	0.1863	4770.16	5515.22	8.884
1100	0.1874	4792.39	5542.10	8.901
1110	0.1886	4814.67	5569.03	8.917
1120	0.1898	4836.99	5596.01	8.934
1130	0.1909	4859.37	5623.04	8.950
1140	0.1921	4881.79	5650.11	8.967
1150	0.1932	4904.27	5677.23	8.983
1160	0.1944	4926.78	5704.40	8.999
1170	0.1956	4949.35	5731.61	9.015
1180	0.1967	4971.96	5758.87	9.031
1190	0.1979	4994.62	5786.17	9.047
1200	0.1991	5017.32	5813.52	9.063
1210	0.2002	5040.06	5840.91	9.079
1220	0.2014	5062.85	5868.35	9.095
1230	0.2025	5085.69	5895.83	9.110
1240	0.2037	5108.57	5923.35	9.126
1250	0.2049	5131.49	5950.92	9.142
1260	0.2060	5154.46	5978.53	9.157
1270	0.2072	5177.47	6006.18	9.173
1280	0.2083	5200.52	6033.87	9.188
1290	0.2095	5223.62	6061.61	9.203
1300	0.2107	5246.75	6089.39	9.219

Table 3(31) Superheated Steam Properties (Pressure Base) Pressure = 42.0 bar and saturation temperature, Tsat = 253.264°C

t (°C)	v (m^3/kg)	u (kJ/kg)	h (kJ/kg)	s (kJ/kg K)
Sat Vapour (g)	0.0473	2601.01	2799.81	6.049
10	0.0493	2630.86	2838.09	6.121
20	0.0512	2657.62	2872.62	6.185
30	0.0529	2682.43	2904.73	6.243
40	0.0546	2705.84	2935.11	6.297
50	0.0562	2728.20	2964.14	6.348
60	0.0577	2749.72	2992.12	6.396
70	0.0592	2770.57	3019.24	6.442
80	0.0607	2790.89	3045.66	6.486
90	0.0621	2810.76	3071.50	6.528
100	0.0635	2830.26	3096.87	6.569
110	0.0648	2849.46	3121.83	6.609
120	0.0662	2868.41	3146.44	6.647
130	0.0675	2887.15	3170.78	6.684
140	0.0688	2905.72	3194.86	6.721
150	0.0701	2924.14	3218.75	6.756
160	0.0714	2942.44	3242.45	6.791
170	0.0727	2960.64	3266.01	6.825
180	0.0740	2978.76	3289.44	6.859
190	0.0752	2996.82	3312.76	6.891
200	0.0765	3014.82	3335.99	6.924
210	0.0777	3032.78	3359.15	6.955
220	0.0789	3050.71	3382.24	6.986
230	0.0802	3068.62	3405.29	7.017
240	0.0814	3086.52	3428.29	7.047
250	0.0826	3104.41	3451.27	7.077
260	0.0838	3122.30	3474.22	7.107
270	0.0850	3140.20	3497.16	7.136
280	0.0862	3158.11	3520.09	7.164
290	0.0874	3176.04	3543.02	7.192
300	0.0886	3193.98	3565.95	7.220
310	0.0897	3211.95	3588.88	7.248
320	0.0909	3229.95	3611.83	7.275
330	0.0921	3247.98	3634.80	7.302
340	0.0933	3266.04	3657.78	7.329
350	0.0944	3284.14	3680.79	7.355
360	0.0956	3302.27	3703.83	7.381
370	0.0968	3320.45	3726.89	7.407
380	0.0979	3338.66	3749.99	7.433
390	0.0991	3356.93	3773.12	7.458
400	0.1003	3375.23	3796.28	7.483
410	0.1014	3393.59	3819.49	7.508
420	0.1026	3411.99	3842.73	7.533
430	0.1037	3430.45	3866.02	7.557
440	0.1049	3448.95	3889.35	7.582

t (°C)	v (m^3/kg)	u (kJ/kg)	h (kJ/kg)	s (kJ/kg K)
Sat Vapour (g)	0.0473	2601.01	2799.81	6.049
450	0.1060	3467.51	3912.73	7.606
460	0.1071	3486.12	3936.15	7.630
470	0.1083	3504.79	3959.62	7.653
480	0.1094	3523.51	3983.14	7.677
490	0.1106	3542.28	4006.71	7.700
500	0.1117	3561.12	4030.33	7.723
510	0.1129	3580.01	4054.00	7.746
520	0.1140	3598.96	4077.72	7.769
530	0.1151	3617.96	4101.50	7.792
540	0.1163	3637.02	4125.33	7.814
550	0.1174	3656.15	4149.21	7.836
560	0.1185	3675.33	4173.15	7.858
570	0.1197	3694.57	4197.15	7.880
580	0.1208	3713.88	4221.20	7.902
590	0.1219	3733.24	4245.30	7.924
600	0.1230	3752.66	4269.47	7.946
610	0.1242	3772.14	4293.69	7.967
620	0.1253	3791.69	4317.97	7.988
630	0.1264	3811.29	4342.30	8.009
640	0.1276	3830.96	4366.69	8.030
650	0.1287	3850.68	4391.15	8.051
660	0.1298	3870.47	4415.65	8.072
670	0.1309	3890.32	4440.22	8.093
680	0.1321	3910.23	4464.85	8.113
690	0.1332	3930.20	4489.53	8.133
700	0.1343	3950.23	4514.27	8.154
710	0.1354	3970.32	4539.07	8.174
720	0.1365	3990.47	4563.92	8.194
730	0.1377	4010.68	4588.84	8.214
740	0.1388	4030.95	4613.81	8.234
750	0.1399	4051.28	4638.84	8.253
760	0.1410	4071.67	4663.93	8.273
770	0.1421	4092.12	4689.08	8.292
780	0.1432	4112.63	4714.28	8.312
790	0.1444	4133.20	4739.54	8.331
800	0.1455	4153.83	4764.86	8.350
810	0.1466	4174.52	4790.23	8.369
820	0.1477	4195.26	4815.66	8.388
830	0.1488	4216.07	4841.15	8.407
840	0.1499	4236.93	4866.69	8.426
850	0.1511	4257.85	4892.29	8.444
860	0.1522	4278.82	4917.95	8.463
870	0.1533	4299.85	4943.66	8.481

t (°C)	v (m^3/kg)	u (kJ/kg)	h (kJ/kg)	s (kJ/kg K)
Sat Vapour (g)	0.0473	2601.01	2799.81	6.049
880	0.1544	4320.94	4969.43	8.500
890	0.1555	4342.09	4995.25	8.518
900	0.1566	4363.29	5021.13	8.536
910	0.1577	4384.55	5047.06	8.554
920	0.1589	4405.86	5073.04	8.572
930	0.1600	4427.23	5099.08	8.590
940	0.1611	4448.65	5125.17	8.608
950	0.1622	4470.13	5151.32	8.626
960	0.1633	4491.66	5177.51	8.644
970	0.1644	4513.25	5203.77	8.661
980	0.1655	4534.88	5230.07	8.679
990	0.1666	4556.57	5256.42	8.696
1000	0.1677	4578.32	5282.83	8.714
1010	0.1689	4600.11	5309.29	8.731
1020	0.1700	4621.96	5335.80	8.748
1030	0.1711	4643.86	5362.35	8.765
1040	0.1722	4665.81	5388.96	8.782
1050	0.1733	4687.81	5415.62	8.799
1060	0.1744	4709.86	5442.33	8.816
1070	0.1755	4731.97	5469.09	8.833
1080	0.1766	4754.12	5495.90	8.850
1090	0.1777	4776.32	5522.75	8.866
1100	0.1788	4798.57	5549.66	8.883
1110	0.1799	4820.86	5576.61	8.899
1120	0.1810	4843.21	5603.61	8.916
1130	0.1822	4865.60	5630.65	8.932
1140	0.1833	4888.05	5657.75	8.949
1150	0.1844	4910.53	5684.88	8.965
1160	0.1855	4933.07	5712.07	8.981
1170	0.1866	4955.65	5739.30	8.997
1180	0.1877	4978.28	5766.58	9.013
1190	0.1888	5000.95	5793.90	9.029
1200	0.1899	5023.67	5821.26	9.045
1210	0.1910	5046.43	5848.67	9.061
1220	0.1921	5069.24	5876.13	9.077
1230	0.1932	5092.09	5903.62	9.092
1240	0.1943	5114.99	5931.16	9.108
1250	0.1954	5137.93	5958.75	9.124
1260	0.1965	5160.91	5986.37	9.139
1270	0.1976	5183.93	6014.04	9.154
1280	0.1988	5207.00	6041.75	9.170
1290	0.1999	5230.11	6069.51	9.185
1300	0.2010	5253.26	6097.30	9.200

Table 3(32) Superheated Steam Properties (Pressure Base) Pressure = 44.0 bar and saturation temperature, Tsat = 256.07°C

t (°C)	v $\left(\frac{m^3}{kg}\right)$	u $\left(\frac{kJ}{kg}\right)$	h $\left(\frac{kJ}{kg}\right)$	s $\left(\frac{kJ}{kg\,K}\right)$
Sat Vapour (g)	0.0451	2600.16	2798.62	6.029
10	0.0471	2630.58	2837.62	6.102
20	0.0488	2657.78	2872.71	6.167
30	0.0505	2682.95	2905.29	6.226
40	0.0521	2706.67	2936.05	6.280
50	0.0537	2729.29	2965.43	6.331
60	0.0552	2751.04	2993.70	6.380
70	0.0566	2772.09	3021.08	6.426
80	0.0580	2792.58	3047.73	6.470
90	0.0594	2812.62	3073.79	6.512
100	0.0607	2832.27	3099.35	6.553
110	0.0620	2851.61	3124.48	6.593
120	0.0633	2870.68	3149.26	6.631
130	0.0646	2889.54	3173.75	6.669
140	0.0659	2908.21	3197.97	6.705
150	0.0671	2926.74	3221.99	6.741
160	0.0683	2945.13	3245.82	6.776
170	0.0696	2963.42	3269.49	6.810
180	0.0708	2981.63	3293.03	6.843
190	0.0720	2999.76	3316.45	6.876
200	0.0732	3017.84	3339.78	6.908
210	0.0744	3035.88	3363.04	6.940
220	0.0755	3053.88	3386.22	6.971
230	0.0767	3071.86	3409.35	7.002
240	0.0779	3089.82	3432.44	7.032
250	0.0790	3107.77	3455.50	7.062
260	0.0802	3125.73	3478.53	7.091
270	0.0813	3143.68	3501.54	7.120
280	0.0825	3161.65	3524.54	7.149
290	0.0836	3179.63	3547.53	7.177
300	0.0848	3197.63	3570.53	7.205
310	0.0859	3215.65	3593.53	7.232
320	0.0870	3233.70	3616.54	7.260
330	0.0881	3251.77	3639.57	7.287
340	0.0893	3269.88	3662.61	7.313
350	0.0904	3288.02	3685.68	7.340
360	0.0915	3306.20	3708.77	7.366
370	0.0926	3324.42	3731.89	7.392
380	0.0937	3342.68	3755.04	7.417
390	0.0948	3360.99	3778.22	7.443
400	0.0959	3379.34	3801.44	7.468
410	0.0970	3397.73	3824.69	7.493
420	0.0981	3416.18	3847.99	7.517
430	0.0992	3434.67	3871.32	7.542
440	0.1003	3453.21	3894.70	7.566

t (°C)	v $\left(\frac{m^3}{kg}\right)$	u $\left(\frac{kJ}{kg}\right)$	h $\left(\frac{kJ}{kg}\right)$	s $\left(\frac{kJ}{kg\,K}\right)$
Sat Vapour (g)	0.0451	2600.16	2798.62	6.029
450	0.1014	3471.81	3918.12	7.590
460	0.1025	3490.46	3941.59	7.614
470	0.1036	3509.16	3965.10	7.638
480	0.1047	3527.92	3988.66	7.661
490	0.1058	3546.73	4012.27	7.684
500	0.1069	3565.59	4035.93	7.707
510	0.1080	3584.52	4059.64	7.730
520	0.1091	3603.50	4083.41	7.753
530	0.1102	3622.54	4107.22	7.776
540	0.1112	3641.64	4131.09	7.798
550	0.1123	3660.79	4155.01	7.820
560	0.1134	3680.01	4178.99	7.843
570	0.1145	3699.28	4203.02	7.864
580	0.1156	3718.61	4227.11	7.886
590	0.1166	3738.01	4251.25	7.908
600	0.1177	3757.46	4275.45	7.929
610	0.1188	3776.97	4299.71	7.951
620	0.1199	3796.55	4324.02	7.972
630	0.1210	3816.18	4348.39	7.993
640	0.1220	3835.87	4372.82	8.014
650	0.1231	3855.63	4397.30	8.035
660	0.1242	3875.44	4421.84	8.056
670	0.1253	3895.32	4446.44	8.076
680	0.1263	3915.26	4471.10	8.097
690	0.1274	3935.25	4495.81	8.117
700	0.1285	3955.31	4520.58	8.137
710	0.1295	3975.43	4545.41	8.158
720	0.1306	3995.60	4570.30	8.178
730	0.1317	4015.84	4595.24	8.197
740	0.1328	4036.14	4620.24	8.217
750	0.1338	4056.50	4645.30	8.237
760	0.1349	4076.91	4670.42	8.256
770	0.1360	4097.39	4695.59	8.276
780	0.1370	4117.92	4720.82	8.295
790	0.1381	4138.52	4746.11	8.315
800	0.1392	4159.17	4771.46	8.334
810	0.1402	4179.88	4796.86	8.353
820	0.1413	4200.65	4822.32	8.372
830	0.1424	4221.47	4847.83	8.390
840	0.1434	4242.36	4873.40	8.409
850	0.1445	4263.30	4899.02	8.428
860	0.1455	4284.30	4924.71	8.446
870	0.1466	4305.35	4950.44	8.465

t (°C)	v $\left(\frac{m^3}{kg}\right)$	u $\left(\frac{kJ}{kg}\right)$	h $\left(\frac{kJ}{kg}\right)$	s $\left(\frac{kJ}{kg\,K}\right)$
Sat Vapour (g)	0.0451	2600.16	2798.62	6.029
880	0.1477	4326.46	4976.23	8.483
890	0.1487	4347.63	5002.08	8.502
900	0.1498	4368.86	5027.98	8.520
910	0.1509	4390.13	5053.93	8.538
920	0.1519	4411.47	5079.94	8.556
930	0.1530	4432.86	5106.00	8.574
940	0.1540	4454.30	5132.12	8.592
950	0.1551	4475.80	5158.29	8.609
960	0.1562	4497.35	5184.51	8.627
970	0.1572	4518.95	5210.78	8.645
980	0.1583	4540.61	5237.10	8.662
990	0.1594	4562.32	5263.48	8.679
1000	0.1604	4584.09	5289.91	8.697
1010	0.1615	4605.90	5316.39	8.714
1020	0.1625	4627.77	5342.92	8.731
1030	0.1636	4649.69	5369.50	8.748
1040	0.1647	4671.66	5396.13	8.765
1050	0.1657	4693.68	5422.81	8.782
1060	0.1668	4715.75	5449.54	8.799
1070	0.1678	4737.87	5476.31	8.816
1080	0.1689	4760.03	5503.14	8.833
1090	0.1699	4782.25	5530.02	8.849
1100	0.1710	4804.52	5556.94	8.866
1110	0.1721	4826.83	5583.91	8.882
1120	0.1731	4849.20	5610.93	8.899
1130	0.1742	4871.61	5637.99	8.915
1140	0.1752	4894.07	5665.10	8.932
1150	0.1763	4916.57	5692.26	8.948
1160	0.1774	4939.12	5719.46	8.964
1170	0.1784	4961.72	5746.71	8.980
1180	0.1795	4984.37	5774.01	8.996
1190	0.1805	5007.05	5801.34	9.012
1200	0.1816	5029.79	5828.73	9.028
1210	0.1826	5052.57	5856.15	9.044
1220	0.1837	5075.39	5883.62	9.059
1230	0.1847	5098.26	5911.14	9.075
1240	0.1858	5121.17	5938.70	9.091
1250	0.1869	5144.12	5966.30	9.106
1260	0.1879	5167.12	5993.94	9.122
1270	0.1890	5190.16	6021.62	9.137
1280	0.1900	5213.24	6049.35	9.153
1290	0.1911	5236.37	6077.12	9.168
1300	0.1921	5259.53	6104.93	9.183

Table 3(33) Superheated Steam Properties (Pressure Base) Pressure = 46.0 bar and saturation temperature, Tsat = 258.78°C

t (°C)	v $\left(\frac{m^3}{kg}\right)$	u $\left(\frac{kJ}{kg}\right)$	h $\left(\frac{kJ}{kg}\right)$	s $\left(\frac{kJ}{kg\,K}\right)$
Sat Vapour (g)	0.0431	2599.21	2797.29	6.010
10	0.0450	2630.19	2837.01	6.084
20	0.0467	2657.83	2872.67	6.149
30	0.0483	2683.36	2905.71	6.209
40	0.0499	2707.38	2936.86	6.264
50	0.0514	2730.26	2966.57	6.315
60	0.0528	2752.23	2995.13	6.364
70	0.0542	2773.49	3022.77	6.410
80	0.0555	2794.16	3049.66	6.455
90	0.0569	2814.35	3075.92	6.497
100	0.0582	2834.15	3101.67	6.538
110	0.0594	2853.62	3126.98	6.578
120	0.0607	2872.82	3151.92	6.616
130	0.0619	2891.80	3176.55	6.654
140	0.0631	2910.58	3200.92	6.691
150	0.0643	2929.20	3225.06	6.726
160	0.0655	2947.69	3249.01	6.761
170	0.0667	2966.07	3272.80	6.795
180	0.0678	2984.36	3296.45	6.829
190	0.0690	3002.57	3319.98	6.861
200	0.0701	3020.73	3343.40	6.894
210	0.0713	3038.83	3366.75	6.925
220	0.0724	3056.91	3390.02	6.957
230	0.0735	3074.95	3413.24	6.987
240	0.0747	3092.98	3436.41	7.017
250	0.0758	3110.99	3459.54	7.047
260	0.0769	3129.00	3482.65	7.077
270	0.0780	3147.02	3505.73	7.106
280	0.0791	3165.04	3528.80	7.134
290	0.0802	3183.07	3551.86	7.162
300	0.0813	3201.12	3574.93	7.190
310	0.0823	3219.19	3597.99	7.218
320	0.0834	3237.29	3621.06	7.245
330	0.0845	3255.41	3644.15	7.272
340	0.0856	3273.57	3667.25	7.299
350	0.0867	3291.75	3690.37	7.325
360	0.0877	3309.98	3713.52	7.351
370	0.0888	3328.24	3736.69	7.377
380	0.0899	3346.54	3759.89	7.402
390	0.0909	3364.89	3783.12	7.428
400	0.0920	3383.28	3806.39	7.453
410	0.0930	3401.72	3829.69	7.478
420	0.0941	3420.20	3853.04	7.502
430	0.0951	3438.73	3876.42	7.527
440	0.0962	3457.31	3899.84	7.551

t (°C)	v $\left(\frac{m^3}{kg}\right)$	u $\left(\frac{kJ}{kg}\right)$	h $\left(\frac{kJ}{kg}\right)$	s $\left(\frac{kJ}{kg\,K}\right)$
Sat Vapour (g)	0.0431	2599.21	2797.29	6.010
450	0.0973	3475.94	3923.31	7.575
460	0.0983	3494.63	3946.82	7.599
470	0.0993	3513.37	3970.37	7.623
480	0.1004	3532.16	3993.98	7.646
490	0.1014	3551.00	4017.63	7.669
500	0.1025	3569.90	4041.33	7.692
510	0.1035	3588.86	4065.08	7.715
520	0.1046	3607.88	4088.88	7.738
530	0.1056	3626.95	4112.74	7.761
540	0.1066	3646.08	4136.64	7.783
550	0.1077	3665.26	4160.60	7.805
560	0.1087	3684.51	4184.62	7.827
570	0.1098	3703.81	4208.69	7.849
580	0.1108	3723.18	4232.81	7.871
590	0.1118	3742.60	4256.99	7.893
600	0.1129	3762.08	4281.22	7.914
610	0.1139	3781.63	4305.51	7.936
620	0.1149	3801.23	4329.86	7.957
630	0.1159	3820.89	4354.26	7.978
640	0.1170	3840.61	4378.72	7.999
650	0.1180	3860.40	4403.23	8.020
660	0.1190	3880.24	4427.80	8.040
670	0.1201	3900.14	4452.43	8.061
680	0.1211	3920.10	4477.12	8.081
690	0.1221	3940.13	4501.87	8.102
700	0.1231	3960.21	4526.67	8.122
710	0.1242	3980.35	4551.53	8.142
720	0.1252	4000.56	4576.44	8.162
730	0.1262	4020.82	4601.41	8.182
740	0.1272	4041.14	4626.45	8.202
750	0.1283	4061.52	4651.53	8.221
760	0.1293	4081.96	4676.68	8.241
770	0.1303	4102.46	4701.88	8.260
780	0.1313	4123.02	4727.14	8.280
790	0.1324	4143.64	4752.45	8.299
800	0.1334	4164.31	4777.82	8.318
810	0.1344	4185.05	4803.25	8.337
820	0.1354	4205.84	4828.73	8.356
830	0.1364	4226.69	4854.27	8.375
840	0.1375	4247.59	4879.87	8.393
850	0.1385	4268.56	4905.52	8.412
860	0.1395	4289.58	4931.22	8.431
870	0.1405	4310.66	4956.98	8.449

t (°C)	v $\left(\frac{m^3}{kg}\right)$	u $\left(\frac{kJ}{kg}\right)$	h $\left(\frac{kJ}{kg}\right)$	s $\left(\frac{kJ}{kg\,K}\right)$
Sat Vapour (g)	0.0431	2599.21	2797.29	6.010
880	0.1415	4331.79	4982.80	8.467
890	0.1425	4352.98	5008.67	8.486
900	0.1436	4374.22	5034.59	8.504
910	0.1446	4395.52	5060.57	8.522
920	0.1456	4416.88	5086.60	8.540
930	0.1466	4438.29	5112.69	8.558
940	0.1476	4459.75	5138.82	8.576
950	0.1486	4481.27	5165.01	8.593
960	0.1497	4502.84	5191.26	8.611
970	0.1507	4524.46	5217.55	8.629
980	0.1517	4546.14	5243.90	8.646
990	0.1527	4567.87	5270.29	8.663
1000	0.1537	4589.65	5296.74	8.681
1010	0.1547	4611.49	5323.24	8.698
1020	0.1557	4633.37	5349.79	8.715
1030	0.1568	4655.31	5376.39	8.732
1040	0.1578	4677.30	5403.04	8.749
1050	0.1588	4699.33	5429.74	8.766
1060	0.1598	4721.42	5456.49	8.783
1070	0.1608	4743.56	5483.29	8.800
1080	0.1618	4765.75	5510.13	8.817
1090	0.1628	4787.98	5537.03	8.833
1100	0.1638	4810.27	5563.97	8.850
1110	0.1649	4832.60	5590.96	8.866
1120	0.1659	4854.98	5617.99	8.883
1130	0.1669	4877.41	5645.08	8.899
1140	0.1679	4899.88	5672.21	8.915
1150	0.1689	4922.40	5699.38	8.931
1160	0.1699	4944.97	5726.60	8.948
1170	0.1709	4967.58	5753.87	8.964
1180	0.1719	4990.24	5781.18	8.980
1190	0.1730	5012.95	5808.53	8.996
1200	0.1740	5035.70	5835.93	9.011
1210	0.1750	5058.49	5863.38	9.027
1220	0.1760	5081.33	5890.87	9.043
1230	0.1770	5104.21	5918.40	9.059
1240	0.1780	5127.14	5945.97	9.074
1250	0.1790	5150.11	5973.59	9.090
1260	0.1800	5173.12	6001.24	9.105
1270	0.1810	5196.17	6028.94	9.121
1280	0.1820	5219.27	6056.69	9.136
1290	0.1831	5242.41	6084.47	9.151
1300	0.1841	5265.59	6112.29	9.167

Table 3(34) Superheated Steam Properties (Pressure Base) Pressure = 48.0 bar and saturation temperature, Tsat = 261.401°C

t (°C)	$v\left(\frac{m^3}{kg}\right)$	$u\left(\frac{kJ}{kg}\right)$	$h\left(\frac{kJ}{kg}\right)$	$s\left(\frac{kJ}{kg\,K}\right)$	t (°C)	$v\left(\frac{m^3}{kg}\right)$	$u\left(\frac{kJ}{kg}\right)$	$h\left(\frac{kJ}{kg}\right)$	$s\left(\frac{kJ}{kg\,K}\right)$	t (°C)	$v\left(\frac{m^3}{kg}\right)$	$u\left(\frac{kJ}{kg}\right)$	$h\left(\frac{kJ}{kg}\right)$	$s\left(\frac{kJ}{kg\,K}\right)$
Sat Vapour (g)	0.0412	2598.15	2795.82	5.992	Sat Vapour (g)	0.0412	2598.15	2795.82	5.992	Sat Vapour (g)	0.0412	2598.15	2795.82	5.992
10	0.0430	2629.70	2836.27	6.067	450	0.0934	3479.93	3928.30	7.561	880	0.1359	4336.93	4989.14	8.452
20	0.0447	2657.79	2872.49	6.133	460	0.0944	3498.65	3951.86	7.585	890	0.1369	4358.14	5015.04	8.470
30	0.0463	2683.68	2905.99	6.193	470	0.0954	3517.42	3975.45	7.608	900	0.1378	4379.41	5040.98	8.489
40	0.0478	2708.00	2937.53	6.248	480	0.0964	3536.25	3999.10	7.632	910	0.1388	4400.73	5066.98	8.507
50	0.0493	2731.13	2967.57	6.300	490	0.0974	3555.12	4022.79	7.655	920	0.1398	4422.10	5093.04	8.525
60	0.0506	2753.33	2996.42	6.349	500	0.0984	3574.06	4046.53	7.678	930	0.1408	4443.53	5119.14	8.543
70	0.0520	2774.78	3024.32	6.395	510	0.0994	3593.05	4070.32	7.701	940	0.1417	4465.02	5145.30	8.560
80	0.0533	2795.63	3051.44	6.440	520	0.1004	3612.10	4094.16	7.724	950	0.1427	4486.55	5171.52	8.578
90	0.0546	2815.98	3077.91	6.482	530	0.1014	3631.20	4118.06	7.746	960	0.1437	4508.14	5197.78	8.596
100	0.0558	2835.92	3103.84	6.524	540	0.1024	3650.36	4142.00	7.768	970	0.1446	4529.79	5224.09	8.613
110	0.0570	2855.53	3129.33	6.563	550	0.1034	3669.58	4166.00	7.791	980	0.1456	4551.48	5250.46	8.631
120	0.0582	2874.85	3154.43	6.602	560	0.1044	3688.85	4190.05	7.813	990	0.1466	4573.23	5276.88	8.648
130	0.0594	2893.94	3179.21	6.640	570	0.1054	3708.19	4214.15	7.835	1000	0.1476	4595.03	5303.35	8.665
140	0.0606	2912.82	3203.71	6.676	580	0.1064	3727.58	4238.31	7.856	1010	0.1485	4616.88	5329.87	8.683
150	0.0618	2931.54	3227.98	6.712	590	0.1074	3747.03	4262.52	7.878	1020	0.1495	4638.79	5356.44	8.700
160	0.0629	2950.13	3252.05	6.747	600	0.1084	3766.55	4286.79	7.900	1030	0.1505	4660.74	5383.06	8.717
170	0.0640	2968.59	3275.95	6.781	610	0.1094	3786.12	4311.11	7.921	1040	0.1515	4682.75	5409.73	8.734
180	0.0652	2986.96	3299.71	6.815	620	0.1104	3805.75	4335.49	7.942	1050	0.1524	4704.80	5436.45	8.751
190	0.0663	3005.26	3323.34	6.847	630	0.1114	3825.44	4359.92	7.963	1060	0.1534	4726.91	5463.22	8.768
200	0.0674	3023.49	3346.86	6.880	640	0.1123	3845.19	4384.41	7.984	1070	0.1544	4749.06	5490.03	8.784
210	0.0685	3041.66	3370.30	6.911	650	0.1133	3865.00	4408.96	8.005	1080	0.1553	4771.27	5516.90	8.801
220	0.0696	3059.80	3393.66	6.943	660	0.1143	3884.87	4433.56	8.026	1090	0.1563	4793.52	5543.81	8.818
230	0.0706	3077.91	3416.96	6.973	670	0.1153	3904.80	4458.22	8.046	1100	0.1573	4815.82	5570.77	8.834
240	0.0717	3096.00	3440.21	7.003	680	0.1163	3924.78	4482.94	8.067	1110	0.1583	4838.17	5597.78	8.851
250	0.0728	3114.07	3463.42	7.033	690	0.1173	3944.83	4507.71	8.087	1120	0.1592	4860.56	5624.83	8.867
260	0.0738	3132.14	3486.60	7.063	700	0.1183	3964.94	4532.54	8.107	1130	0.1602	4883.01	5651.93	8.883
270	0.0749	3150.21	3509.75	7.091	710	0.1192	3985.11	4557.43	8.127	1140	0.1612	4905.50	5679.08	8.900
280	0.0760	3168.29	3532.89	7.120	720	0.1202	4005.34	4582.38	8.147	1150	0.1621	4928.04	5706.27	8.916
290	0.0770	3186.37	3556.02	7.148	730	0.1212	4025.62	4607.38	8.167	1160	0.1631	4950.62	5733.51	8.932
300	0.0781	3204.48	3579.15	7.176	740	0.1222	4045.97	4632.44	8.187	1170	0.1641	4973.25	5760.79	8.948
310	0.0791	3222.60	3602.27	7.204	750	0.1232	4066.38	4657.55	8.206	1180	0.1650	4995.92	5788.12	8.964
320	0.0801	3240.74	3625.41	7.231	760	0.1241	4086.84	4682.72	8.226	1190	0.1660	5018.64	5815.49	8.980
330	0.0812	3258.91	3648.55	7.258	770	0.1251	4107.36	4707.95	8.245	1200	0.1670	5041.41	5842.90	8.996
340	0.0822	3277.11	3671.71	7.285	780	0.1261	4127.95	4733.24	8.265	1210	0.1679	5064.22	5870.36	9.012
350	0.0832	3295.34	3694.89	7.311	790	0.1271	4148.59	4758.58	8.284	1220	0.1689	5087.07	5897.87	9.027
360	0.0843	3313.61	3718.09	7.337	800	0.1281	4169.28	4783.97	8.303	1230	0.1699	5109.97	5925.41	9.043
370	0.0853	3331.92	3741.31	7.363	810	0.1290	4190.04	4809.43	8.322	1240	0.1709	5132.91	5953.00	9.059
380	0.0863	3350.26	3764.56	7.388	820	0.1300	4210.85	4834.93	8.341	1250	0.1718	5155.89	5980.63	9.074
390	0.0873	3368.65	3787.85	7.414	830	0.1310	4231.73	4860.50	8.360	1260	0.1728	5178.92	6008.31	9.090
400	0.0884	3387.08	3811.16	7.439	840	0.1320	4252.65	4886.12	8.378	1270	0.1738	5201.98	6036.02	9.105
410	0.0894	3405.55	3834.51	7.464	850	0.1329	4273.64	4911.79	8.397	1280	0.1747	5225.09	6063.78	9.120
420	0.0904	3424.07	3857.90	7.488	860	0.1339	4294.68	4937.52	8.416	1290	0.1757	5248.25	6091.58	9.136
430	0.0914	3442.64	3881.33	7.513	870	0.1349	4315.78	4963.31	8.434	1300	0.1767	5271.44	6119.42	9.151
440	0.0924	3461.26	3904.79	7.537										

Table 3(35) Superheated Steam Properties (Pressure Base) Pressure = 50.0 bar and saturation temperature, Tsat = 263.94°C

t (°C)	v (m^3/kg)	u (kJ/kg)	h (kJ/kg)	s (kJ/kg K)
Sat Vapour (g)	0.0394	2596.99	2794.23	5.974
10	0.0413	2629.12	2835.41	6.050
20	0.0429	2657.65	2872.20	6.116
30	0.0445	2683.89	2906.16	6.177
40	0.0459	2708.52	2938.08	6.233
50	0.0473	2731.90	2968.45	6.285
60	0.0487	2754.32	2997.59	6.334
70	0.0500	2775.97	3025.74	6.381
80	0.0512	2797.00	3053.09	6.425
90	0.0525	2817.51	3079.76	6.468
100	0.0537	2837.59	3105.88	6.510
110	0.0548	2857.33	3131.54	6.550
120	0.0560	2876.77	3156.80	6.588
130	0.0572	2895.97	3181.72	6.626
140	0.0583	2914.96	3206.36	6.663
150	0.0594	2933.77	3230.76	6.698
160	0.0605	2952.45	3254.95	6.733
170	0.0616	2971.00	3278.96	6.768
180	0.0627	2989.45	3302.82	6.801
190	0.0637	3007.82	3326.55	6.834
200	0.0648	3026.13	3350.17	6.866
210	0.0659	3044.37	3373.69	6.898
220	0.0669	3062.58	3397.14	6.929
230	0.0680	3080.75	3420.53	6.960
240	0.0690	3098.90	3443.85	6.990
250	0.0700	3117.04	3467.14	7.020
260	0.0710	3135.16	3490.39	7.049
270	0.0721	3153.29	3513.62	7.078
280	0.0731	3171.42	3536.82	7.107
290	0.0741	3189.55	3560.02	7.135
300	0.0751	3207.70	3583.20	7.163
310	0.0761	3225.87	3606.39	7.190
320	0.0771	3244.06	3629.59	7.217
330	0.0781	3262.28	3652.79	7.244
340	0.0791	3280.52	3676.00	7.271
350	0.0801	3298.80	3699.23	7.297
360	0.0811	3317.11	3722.49	7.323
370	0.0821	3335.46	3745.76	7.349
380	0.0830	3353.84	3769.06	7.375
390	0.0840	3372.27	3792.40	7.400
400	0.0850	3390.74	3815.76	7.425
410	0.0860	3409.25	3839.16	7.450
420	0.0870	3427.81	3862.59	7.475
430	0.0879	3446.41	3886.06	7.499
440	0.0889	3465.07	3909.57	7.523

t (°C)	v (m^3/kg)	u (kJ/kg)	h (kJ/kg)	s (kJ/kg K)
Sat Vapour (g)	0.0394	2596.99	2794.23	5.974
450	0.0899	3483.77	3933.13	7.547
460	0.0908	3502.53	3956.72	7.571
470	0.0918	3521.33	3980.36	7.595
480	0.0928	3540.19	4004.05	7.618
490	0.0937	3559.10	4027.78	7.641
500	0.0947	3578.07	4051.56	7.664
510	0.0957	3597.09	4075.39	7.687
520	0.0966	3616.17	4099.26	7.710
530	0.0976	3635.31	4123.19	7.732
540	0.0985	3654.50	4147.17	7.755
550	0.0995	3673.75	4171.21	7.777
560	0.1004	3693.05	4195.29	7.799
570	0.1014	3712.41	4219.43	7.821
580	0.1024	3731.84	4243.62	7.843
590	0.1033	3751.32	4267.87	7.864
600	0.1043	3770.86	4292.17	7.886
610	0.1052	3790.46	4316.52	7.907
620	0.1062	3810.11	4340.93	7.928
630	0.1071	3829.83	4365.40	7.949
640	0.1081	3849.61	4389.92	7.970
650	0.1090	3869.44	4414.50	7.991
660	0.1100	3889.34	4439.13	8.012
670	0.1109	3909.29	4463.82	8.032
680	0.1119	3929.31	4488.57	8.053
690	0.1128	3949.38	4513.37	8.073
700	0.1137	3969.52	4538.23	8.093
710	0.1147	3989.71	4563.15	8.113
720	0.1156	4009.96	4588.12	8.133
730	0.1166	4030.27	4613.15	8.153
740	0.1175	4050.64	4638.23	8.173
750	0.1185	4071.07	4663.37	8.192
760	0.1194	4091.56	4688.57	8.212
770	0.1203	4112.11	4713.83	8.231
780	0.1213	4132.71	4739.14	8.250
790	0.1222	4153.37	4764.50	8.270
800	0.1232	4174.09	4789.92	8.289
810	0.1241	4194.87	4815.40	8.308
820	0.1250	4215.71	4840.93	8.326
830	0.1260	4236.60	4866.52	8.345
840	0.1269	4257.55	4892.17	8.364
850	0.1279	4278.56	4917.87	8.383
860	0.1288	4299.62	4943.62	8.401
870	0.1297	4320.74	4969.43	8.419

t (°C)	v (m^3/kg)	u (kJ/kg)	h (kJ/kg)	s (kJ/kg K)
Sat Vapour (g)	0.0394	2596.99	2794.23	5.974
880	0.1307	4341.91	4995.29	8.438
890	0.1316	4363.14	5021.20	8.456
900	0.1325	4384.43	5047.17	8.474
910	0.1335	4405.77	5073.19	8.492
920	0.1344	4427.16	5099.27	8.510
930	0.1354	4448.61	5125.40	8.528
940	0.1363	4470.11	5151.58	8.546
950	0.1372	4491.67	5177.81	8.563
960	0.1382	4513.28	5204.09	8.581
970	0.1391	4534.94	5230.43	8.599
980	0.1400	4556.65	5256.82	8.616
990	0.1410	4578.42	5283.26	8.633
1000	0.1419	4600.24	5309.75	8.651
1010	0.1428	4622.11	5336.29	8.668
1020	0.1438	4644.03	5362.87	8.685
1030	0.1447	4666.00	5389.51	8.702
1040	0.1456	4688.02	5416.20	8.719
1050	0.1466	4710.09	5442.94	8.736
1060	0.1475	4732.22	5469.73	8.753
1070	0.1484	4754.39	5496.56	8.770
1080	0.1494	4776.61	5523.44	8.786
1090	0.1503	4798.88	5550.37	8.803
1100	0.1512	4821.19	5577.35	8.819
1110	0.1522	4843.56	5604.38	8.836
1120	0.1531	4865.97	5631.45	8.852
1130	0.1540	4888.43	5658.56	8.869
1140	0.1550	4910.94	5685.73	8.885
1150	0.1559	4933.49	5712.94	8.901
1160	0.1568	4956.09	5740.19	8.917
1170	0.1578	4978.73	5767.49	8.933
1180	0.1587	5001.42	5794.83	8.949
1190	0.1596	5024.16	5822.22	8.965
1200	0.1605	5046.94	5849.65	8.981
1210	0.1615	5069.76	5877.13	8.997
1220	0.1624	5092.63	5904.65	9.012
1230	0.1633	5115.54	5932.21	9.028
1240	0.1643	5138.49	5959.82	9.044
1250	0.1652	5161.49	5987.46	9.059
1260	0.1661	5184.53	6015.15	9.074
1270	0.1671	5207.61	6042.88	9.090
1280	0.1680	5230.73	6070.65	9.105
1290	0.1689	5253.90	6098.46	9.120
1300	0.1698	5277.10	6126.32	9.136

Table 3(36) Superheated Steam Properties (Pressure Base) Pressure = 52.0 bar and saturation temperature, Tsat = 266.403°C

t (°C)	*v* $\left(\frac{m^3}{kg}\right)$	*u* $\left(\frac{kJ}{kg}\right)$	*h* $\left(\frac{kJ}{kg}\right)$	*s* $\left(\frac{kJ}{kg\,K}\right)$	*t* (°C)	*v* $\left(\frac{m^3}{kg}\right)$	*u* $\left(\frac{kJ}{kg}\right)$	*h* $\left(\frac{kJ}{kg}\right)$	*s* $\left(\frac{kJ}{kg\,K}\right)$	*t* (°C)	*v* $\left(\frac{m^3}{kg}\right)$	*u* $\left(\frac{kJ}{kg}\right)$	*h* $\left(\frac{kJ}{kg}\right)$	*s* $\left(\frac{kJ}{kg\,K}\right)$
Sat Vapour (g)	0.0378	2595.74	2792.52	5.956	Sat Vapour (g)	0.0378	2595.74	2792.52	5.956	Sat Vapour (g)	0.0378	2595.74	2792.52	5.956
10	0.0396	2628.45	2834.44	6.033	450	0.0866	3487.49	3937.78	7.534	880	0.1259	4346.73	5001.24	8.424
20	0.0412	2657.42	2871.79	6.101	460	0.0875	3506.27	3961.42	7.558	890	0.1268	4367.98	5027.18	8.442
30	0.0427	2684.03	2906.21	6.162	470	0.0885	3525.11	3985.10	7.581	900	0.1277	4389.29	5053.17	8.460
40	0.0441	2708.95	2938.52	6.218	480	0.0894	3544.01	4008.83	7.605	910	0.1286	4410.65	5079.21	8.478
50	0.0455	2732.59	2969.22	6.270	490	0.0903	3562.95	4032.60	7.628	920	0.1295	4432.06	5105.31	8.496
60	0.0468	2755.23	2998.65	6.320	500	0.0912	3581.95	4056.42	7.651	930	0.1304	4453.53	5131.46	8.514
70	0.0481	2777.07	3027.05	6.367	510	0.0922	3601.00	4080.28	7.674	940	0.1313	4475.05	5157.66	8.532
80	0.0493	2798.27	3054.62	6.412	520	0.0931	3620.11	4104.20	7.696	950	0.1322	4496.62	5183.91	8.549
90	0.0505	2818.93	3081.50	6.455	530	0.0940	3639.28	4128.16	7.719	960	0.1331	4518.25	5210.22	8.567
100	0.0517	2839.16	3107.80	6.496	540	0.0949	3658.50	4152.18	7.741	970	0.1340	4539.93	5236.57	8.585
110	0.0528	2859.03	3133.62	6.536	550	0.0959	3677.78	4176.25	7.764	980	0.1349	4561.66	5262.98	8.602
120	0.0539	2878.59	3159.04	6.575	560	0.0968	3697.11	4200.36	7.786	990	0.1358	4583.45	5289.44	8.619
130	0.0550	2897.89	3184.11	6.613	570	0.0977	3716.50	4224.54	7.807	1000	0.1367	4605.28	5315.95	8.637
140	0.0561	2916.99	3208.88	6.649	580	0.0986	3735.95	4248.76	7.829	1010	0.1376	4627.17	5342.50	8.654
150	0.0572	2935.90	3233.40	6.685	590	0.0995	3755.46	4273.04	7.851	1020	0.1385	4649.11	5369.11	8.671
160	0.0583	2954.67	3257.71	6.720	600	0.1005	3775.03	4297.37	7.872	1030	0.1394	4671.10	5395.77	8.688
170	0.0593	2973.30	3281.83	6.755	610	0.1014	3794.66	4321.76	7.894	1040	0.1403	4693.13	5422.48	8.705
180	0.0604	2991.84	3305.79	6.788	620	0.1023	3814.34	4346.20	7.915	1050	0.1412	4715.22	5449.23	8.722
190	0.0614	3010.28	3329.62	6.821	630	0.1032	3834.08	4370.70	7.936	1060	0.1421	4737.36	5476.04	8.739
200	0.0624	3028.66	3353.34	6.853	640	0.1041	3853.89	4395.25	7.957	1070	0.1430	4759.55	5502.89	8.755
210	0.0635	3046.97	3376.95	6.885	650	0.1050	3873.75	4419.86	7.977	1080	0.1438	4781.79	5529.79	8.772
220	0.0645	3065.24	3400.48	6.916	660	0.1059	3893.67	4444.52	7.998	1090	0.1447	4804.07	5556.74	8.789
230	0.0655	3083.48	3423.95	6.947	670	0.1068	3913.65	4469.24	8.019	1100	0.1456	4826.40	5583.73	8.805
240	0.0665	3101.69	3447.36	6.977	680	0.1078	3933.69	4494.02	8.039	1110	0.1465	4848.78	5610.77	8.822
250	0.0675	3119.88	3470.72	7.007	690	0.1087	3953.79	4518.85	8.059	1120	0.1474	4871.21	5637.86	8.838
260	0.0685	3138.07	3494.04	7.036	700	0.1096	3973.95	4543.73	8.079	1130	0.1483	4893.69	5665.00	8.854
270	0.0694	3156.24	3517.33	7.065	710	0.1105	3994.16	4568.68	8.099	1140	0.1492	4916.21	5692.18	8.871
280	0.0704	3174.43	3540.60	7.094	720	0.1114	4014.44	4593.68	8.119	1150	0.1501	4938.77	5719.40	8.887
290	0.0714	3192.61	3563.86	7.122	730	0.1123	4034.77	4618.73	8.139	1160	0.1510	4961.39	5746.67	8.903
300	0.0724	3210.81	3587.11	7.150	740	0.1132	4055.17	4643.84	8.159	1170	0.1519	4984.05	5773.99	8.919
310	0.0733	3229.03	3610.36	7.177	750	0.1141	4075.62	4669.01	8.178	1180	0.1528	5006.75	5801.35	8.935
320	0.0743	3247.27	3633.61	7.205	760	0.1150	4096.13	4694.24	8.198	1190	0.1537	5029.50	5828.75	8.951
330	0.0753	3265.53	3656.87	7.231	770	0.1159	4116.70	4719.52	8.217	1200	0.1546	5052.29	5856.20	8.966
340	0.0762	3283.82	3680.14	7.258	780	0.1168	4137.32	4744.85	8.237	1210	0.1555	5075.13	5883.69	8.982
350	0.0772	3302.14	3703.43	7.284	790	0.1177	4158.01	4770.24	8.256	1220	0.1564	5098.01	5911.23	8.998
360	0.0781	3320.49	3726.73	7.310	800	0.1186	4178.75	4795.69	8.275	1230	0.1573	5120.94	5938.80	9.014
370	0.0791	3338.88	3750.06	7.336	810	0.1195	4199.55	4821.19	8.294	1240	0.1582	5143.90	5966.42	9.029
380	0.0800	3357.30	3773.41	7.362	820	0.1205	4220.41	4846.75	8.313	1250	0.1591	5166.92	5994.08	9.045
390	0.0810	3375.76	3796.79	7.387	830	0.1214	4241.32	4872.36	8.331	1260	0.1600	5189.97	6021.79	9.060
400	0.0819	3394.27	3820.20	7.412	840	0.1223	4262.29	4898.03	8.350	1270	0.1609	5213.06	6049.53	9.075
410	0.0829	3412.82	3843.64	7.437	850	0.1232	4283.32	4923.75	8.369	1280	0.1618	5236.20	6077.32	9.091
420	0.0838	3431.42	3867.12	7.461	860	0.1241	4304.40	4949.52	8.387	1290	0.1626	5259.38	6105.14	9.106
430	0.0847	3450.06	3890.64	7.486	870	0.1250	4325.54	4975.35	8.406	1300	0.1635	5282.60	6133.01	9.121
440	0.0857	3468.75	3914.19	7.510										

Table 3(37) Superheated Steam Properties (Pressure Base) Pressure = 54.0 bar and saturation temperature, Tsat = 268.794°C

t (°C)	v $\left(\frac{m^3}{kg}\right)$	u $\left(\frac{kJ}{kg}\right)$	h $\left(\frac{kJ}{kg}\right)$	s $\left(\frac{kJ}{kg\,K}\right)$	t (°C)	v $\left(\frac{m^3}{kg}\right)$	u $\left(\frac{kJ}{kg}\right)$	h $\left(\frac{kJ}{kg}\right)$	s $\left(\frac{kJ}{kg\,K}\right)$	t (°C)	v $\left(\frac{m^3}{kg}\right)$	u $\left(\frac{kJ}{kg}\right)$	h $\left(\frac{kJ}{kg}\right)$	s $\left(\frac{kJ}{kg\,K}\right)$
Sat Vapour (g)	0.0364	2594.41	2790.70	5.939	Sat Vapour (g)	0.0364	2594.41	2790.70	5.939	Sat Vapour (g)	0.0364	2594.41	2790.70	5.939
10	0.0381	2627.70	2833.36	6.017	450	0.0836	3491.08	3942.29	7.521	880	0.1214	4351.41	5007.01	8.411
20	0.0397	2657.11	2871.28	6.085	460	0.0845	3509.90	3965.97	7.545	890	0.1223	4372.68	5032.97	8.429
30	0.0411	2684.08	2906.16	6.147	470	0.0854	3528.77	3989.69	7.569	900	0.1231	4394.00	5058.98	8.447
40	0.0425	2709.29	2938.85	6.203	480	0.0863	3547.70	4013.45	7.592	910	0.1240	4415.38	5085.05	8.465
50	0.0438	2733.19	2969.88	6.256	490	0.0871	3566.67	4037.26	7.615	920	0.1249	4436.81	5111.17	8.483
60	0.0451	2756.05	2999.59	6.306	500	0.0880	3585.70	4061.12	7.638	930	0.1257	4458.30	5137.34	8.501
70	0.0463	2778.09	3028.24	6.353	510	0.0889	3604.79	4085.02	7.661	940	0.1266	4479.84	5163.56	8.518
80	0.0475	2799.45	3056.04	6.398	520	0.0898	3623.93	4108.97	7.684	950	0.1275	4501.43	5189.83	8.536
90	0.0487	2820.27	3083.12	6.442	530	0.0907	3643.12	4132.97	7.706	960	0.1283	4523.07	5216.15	8.554
100	0.0498	2840.64	3109.61	6.483	540	0.0916	3662.37	4157.02	7.729	970	0.1292	4544.77	5242.53	8.571
110	0.0509	2860.64	3135.59	6.523	550	0.0925	3681.68	4181.13	7.751	980	0.1301	4566.52	5268.96	8.589
120	0.0520	2880.31	3161.16	6.562	560	0.0934	3701.04	4205.28	7.773	990	0.1309	4588.32	5295.43	8.606
130	0.0531	2899.73	3186.37	6.600	570	0.0943	3720.46	4229.48	7.795	1000	0.1318	4610.18	5321.96	8.623
140	0.0541	2918.92	3211.28	6.637	580	0.0951	3739.94	4253.74	7.816	1010	0.1327	4632.08	5348.54	8.640
150	0.0552	2937.93	3235.92	6.673	590	0.0960	3759.48	4278.05	7.838	1020	0.1335	4654.03	5375.17	8.657
160	0.0562	2956.79	3260.35	6.708	600	0.0969	3779.07	4302.42	7.859	1030	0.1344	4676.04	5401.84	8.674
170	0.0572	2975.51	3284.58	6.742	610	0.0978	3798.72	4326.83	7.881	1040	0.1353	4698.09	5428.57	8.691
180	0.0582	2994.12	3308.64	6.776	620	0.0987	3818.43	4351.31	7.902	1050	0.1361	4720.20	5455.34	8.708
190	0.0592	3012.64	3332.57	6.808	630	0.0996	3838.20	4375.83	7.923	1060	0.1370	4742.35	5482.16	8.725
200	0.0602	3031.09	3356.37	6.841	640	0.1004	3858.03	4400.42	7.944	1070	0.1379	4764.56	5509.03	8.742
210	0.0612	3049.47	3380.08	6.873	650	0.1013	3877.92	4425.05	7.964	1080	0.1387	4786.81	5535.95	8.758
220	0.0622	3067.81	3403.69	6.904	660	0.1022	3897.87	4449.75	7.985	1090	0.1396	4809.11	5562.92	8.775
230	0.0632	3086.11	3427.24	6.934	670	0.1031	3917.87	4474.49	8.006	1100	0.1405	4831.46	5589.93	8.792
240	0.0641	3104.38	3450.72	6.965	680	0.1040	3937.93	4499.30	8.026	1110	0.1413	4853.85	5616.99	8.808
250	0.0651	3122.62	3474.16	6.994	690	0.1048	3958.06	4524.16	8.046	1120	0.1422	4876.30	5644.09	8.824
260	0.0661	3140.86	3497.55	7.024	700	0.1057	3978.24	4549.07	8.066	1130	0.1430	4898.78	5671.24	8.841
270	0.0670	3159.09	3520.91	7.053	710	0.1066	3998.48	4574.04	8.086	1140	0.1439	4921.32	5698.44	8.857
280	0.0679	3177.33	3544.25	7.081	720	0.1075	4018.78	4599.07	8.106	1150	0.1448	4943.90	5725.68	8.873
290	0.0689	3195.56	3567.57	7.110	730	0.1083	4039.13	4624.15	8.126	1160	0.1456	4966.53	5752.97	8.889
300	0.0698	3213.81	3590.88	7.137	740	0.1092	4059.55	4649.29	8.146	1170	0.1465	4989.20	5780.30	8.905
310	0.0708	3232.08	3614.19	7.165	750	0.1101	4080.03	4674.48	8.165	1180	0.1474	5011.92	5807.67	8.921
320	0.0717	3250.36	3637.50	7.192	760	0.1110	4100.56	4699.73	8.185	1190	0.1482	5034.69	5835.09	8.937
330	0.0726	3268.66	3660.81	7.219	770	0.1118	4121.15	4725.04	8.204	1200	0.1491	5057.49	5862.55	8.953
340	0.0735	3287.00	3684.14	7.246	780	0.1127	4141.80	4750.40	8.223	1210	0.1499	5080.34	5890.06	8.968
350	0.0745	3305.36	3707.47	7.272	790	0.1136	4162.50	4775.81	8.243	1220	0.1508	5103.24	5917.61	8.984
360	0.0754	3323.75	3730.83	7.298	800	0.1144	4183.27	4801.28	8.262	1230	0.1517	5126.18	5945.20	9.000
370	0.0763	3342.18	3754.21	7.324	810	0.1153	4204.09	4826.81	8.281	1240	0.1525	5149.16	5972.83	9.015
380	0.0772	3360.64	3777.61	7.349	820	0.1162	4224.96	4852.39	8.299	1250	0.1534	5172.18	6000.51	9.031
390	0.0781	3379.14	3801.03	7.374	830	0.1171	4245.90	4878.02	8.318	1260	0.1543	5195.25	6028.23	9.046
400	0.0790	3397.69	3824.49	7.400	840	0.1179	4266.89	4903.71	8.337	1270	0.1551	5218.36	6055.99	9.062
410	0.0799	3416.27	3847.98	7.424	850	0.1188	4287.94	4929.46	8.355	1280	0.1560	5241.51	6083.79	9.077
420	0.0809	3434.91	3871.50	7.449	860	0.1197	4309.04	4955.26	8.374	1290	0.1568	5264.70	6111.63	9.092
430	0.0818	3453.58	3895.06	7.473	870	0.1205	4330.20	4981.11	8.392	1300	0.1577	5287.93	6139.51	9.107
440	0.0827	3472.31	3918.66	7.497										

Table 3(38) Superheated Steam Properties (Pressure Base) Pressure = 56.0 bar and saturation temperature, Tsat = 271.119°C

t (°C)	v ($\frac{m^3}{kg}$)	u ($\frac{kJ}{kg}$)	h ($\frac{kJ}{kg}$)	s ($\frac{kJ}{kg\ K}$)	t (°C)	v ($\frac{m^3}{kg}$)	u ($\frac{kJ}{kg}$)	h ($\frac{kJ}{kg}$)	s ($\frac{kJ}{kg\ K}$)	t (°C)	v ($\frac{m^3}{kg}$)	u ($\frac{kJ}{kg}$)	h ($\frac{kJ}{kg}$)	s ($\frac{kJ}{kg\ K}$)
Sat Vapour (g)	*0.0350*	*2592.99*	*2788.77*	*5.922*	*Sat Vapour (g)*	*0.0350*	*2592.99*	*2788.77*	*5.922*	*Sat Vapour (g)*	*0.0350*	*2592.99*	*2788.77*	*5.922*
10	0.0367	2626.87	2832.18	6.001	450	0.0807	3494.56	3946.66	7.509	880	0.1173	4355.95	5012.62	8.398
20	0.0382	2656.73	2870.67	6.070	460	0.0816	3513.41	3970.38	7.533	890	0.1181	4377.24	5038.60	8.416
30	0.0396	2684.05	2906.01	6.132	470	0.0825	3532.32	3994.13	7.557	900	0.1189	4398.58	5064.63	8.434
40	0.0410	2709.56	2939.09	6.190	480	0.0833	3551.27	4017.94	7.580	910	0.1198	4419.98	5090.72	8.452
50	0.0423	2733.71	2970.44	6.243	490	0.0842	3570.28	4041.78	7.603	920	0.1206	4441.43	5116.86	8.470
60	0.0435	2756.79	3000.43	6.293	500	0.0851	3589.34	4065.68	7.626	930	0.1214	4462.93	5143.05	8.488
70	0.0447	2779.02	3029.34	6.340	510	0.0859	3608.46	4089.61	7.649	940	0.1223	4484.49	5169.29	8.505
80	0.0459	2800.56	3057.35	6.385	520	0.0868	3627.63	4113.60	7.671	950	0.1231	4506.10	5195.58	8.523
90	0.0470	2821.53	3084.63	6.429	530	0.0876	3646.85	4137.64	7.694	960	0.1240	4527.76	5221.92	8.541
100	0.0481	2842.04	3111.30	6.471	540	0.0885	3666.13	4161.72	7.716	970	0.1248	4549.47	5248.32	8.558
110	0.0492	2862.16	3137.46	6.511	550	0.0894	3685.46	4185.86	7.738	980	0.1256	4571.24	5274.77	8.576
120	0.0502	2881.95	3163.18	6.550	560	0.0902	3704.85	4210.04	7.760	990	0.1265	4593.06	5301.26	8.593
130	0.0513	2901.48	3188.53	6.588	570	0.0911	3724.30	4234.28	7.782	1000	0.1273	4614.93	5327.81	8.610
140	0.0523	2920.77	3213.56	6.625	580	0.0919	3743.81	4258.57	7.804	1010	0.1281	4636.85	5354.40	8.627
150	0.0533	2939.87	3238.33	6.661	590	0.0928	3763.37	4282.91	7.826	1020	0.1290	4658.82	5381.05	8.644
160	0.0543	2958.81	3262.87	6.696	600	0.0936	3782.99	4307.31	7.847	1030	0.1298	4680.84	5407.74	8.661
170	0.0553	2977.62	3287.21	6.730	610	0.0945	3802.67	4331.76	7.868	1040	0.1306	4702.91	5434.49	8.678
180	0.0563	2996.31	3311.37	6.764	620	0.0953	3822.41	4356.26	7.889	1050	0.1315	4725.03	5461.28	8.695
190	0.0572	3014.90	3335.40	6.796	630	0.0962	3842.20	4380.82	7.910	1060	0.1323	4747.20	5488.12	8.712
200	0.0582	3033.42	3359.29	6.829	640	0.0970	3862.05	4405.43	7.931	1070	0.1331	4769.42	5515.00	8.729
210	0.0591	3051.87	3383.08	6.861	650	0.0979	3881.97	4430.09	7.952	1080	0.1340	4791.69	5541.94	8.745
220	0.0601	3070.27	3406.78	6.892	660	0.0987	3901.94	4454.82	7.973	1090	0.1348	4814.00	5568.92	8.762
230	0.0610	3088.63	3430.40	6.922	670	0.0996	3921.96	4479.59	7.993	1100	0.1356	4836.37	5595.95	8.778
240	0.0620	3106.96	3453.96	6.953	680	0.1004	3942.05	4504.42	8.013	1110	0.1365	4858.78	5623.02	8.795
250	0.0629	3125.27	3477.47	6.982	690	0.1013	3962.20	4529.31	8.034	1120	0.1373	4881.23	5650.14	8.811
260	0.0638	3143.56	3500.93	7.012	700	0.1021	3982.40	4554.25	8.054	1130	0.1381	4903.74	5677.31	8.827
270	0.0647	3161.84	3524.36	7.041	710	0.1030	4002.67	4579.25	8.074	1140	0.1390	4926.29	5704.52	8.844
280	0.0657	3180.13	3547.76	7.069	720	0.1038	4022.99	4604.30	8.094	1150	0.1398	4948.88	5731.78	8.860
290	0.0666	3198.41	3571.15	7.098	730	0.1047	4043.37	4629.41	8.113	1160	0.1406	4971.53	5759.08	8.876
300	0.0675	3216.71	3594.52	7.125	740	0.1055	4063.81	4654.57	8.133	1170	0.1415	4994.21	5786.43	8.892
310	0.0684	3235.02	3617.89	7.153	750	0.1063	4084.30	4679.79	8.153	1180	0.1423	5016.95	5813.82	8.908
320	0.0693	3253.35	3641.25	7.180	760	0.1072	4104.86	4705.06	8.172	1190	0.1431	5039.72	5841.25	8.924
330	0.0702	3271.69	3664.62	7.207	770	0.1080	4125.47	4730.39	8.192	1200	0.1440	5062.54	5868.73	8.939
340	0.0711	3290.07	3688.00	7.234	780	0.1089	4146.14	4755.78	8.211	1210	0.1448	5085.41	5896.25	8.955
350	0.0719	3308.47	3711.39	7.260	790	0.1097	4166.87	4781.22	8.230	1220	0.1456	5108.32	5923.81	8.971
360	0.0728	3326.90	3734.79	7.286	800	0.1105	4187.65	4806.71	8.249	1230	0.1465	5131.27	5951.42	8.986
370	0.0737	3345.37	3758.22	7.312	810	0.1114	4208.49	4832.26	8.268	1240	0.1473	5154.26	5979.07	9.002
380	0.0746	3363.87	3781.67	7.337	820	0.1122	4229.39	4857.86	8.287	1250	0.1481	5177.30	6006.76	9.017
390	0.0755	3382.41	3805.14	7.362	830	0.1131	4250.34	4883.52	8.305	1260	0.1489	5200.38	6034.49	9.033
400	0.0764	3400.99	3828.64	7.387	840	0.1139	4271.35	4909.23	8.324	1270	0.1498	5223.50	6062.26	9.048
410	0.0772	3419.62	3852.18	7.412	850	0.1147	4292.42	4935.00	8.343	1280	0.1506	5246.66	6090.07	9.063
420	0.0781	3438.28	3875.74	7.437	860	0.1156	4313.54	4960.82	8.361	1290	0.1514	5269.86	6117.93	9.079
430	0.0790	3456.99	3899.34	7.461	870	0.1164	4334.72	4986.69	8.379	1300	0.1523	5293.11	6145.82	9.094
440	0.0799	3475.75	3922.98	7.485										

Table 3(39) Superheated Steam Properties (Pressure Base) Pressure = 58.0 bar and saturation temperature, Tsat = 273.382°C

t (°C)	v $\left(\frac{m^3}{kg}\right)$	u $\left(\frac{kJ}{kg}\right)$	h $\left(\frac{kJ}{kg}\right)$	s $\left(\frac{kJ}{kg\,K}\right)$	t (°C)	v $\left(\frac{m^3}{kg}\right)$	u $\left(\frac{kJ}{kg}\right)$	h $\left(\frac{kJ}{kg}\right)$	s $\left(\frac{kJ}{kg\,K}\right)$	t (°C)	v $\left(\frac{m^3}{kg}\right)$	u $\left(\frac{kJ}{kg}\right)$	h $\left(\frac{kJ}{kg}\right)$	s $\left(\frac{kJ}{kg\,K}\right)$
Sat Vapour (g)	0.0337	2591.49	2786.74	5.906	Sat Vapour (g)	0.0337	2591.49	2786.74	5.906	Sat Vapour (g)	0.0337	2591.49	2786.74	5.906
10	0.0353	2625.97	2830.91	5.986	450	0.0781	3497.93	3950.89	7.498	880	0.1134	4360.36	5018.07	8.385
20	0.0368	2656.28	2869.98	6.056	460	0.0789	3516.82	3974.65	7.521	890	0.1142	4381.67	5044.07	8.404
30	0.0382	2683.96	2905.78	6.118	470	0.0798	3535.76	3998.45	7.545	900	0.1150	4403.03	5070.13	8.422
40	0.0396	2709.77	2939.23	6.176	480	0.0806	3554.74	4022.29	7.568	910	0.1158	4424.44	5096.23	8.440
50	0.0408	2734.16	2970.91	6.230	490	0.0814	3573.78	4046.17	7.591	920	0.1166	4445.91	5122.39	8.457
60	0.0420	2757.46	3001.18	6.280	500	0.0823	3592.87	4070.10	7.614	930	0.1174	4467.43	5148.60	8.475
70	0.0432	2779.88	3030.34	6.328	510	0.0831	3612.02	4094.07	7.637	940	0.1183	4489.01	5174.86	8.493
80	0.0443	2801.59	3058.57	6.373	520	0.0839	3631.21	4118.09	7.660	950	0.1191	4510.63	5201.17	8.511
90	0.0454	2822.71	3086.05	6.417	530	0.0848	3650.47	4142.16	7.682	960	0.1199	4532.31	5227.54	8.528
100	0.0465	2843.36	3112.90	6.458	540	0.0856	3669.77	4166.28	7.704	970	0.1207	4554.05	5253.95	8.546
110	0.0475	2863.60	3139.21	6.499	550	0.0864	3689.14	4190.45	7.727	980	0.1215	4575.83	5280.41	8.563
120	0.0485	2883.51	3165.09	6.538	560	0.0873	3708.55	4214.67	7.749	990	0.1223	4597.66	5306.93	8.580
130	0.0496	2903.14	3190.58	6.576	570	0.0881	3728.03	4238.94	7.770	1000	0.1231	4619.55	5333.49	8.598
140	0.0506	2922.54	3215.74	6.613	580	0.0889	3747.56	4263.26	7.792	1010	0.1239	4641.49	5360.11	8.615
150	0.0515	2941.73	3240.63	6.649	590	0.0897	3767.15	4287.63	7.814	1020	0.1247	4663.47	5386.77	8.632
160	0.0525	2960.76	3265.28	6.684	600	0.0906	3786.80	4312.06	7.835	1030	0.1255	4685.51	5413.48	8.649
170	0.0535	2979.65	3289.72	6.718	610	0.0914	3806.50	4336.54	7.856	1040	0.1263	4707.60	5440.24	8.666
180	0.0544	2998.41	3313.99	6.752	620	0.0922	3826.26	4361.07	7.877	1050	0.1271	4729.73	5467.05	8.683
190	0.0554	3017.08	3338.11	6.785	630	0.0930	3846.08	4385.66	7.898	1060	0.1279	4751.92	5493.91	8.699
200	0.0563	3035.67	3362.10	6.817	640	0.0939	3865.96	4410.30	7.919	1070	0.1287	4774.15	5520.81	8.716
210	0.0572	3054.18	3385.97	6.849	650	0.0947	3885.90	4434.99	7.940	1080	0.1295	4796.43	5547.76	8.733
220	0.0581	3072.65	3409.75	6.880	660	0.0955	3905.89	4459.74	7.961	1090	0.1303	4818.76	5574.76	8.749
230	0.0590	3091.07	3433.46	6.911	670	0.0963	3925.94	4484.54	7.981	1100	0.1311	4841.14	5601.80	8.766
240	0.0599	3109.45	3457.09	6.941	680	0.0971	3946.05	4509.40	8.001	1110	0.1320	4863.57	5628.89	8.782
250	0.0608	3127.82	3480.67	6.971	690	0.0979	3966.22	4534.32	8.022	1120	0.1328	4886.04	5656.03	8.798
260	0.0617	3146.16	3504.20	7.000	700	0.0988	3986.45	4559.28	8.042	1130	0.1336	4908.56	5683.21	8.815
270	0.0626	3164.50	3527.69	7.029	710	0.0996	4006.74	4584.31	8.062	1140	0.1344	4931.12	5710.44	8.831
280	0.0635	3182.83	3551.16	7.058	720	0.1004	4027.08	4609.39	8.082	1150	0.1352	4953.73	5737.71	8.847
290	0.0644	3201.16	3574.61	7.086	730	0.1012	4047.48	4634.52	8.101	1160	0.1360	4976.39	5765.03	8.863
300	0.0653	3219.51	3598.04	7.114	740	0.1020	4067.94	4659.71	8.121	1170	0.1368	4999.09	5792.39	8.879
310	0.0661	3237.86	3621.46	7.141	750	0.1028	4088.46	4684.95	8.141	1180	0.1376	5021.83	5819.80	8.895
320	0.0670	3256.23	3644.88	7.169	760	0.1037	4109.03	4710.25	8.160	1190	0.1384	5044.62	5847.25	8.911
330	0.0679	3274.63	3668.31	7.195	770	0.1045	4129.67	4735.60	8.179	1200	0.1392	5067.46	5874.74	8.927
340	0.0687	3293.04	3691.74	7.222	780	0.1053	4150.36	4761.01	8.199	1210	0.1400	5090.33	5902.27	8.942
350	0.0696	3311.49	3715.18	7.248	790	0.1061	4171.11	4786.47	8.218	1220	0.1408	5113.25	5929.85	8.958
360	0.0705	3329.96	3738.63	7.274	800	0.1069	4191.91	4811.99	8.237	1230	0.1416	5136.22	5957.47	8.974
370	0.0713	3348.46	3762.11	7.300	810	0.1077	4212.77	4837.56	8.256	1240	0.1424	5159.23	5985.13	8.989
380	0.0722	3367.00	3785.60	7.326	820	0.1085	4233.69	4863.19	8.274	1250	0.1432	5182.27	6012.84	9.005
390	0.0730	3385.58	3809.12	7.351	830	0.1093	4254.66	4888.87	8.293	1260	0.1440	5205.37	6040.58	9.020
400	0.0739	3404.20	3832.67	7.376	840	0.1102	4275.69	4914.60	8.312	1270	0.1448	5228.50	6068.37	9.035
410	0.0747	3422.86	3856.24	7.401	850	0.1110	4296.78	4940.39	8.330	1280	0.1456	5251.67	6096.19	9.051
420	0.0756	3441.56	3879.85	7.425	860	0.1118	4317.92	4966.23	8.349	1290	0.1464	5274.89	6124.06	9.066
430	0.0764	3460.30	3903.50	7.449	870	0.1126	4339.11	4992.13	8.367	1300	0.1472	5298.14	6151.97	9.081
440	0.0773	3479.09	3927.17	7.474										

Table 3(40) Superheated Steam Properties (Pressure Base) Pressure = 60.0 bar and saturation temperature, Tsat = 275.585°C

t (°C)	v $\left(\frac{m^3}{kg}\right)$	u $\left(\frac{kJ}{kg}\right)$	h $\left(\frac{kJ}{kg}\right)$	s $\left(\frac{kJ}{kg\,K}\right)$	t (°C)	v $\left(\frac{m^3}{kg}\right)$	u $\left(\frac{kJ}{kg}\right)$	h $\left(\frac{kJ}{kg}\right)$	s $\left(\frac{kJ}{kg\,K}\right)$	t (°C)	v $\left(\frac{m^3}{kg}\right)$	u $\left(\frac{kJ}{kg}\right)$	h $\left(\frac{kJ}{kg}\right)$	s $\left(\frac{kJ}{kg\,K}\right)$
Sat Vapour (g)	0.0324	2589.92	2784.61	5.890	Sat Vapour (g)	0.0324	2589.92	2784.61	5.890	Sat Vapour (g)	0.0324	2589.92	2784.61	5.890
10	0.0341	2625.00	2829.55	5.971	450	0.0756	3501.21	3955.00	7.486	880	0.1098	4364.66	5023.38	8.373
20	0.0356	2655.76	2869.19	6.042	460	0.0764	3520.13	3978.79	7.510	890	0.1106	4385.98	5049.40	8.392
30	0.0369	2683.79	2905.46	6.105	470	0.0773	3539.09	4002.63	7.533	900	0.1114	4407.36	5075.47	8.410
40	0.0382	2709.90	2939.29	6.163	480	0.0781	3558.11	4026.51	7.557	910	0.1121	4428.79	5101.60	8.428
50	0.0395	2734.55	2971.29	6.217	490	0.0789	3577.18	4050.43	7.580	920	0.1129	4450.28	5127.78	8.446
60	0.0406	2758.06	3001.85	6.267	500	0.0797	3596.30	4074.39	7.603	930	0.1137	4471.81	5154.01	8.463
70	0.0418	2780.67	3031.24	6.315	510	0.0805	3615.47	4098.40	7.626	940	0.1145	4493.41	5180.29	8.481
80	0.0429	2802.54	3059.70	6.361	520	0.0813	3634.70	4122.46	7.648	950	0.1153	4515.05	5206.62	8.499
90	0.0439	2823.82	3087.37	6.405	530	0.0821	3653.98	4146.56	7.671	960	0.1160	4536.75	5233.00	8.516
100	0.0450	2844.60	3114.40	6.447	540	0.0829	3673.31	4170.71	7.693	970	0.1160	4558.49	5259.43	8.534
110	0.0460	2864.97	3140.88	6.487	550	0.0837	3692.70	4194.91	7.715	980	0.1176	4580.29	5285.91	8.551
120	0.0470	2885.00	3166.90	6.520	560	0.0845	3712.15	4219.17	7.737	990	0.1184	4602.14	5312.45	8.568
130	0.0480	2904.73	3192.53	6.564	570	0.0853	3731.65	4243.47	7.759	1000	0.1192	4624.05	5339.03	8.586
140	0.0489	2924.22	3217.82	6.601	580	0.0861	3751.21	4267.82	7.781	1010	0.1199	4646.00	5365.66	8.603
150	0.0499	2943.51	3242.83	6.637	590	0.0869	3770.82	4292.22	7.802	1020	0.1207	4668.00	5392.34	8.620
160	0.0508	2962.62	3267.59	6.673	600	0.0877	3790.49	4316.68	7.824	1030	0.1215	4690.05	5419.07	8.637
170	0.0518	2981.59	3292.14	6.707	610	0.0885	3810.22	4341.19	7.845	1040	0.1223	4712.15	5445.85	8.654
180	0.0527	3000.43	3316.51	6.741	620	0.0893	3830.01	4365.75	7.866	1050	0.1231	4734.31	5472.67	8.670
190	0.0536	3019.17	3340.72	6.774	630	0.0901	3849.85	4390.36	7.887	1060	0.1238	4756.51	5499.54	8.687
200	0.0545	3037.83	3364.80	6.806	640	0.0909	3869.76	4415.03	7.908	1070	0.1246	4778.75	5526.46	8.704
210	0.0554	3056.41	3388.76	6.838	650	0.0917	3889.72	4439.75	7.928	1080	0.1254	4801.05	5553.43	8.721
220	0.0563	3074.94	3412.62	6.869	660	0.0925	3909.73	4464.53	7.949	1090	0.1262	4823.39	5580.44	8.737
230	0.0572	3093.42	3436.40	6.900	670	0.0933	3929.81	4489.36	7.969	1100	0.1270	4845.79	5607.50	8.754
240	0.0580	3111.86	3460.10	6.930	680	0.0941	3949.94	4514.25	7.990	1110	0.1277	4868.23	5634.61	8.770
250	0.0589	3130.28	3483.75	6.960	690	0.0948	3970.13	4539.18	8.010	1120	0.1285	4890.71	5661.76	8.786
260	0.0598	3148.68	3507.35	6.989	700	0.0956	3990.38	4564.18	8.030	1130	0.1293	4913.24	5688.96	8.802
270	0.0606	3167.06	3530.91	7.018	710	0.0964	4010.69	4589.23	8.050	1140	0.1301	4935.82	5716.20	8.819
280	0.0615	3185.44	3554.44	7.047	720	0.0972	4031.06	4614.33	8.070	1150	0.1308	4958.45	5743.49	8.835
290	0.0624	3203.83	3577.95	7.075	730	0.0980	4051.48	4639.49	8.090	1160	0.1316	4981.11	5770.82	8.851
300	0.0632	3222.22	3601.44	7.103	740	0.0988	4071.96	4664.70	8.109	1170	0.1324	5003.83	5798.20	8.867
310	0.0641	3240.62	3624.92	7.130	750	0.0996	4092.50	4689.97	8.129	1180	0.1332	5026.59	5825.62	8.883
320	0.0649	3259.03	3648.40	7.157	760	0.1004	4113.10	4715.29	8.148	1190	0.1339	5049.39	5853.08	8.899
330	0.0657	3277.47	3671.87	7.184	770	0.1012	4133.75	4740.67	8.168	1200	0.1347	5072.24	5880.59	8.914
340	0.0666	3295.92	3695.35	7.211	780	0.1019	4154.46	4766.10	8.187	1210	0.1355	5095.13	5908.14	8.930
350	0.0674	3314.41	3718.85	7.237	790	0.1027	4175.23	4791.59	8.206	1220	0.1363	5118.06	5935.73	8.946
360	0.0682	3332.92	3742.35	7.263	800	0.1035	4196.05	4817.13	8.225	1230	0.1371	5141.04	5963.36	8.961
370	0.0691	3351.46	3765.87	7.289	810	0.1043	4216.93	4842.72	8.244	1240	0.1378	5164.06	5991.04	8.977
380	0.0699	3370.04	3789.41	7.314	820	0.1051	4237.87	4868.37	8.263	1250	0.1386	5187.12	6018.75	8.992
390	0.0707	3388.65	3812.98	7.340	830	0.1059	4258.86	4894.07	8.281	1260	0.1394	5210.22	6046.51	9.008
400	0.0715	3407.31	3836.57	7.365	840	0.1067	4279.91	4919.83	8.300	1270	0.1402	5233.37	6074.31	9.023
410	0.0724	3426.00	3860.19	7.389	850	0.1074	4301.02	4945.63	8.318	1280	0.1409	5256.55	6102.15	9.038
420	0.0732	3444.73	3883.84	7.414	860	0.1082	4322.17	4971.50	8.337	1290	0.1417	5279.78	6130.03	9.053
430	0.0740	3463.51	3907.52	7.438	870	0.1090	4343.39	4997.41	8.355	1300	0.1425	5303.05	6157.95	9.069
440	0.0748	3482.34	3931.24	7.462										

Table 3(41) Superheated Steam Properties (Pressure Base) Pressure = 65.0 bar and saturation temperature, Tsat = 280.858°C

t (°C)	v $\left(\frac{m^3}{kg}\right)$	u $\left(\frac{kJ}{kg}\right)$	h $\left(\frac{kJ}{kg}\right)$	s $\left(\frac{kJ}{kg\,K}\right)$
Sat Vapour (g)	0.0297	2586.00	2779.00	5.852
10	0.0313	2622.29	2825.79	5.936
20	0.0327	2654.19	2866.89	6.008
30	0.0340	2683.12	2904.31	6.072
40	0.0353	2709.96	2939.10	6.132
50	0.0364	2735.23	2971.90	6.186
60	0.0375	2759.28	3003.15	6.238
70	0.0386	2782.36	3033.16	6.286
80	0.0396	2804.65	3062.15	6.332
90	0.0406	2826.30	3090.31	6.376
100	0.0416	2847.42	3117.78	6.419
110	0.0425	2868.09	3144.66	6.459
120	0.0435	2888.40	3171.05	6.499
130	0.0444	2908.40	3197.02	6.537
140	0.0453	2928.13	3222.63	6.574
150	0.0462	2947.64	3247.93	6.610
160	0.0471	2966.96	3272.96	6.646
170	0.0479	2986.13	3297.77	6.680
180	0.0488	3005.16	3322.38	6.714
190	0.0497	3024.08	3346.83	6.747
200	0.0505	3042.90	3371.12	6.780
210	0.0513	3061.64	3395.29	6.811
220	0.0522	3080.32	3419.35	6.843
230	0.0530	3098.95	3443.32	6.873
240	0.0538	3117.53	3467.20	6.904
250	0.0546	3136.08	3491.03	6.933
260	0.0554	3154.61	3514.79	6.963
270	0.0562	3173.12	3538.51	6.992
280	0.0570	3191.62	3562.20	7.020
290	0.0578	3210.12	3585.85	7.049
300	0.0586	3228.62	3609.49	7.076
310	0.0594	3247.13	3633.11	7.104
320	0.0602	3265.66	3656.72	7.131
330	0.0609	3284.19	3680.32	7.158
340	0.0617	3302.75	3703.93	7.185
350	0.0625	3321.33	3727.55	7.211
360	0.0633	3339.94	3751.17	7.237
370	0.0640	3358.57	3774.81	7.262
380	0.0648	3377.24	3798.46	7.288
390	0.0656	3395.95	3822.14	7.313
400	0.0663	3414.68	3845.84	7.338
410	0.0671	3433.46	3869.56	7.363
420	0.0679	3452.28	3893.32	7.387
430	0.0686	3471.14	3917.10	7.412
440	0.0694	3490.04	3940.92	7.436

t (°C)	v $\left(\frac{m^3}{kg}\right)$	u $\left(\frac{kJ}{kg}\right)$	h $\left(\frac{kJ}{kg}\right)$	s $\left(\frac{kJ}{kg\,K}\right)$
Sat Vapour (g)	0.0297	2586.00	2779.00	5.852
450	0.0701	3508.99	3964.77	7.460
460	0.0709	3527.99	3988.66	7.483
470	0.0716	3547.03	4012.59	7.507
480	0.0724	3566.12	4036.55	7.530
490	0.0731	3585.26	4060.56	7.553
500	0.0739	3604.46	4084.61	7.576
510	0.0746	3623.70	4108.71	7.599
520	0.0754	3642.99	4132.85	7.621
530	0.0761	3662.34	4157.04	7.644
540	0.0769	3681.75	4181.27	7.666
550	0.0776	3701.20	4205.55	7.688
560	0.0783	3720.71	4229.88	7.710
570	0.0791	3740.28	4254.26	7.732
580	0.0798	3759.90	4278.68	7.754
590	0.0806	3779.58	4303.16	7.775
600	0.0813	3799.31	4327.69	7.796
610	0.0820	3819.10	4352.27	7.818
620	0.0828	3838.94	4376.90	7.839
630	0.0835	3858.85	4401.59	7.860
640	0.0842	3878.81	4426.32	7.880
650	0.0850	3898.83	4451.11	7.901
660	0.0857	3918.90	4475.96	7.922
670	0.0864	3939.03	4500.85	7.942
680	0.0872	3959.22	4525.80	7.962
690	0.0879	3979.47	4550.81	7.983
700	0.0886	3999.77	4575.86	8.003
710	0.0894	4020.13	4600.97	8.023
720	0.0901	4040.55	4626.14	8.042
730	0.0908	4061.03	4651.36	8.062
740	0.0915	4081.56	4676.63	8.082
750	0.0923	4102.15	4701.95	8.101
760	0.0930	4122.79	4727.33	8.121
770	0.0937	4143.50	4752.77	8.140
780	0.0945	4164.26	4778.26	8.159
790	0.0952	4185.07	4803.80	8.178
800	0.0959	4205.94	4829.39	8.197
810	0.0966	4226.87	4855.04	8.216
820	0.0974	4247.85	4880.74	8.235
830	0.0981	4268.89	4906.50	8.253
840	0.0988	4289.99	4932.30	8.272
850	0.0995	4311.14	4958.17	8.291
860	0.1003	4332.34	4984.08	8.309
870	0.1010	4353.60	5010.04	8.327

t (°C)	v $\left(\frac{m^3}{kg}\right)$	u $\left(\frac{kJ}{kg}\right)$	h $\left(\frac{kJ}{kg}\right)$	s $\left(\frac{kJ}{kg\,K}\right)$
Sat Vapour (g)	0.0297	2586.00	2779.00	5.852
880	0.1017	4374.91	5036.06	8.345
890	0.1024	4396.28	5062.13	8.364
900	0.1032	4417.70	5088.25	8.382
910	0.1039	4439.17	5114.43	8.399
920	0.1046	4460.70	5140.65	8.417
930	0.1053	4482.28	5166.93	8.435
940	0.1061	4503.91	5193.25	8.453
950	0.1068	4525.60	5219.63	8.470
960	0.1075	4547.33	5246.06	8.488
970	0.1082	4569.12	5272.53	8.505
980	0.1089	4590.96	5299.06	8.523
990	0.1097	4612.85	5325.63	8.540
1000	0.1104	4634.79	5352.26	8.557
1010	0.1111	4656.78	5378.93	8.574
1020	0.1118	4678.82	5405.66	8.591
1030	0.1125	4700.91	5432.43	8.608
1040	0.1133	4723.05	5459.24	8.625
1050	0.1140	4745.23	5486.11	8.642
1060	0.1147	4767.47	5513.02	8.659
1070	0.1154	4789.75	5539.98	8.675
1080	0.1161	4812.08	5566.99	8.692
1090	0.1169	4834.46	5594.04	8.708
1100	0.1176	4856.89	5621.14	8.725
1110	0.1183	4879.36	5648.28	8.741
1120	0.1190	4901.88	5675.47	8.757
1130	0.1197	4924.45	5702.71	8.774
1140	0.1205	4947.06	5729.99	8.790
1150	0.1212	4969.72	5757.31	8.806
1160	0.1219	4992.42	5784.68	8.822
1170	0.1226	5015.16	5812.09	8.838
1180	0.1233	5037.95	5839.54	8.854
1190	0.1240	5060.79	5867.04	8.870
1200	0.1248	5083.67	5894.58	8.885
1210	0.1255	5106.59	5922.17	8.901
1220	0.1262	5129.55	5949.79	8.917
1230	0.1269	5152.56	5977.46	8.932
1240	0.1276	5175.61	6005.17	8.948
1250	0.1283	5198.70	6032.91	8.963
1260	0.1291	5221.83	6060.70	8.978
1270	0.1298	5245.00	6088.54	8.994
1280	0.1305	5268.22	6116.41	9.009
1290	0.1312	5291.47	6144.32	9.024
1300	0.1319	5314.77	6172.27	9.039

Table 3(42) Superheated Steam Properties (Pressure Base) Pressure = 70.0 bar and saturation temperature, Tsat = 290.536°C

t (°C)	v $\left(\frac{m^3}{kg}\right)$	u $\left(\frac{kJ}{kg}\right)$	h $\left(\frac{kJ}{kg}\right)$	s $\left(\frac{kJ}{kg\,K}\right)$	t (°C)	v $\left(\frac{m^3}{kg}\right)$	u $\left(\frac{kJ}{kg}\right)$	h $\left(\frac{kJ}{kg}\right)$	s $\left(\frac{kJ}{kg\,K}\right)$	t (°C)	v $\left(\frac{m^3}{kg}\right)$	u $\left(\frac{kJ}{kg}\right)$	h $\left(\frac{kJ}{kg}\right)$	s $\left(\frac{kJ}{kg\,K}\right)$
Sat Vapour (g)	0.0274	2580.99	2772.64	5.815	Sat Vapour (g)	0.0274	2580.99	2772.64	5.815	Sat Vapour (g)	0.0274	2580.99	2772.64	5.815
10	0.0289	2619.21	2821.57	5.902	450	0.0654	3516.27	3973.90	7.435	880	0.0948	4384.55	5047.99	8.320
20	0.0303	2652.27	2864.15	5.976	460	0.0661	3535.34	3997.88	7.459	890	0.0955	4405.96	5074.11	8.338
30	0.0315	2682.11	2902.73	6.042	470	0.0668	3554.45	4021.90	7.482	900	0.0961	4427.42	5100.28	8.356
40	0.0327	2709.69	2938.47	6.102	480	0.0675	3573.62	4045.96	7.505	910	0.0968	4448.94	5126.50	8.374
50	0.0338	2735.57	2972.07	6.158	490	0.0682	3592.83	4070.05	7.529	920	0.0975	4470.50	5152.77	8.391
60	0.0348	2760.15	3004.01	6.210	500	0.0689	3612.09	4094.18	7.551	930	0.0981	4492.12	5179.09	8.409
70	0.0358	2783.69	3034.62	6.259	510	0.0696	3631.40	4118.36	7.574	940	0.0988	4513.79	5205.46	8.427
80	0.0368	2806.39	3064.14	6.305	520	0.0703	3650.76	4142.58	7.597	950	0.0995	4535.52	5231.88	8.444
90	0.0378	2828.41	3092.78	6.350	530	0.0710	3670.18	4166.85	7.619	960	0.1002	4557.29	5258.35	8.462
100	0.0387	2849.86	3120.68	6.392	540	0.0716	3689.64	4191.16	7.641	970	0.1008	4579.11	5284.87	8.479
110	0.0396	2870.84	3147.96	6.434	550	0.0723	3709.16	4215.52	7.663	980	0.1015	4600.99	5311.44	8.497
120	0.0405	2891.42	3174.71	6.473	560	0.0730	3728.73	4239.92	7.685	990	0.1022	4622.92	5338.05	8.514
130	0.0413	2911.67	3201.01	6.512	570	0.0737	3748.36	4264.37	7.707	1000	0.1028	4644.89	5364.72	8.531
140	0.0422	2931.64	3226.93	6.549	580	0.0744	3768.04	4288.87	7.729	1010	0.1035	4666.92	5391.43	8.548
150	0.0430	2951.37	3252.51	6.585	590	0.0751	3787.78	4313.42	7.750	1020	0.1042	4688.99	5418.19	8.565
160	0.0438	2970.90	3277.82	6.621	600	0.0758	3807.57	4338.02	7.771	1030	0.1048	4711.12	5445.00	8.582
170	0.0447	2990.26	3302.88	6.655	610	0.0765	3827.42	4362.67	7.793	1040	0.1055	4733.29	5471.86	8.599
180	0.0455	3009.47	3327.73	6.689	620	0.0771	3847.32	4387.37	7.814	1050	0.1062	4755.52	5498.77	8.616
190	0.0463	3028.56	3352.40	6.722	630	0.0778	3867.28	4412.12	7.835	1060	0.1068	4777.79	5525.72	8.632
200	0.0471	3047.55	3376.91	6.755	640	0.0785	3887.30	4436.92	7.855	1070	0.1075	4800.10	5552.71	8.649
210	0.0478	3066.45	3401.28	6.787	650	0.0792	3907.37	4461.78	7.876	1080	0.1082	4822.47	5579.76	8.665
220	0.0486	3085.27	3425.53	6.818	660	0.0799	3927.50	4486.69	7.896	1090	0.1089	4844.88	5606.85	8.682
230	0.0494	3104.04	3449.68	6.849	670	0.0806	3947.69	4511.64	7.917	1100	0.1095	4867.34	5633.98	8.698
240	0.0501	3122.76	3473.74	6.879	680	0.0812	3967.93	4536.66	7.937	1110	0.1102	4889.84	5661.16	8.715
250	0.0509	3141.44	3497.74	6.909	690	0.0819	3988.23	4561.72	7.957	1120	0.1109	4912.40	5688.39	8.731
260	0.0517	3160.09	3521.66	6.939	700	0.0826	4008.58	4586.84	7.977	1130	0.1115	4934.99	5715.66	8.747
270	0.0524	3178.73	3545.54	6.968	710	0.0833	4028.99	4612.01	7.997	1140	0.1122	4957.64	5742.97	8.763
280	0.0531	3197.35	3569.37	6.996	720	0.0840	4049.46	4637.23	8.017	1150	0.1129	4980.32	5770.33	8.779
290	0.0539	3215.96	3593.18	7.024	730	0.0846	4069.99	4662.51	8.037	1160	0.1135	5003.06	5797.73	8.795
300	0.0546	3234.57	3616.95	7.052	740	0.0853	4090.57	4687.84	8.056	1170	0.1142	5025.83	5825.18	8.811
310	0.0554	3253.18	3640.70	7.080	750	0.0860	4111.21	4713.22	8.076	1180	0.1149	5048.65	5852.67	8.827
320	0.0561	3271.81	3664.44	7.107	760	0.0867	4131.90	4738.65	8.095	1190	0.1155	5071.52	5880.20	8.843
330	0.0568	3290.45	3688.18	7.134	770	0.0874	4152.65	4764.14	8.114	1200	0.1162	5094.42	5907.77	8.859
340	0.0575	3309.10	3711.91	7.160	780	0.0880	4173.46	4789.68	8.134	1210	0.1169	5117.37	5935.38	8.874
350	0.0583	3327.78	3735.64	7.186	790	0.0887	4194.32	4815.28	8.153	1220	0.1175	5140.37	5963.04	8.890
360	0.0590	3346.47	3759.38	7.212	800	0.0894	4215.24	4840.93	8.172	1230	0.1182	5163.40	5990.74	8.905
370	0.0597	3365.20	3783.13	7.238	810	0.0901	4236.21	4866.63	8.190	1240	0.1189	5186.48	6018.48	8.921
380	0.0604	3383.96	3806.90	7.264	820	0.0907	4257.24	4892.38	8.209	1250	0.1195	5209.60	6046.26	8.936
390	0.0611	3402.74	3830.68	7.289	830	0.0914	4278.32	4918.19	8.228	1260	0.1202	5232.76	6074.08	8.952
400	0.0618	3421.57	3854.48	7.314	840	0.0921	4299.46	4944.04	8.246	1270	0.1209	5255.96	6101.94	8.967
410	0.0626	3440.43	3878.31	7.338	850	0.0928	4320.65	4969.95	8.265	1280	0.1215	5279.20	6129.84	8.982
420	0.0633	3459.33	3902.16	7.363	860	0.0934	4341.90	4995.92	8.283	1290	0.1222	5302.48	6157.78	8.997
430	0.0640	3478.26	3926.05	7.387	870	0.0941	4363.20	5021.93	8.301	1300	0.1229	5325.80	6185.76	9.012
440	0.0647	3497.24	3949.96	7.411										

Table 3(43) Superheated Steam Properties (Pressure Base) Pressure = 75.0 bar and saturation temperature, Tsat = 290.536°C

t (°C)	$v\left(\frac{m^3}{kg}\right)$	$u\left(\frac{kJ}{kg}\right)$	$h\left(\frac{kJ}{kg}\right)$	$s\left(\frac{kJ}{kg\,K}\right)$	*t* (°C)	$v\left(\frac{m^3}{kg}\right)$	$u\left(\frac{kJ}{kg}\right)$	$h\left(\frac{kJ}{kg}\right)$	$s\left(\frac{kJ}{kg\,K}\right)$	*t* (°C)	$v\left(\frac{m^3}{kg}\right)$	$u\left(\frac{kJ}{kg}\right)$	$h\left(\frac{kJ}{kg}\right)$	$s\left(\frac{kJ}{kg\,K}\right)$
Sat Vapour (g)	*0.0253*	*2576.00*	*2766.00*	*5.779*	*Sat Vapour (g)*	*0.0253*	*2576.00*	*2766.00*	*5.779*	*Sat Vapour (g)*	*0.0253*	*2576.00*	*2766.00*	*5.779*
10	0.0268	2615.80	2816.92	5.869	450	0.0613	3523.10	3982.48	7.412	880	0.0887	4393.65	5059.27	8.296
20	0.0281	2650.03	2861.01	5.945	460	0.0619	3542.24	4006.55	7.436	890	0.0894	4415.10	5085.43	8.314
30	0.0293	2680.79	2900.76	6.013	470	0.0626	3561.42	4030.65	7.459	900	0.0900	4436.60	5111.64	8.332
40	0.0304	2709.11	2937.46	6.074	480	0.0632	3580.66	4054.79	7.483	910	0.0906	4458.15	5137.91	8.349
50	0.0315	2735.61	2971.86	6.131	490	0.0639	3599.93	4078.97	7.506	920	0.0913	4479.76	5164.22	8.367
60	0.0325	2760.72	3004.48	6.183	500	0.0645	3619.26	4103.18	7.529	930	0.0919	4501.42	5190.58	8.385
70	0.0335	2784.72	3035.68	6.233	510	0.0652	3638.64	4127.44	7.551	940	0.0925	4523.12	5216.99	8.403
80	0.0344	2807.82	3065.74	6.280	520	0.0658	3658.06	4151.74	7.574	950	0.0931	4544.88	5243.46	8.420
90	0.0353	2830.20	3094.85	6.325	530	0.0665	3677.54	4176.08	7.596	960	0.0938	4566.69	5269.97	8.438
100	0.0362	2851.97	3123.17	6.368	540	0.0671	3697.07	4200.47	7.618	970	0.0944	4588.55	5296.53	8.455
110	0.0370	2873.25	3150.83	6.409	550	0.0678	3716.65	4224.90	7.640	980	0.0950	4610.47	5323.14	8.472
120	0.0378	2894.10	3177.94	6.449	560	0.0684	3736.28	4249.37	7.662	990	0.0956	4632.43	5349.79	8.489
130	0.0387	2914.60	3204.57	6.488	570	0.0691	3755.97	4273.90	7.684	1000	0.0963	4654.44	5376.50	8.507
140	0.0395	2934.80	3230.79	6.525	580	0.0697	3775.71	4298.47	7.706	1010	0.0969	4676.50	5403.25	8.524
150	0.0403	2954.75	3256.66	6.562	590	0.0703	3795.51	4323.08	7.727	1020	0.0975	4698.61	5430.05	8.541
160	0.0410	2974.48	3282.23	6.598	600	0.0710	3815.35	4347.75	7.748	1030	0.0982	4720.77	5456.90	8.558
170	0.0418	2994.03	3307.54	6.632	610	0.0716	3835.26	4372.47	7.769	1040	0.0988	4742.97	5483.79	8.574
180	0.0426	3013.42	3332.62	6.666	620	0.0723	3855.22	4397.23	7.790	1050	0.0994	4765.23	5510.74	8.591
190	0.0433	3032.68	3357.51	6.699	630	0.0729	3875.23	4422.05	7.811	1060	0.1000	4787.53	5537.72	8.608
200	0.0441	3051.82	3382.22	6.732	640	0.0735	3895.30	4446.92	7.832	1070	0.1007	4809.88	5564.76	8.624
210	0.0448	3070.87	3406.79	6.764	650	0.0742	3915.43	4471.83	7.853	1080	0.1013	4832.28	5591.84	8.641
220	0.0455	3089.84	3431.23	6.795	660	0.0748	3935.61	4496.80	7.873	1090	0.1019	4854.72	5618.96	8.657
230	0.0462	3108.75	3455.56	6.826	670	0.0755	3955.84	4521.82	7.893	1100	0.1025	4877.21	5646.13	8.674
240	0.0470	3127.60	3479.80	6.857	680	0.0761	3976.14	4546.89	7.914	1110	0.1031	4899.75	5673.35	8.690
250	0.0477	3146.41	3503.95	6.886	690	0.0767	3996.48	4572.02	7.934	1120	0.1038	4922.33	5700.61	8.706
260	0.0484	3165.19	3528.04	6.916	700	0.0774	4016.89	4597.19	7.954	1130	0.1044	4944.96	5727.91	8.722
270	0.0491	3183.94	3552.07	6.945	710	0.0780	4037.35	4622.42	7.974	1140	0.1050	4967.63	5755.26	8.739
280	0.0498	3202.67	3576.05	6.973	720	0.0786	4057.87	4647.70	7.994	1150	0.1056	4990.35	5782.65	8.755
290	0.0505	3221.39	3599.99	7.002	730	0.0793	4078.44	4673.03	8.013	1160	0.1063	5013.11	5810.08	8.771
300	0.0512	3240.11	3623.90	7.030	740	0.0799	4099.07	4698.42	8.033	1170	0.1069	5035.92	5837.56	8.786
310	0.0519	3258.83	3647.79	7.057	750	0.0805	4119.75	4723.85	8.052	1180	0.1075	5058.76	5865.08	8.802
320	0.0525	3277.55	3671.65	7.084	760	0.0812	4140.49	4749.34	8.071	1190	0.1081	5081.66	5892.64	8.818
330	0.0532	3296.28	3695.51	7.111	770	0.0818	4161.29	4774.88	8.091	1200	0.1088	5104.59	5920.25	8.834
340	0.0539	3315.03	3719.36	7.138	780	0.0824	4182.14	4800.48	8.110	1210	0.1094	5127.57	5947.89	8.849
350	0.0546	3333.80	3743.21	7.164	790	0.0831	4203.05	4826.12	8.129	1220	0.1100	5150.59	5975.58	8.865
360	0.0553	3352.59	3767.06	7.190	800	0.0837	4224.01	4851.82	8.148	1230	0.1106	5173.65	6003.31	8.880
370	0.0559	3371.40	3790.92	7.215	810	0.0843	4245.02	4877.57	8.167	1240	0.1112	5196.76	6031.08	8.896
380	0.0566	3390.24	3814.80	7.241	820	0.0850	4266.09	4903.37	8.185	1250	0.1119	5219.90	6058.89	8.911
390	0.0573	3409.11	3838.68	7.266	830	0.0856	4287.22	4929.23	8.204	1260	0.1125	5243.09	6086.74	8.927
400	0.0579	3428.02	3862.59	7.291	840	0.0862	4308.40	4955.13	8.222	1270	0.1131	5266.32	6114.63	8.942
410	0.0586	3446.96	3886.51	7.316	850	0.0869	4329.63	4981.09	8.241	1280	0.1137	5289.58	6142.56	8.957
420	0.0593	3465.93	3910.46	7.340	860	0.0875	4350.92	5007.10	8.259	1290	0.1144	5312.89	6170.53	8.972
430	0.0599	3484.95	3934.44	7.364	870	0.0881	4372.26	5033.16	8.277	1300	0.1150	5336.24	6198.53	8.987
440	0.0606	3504.00	3958.44	7.388										

Table 3(44) Superheated Steam Properties (Pressure Base) Pressure = 80.0 bar and saturation temperature, Tsat = 295.008°C

t (°C)	v $\left(\frac{m^3}{kg}\right)$	u $\left(\frac{kJ}{kg}\right)$	h $\left(\frac{kJ}{kg}\right)$	s $\left(\frac{kJ}{kg\,K}\right)$
Sat Vapour (g)	0.0235	2570.48	2758.67	5.745
10	0.0250	2612.08	2811.90	5.838
20	0.0263	2647.52	2857.53	5.916
30	0.0274	2679.20	2898.46	5.985
40	0.0285	2708.26	2936.11	6.048
50	0.0295	2735.38	2971.31	6.105
60	0.0304	2761.01	3004.61	6.158
70	0.0314	2785.47	3036.41	6.209
80	0.0323	2808.97	3066.98	6.256
90	0.0331	2831.71	3096.56	6.301
100	0.0339	2853.80	3125.30	6.345
110	0.0347	2875.37	3153.34	6.386
120	0.0355	2896.49	3180.80	6.427
130	0.0363	2917.23	3207.75	6.466
140	0.0371	2937.66	3234.26	6.503
150	0.0378	2957.82	3260.41	6.540
160	0.0386	2977.75	3286.24	6.576
170	0.0393	2997.48	3311.79	6.611
180	0.0400	3017.04	3337.10	6.645
190	0.0407	3036.47	3362.20	6.678
200	0.0414	3055.77	3387.12	6.711
210	0.0421	3074.97	3411.89	6.743
220	0.0428	3094.08	3436.51	6.774
230	0.0435	3113.12	3461.02	6.805
240	0.0442	3132.10	3485.42	6.835
250	0.0448	3151.04	3509.74	6.865
260	0.0455	3169.93	3533.98	6.895
270	0.0462	3188.80	3558.16	6.924
280	0.0468	3207.64	3582.28	6.952
290	0.0475	3226.47	3606.36	6.981
300	0.0481	3245.29	3630.40	7.008
310	0.0488	3264.11	3654.41	7.036
320	0.0494	3282.93	3678.41	7.063
330	0.0501	3301.76	3702.38	7.090
340	0.0507	3320.60	3726.35	7.116
350	0.0514	3339.45	3750.31	7.143
360	0.0520	3358.33	3774.27	7.169
370	0.0526	3377.22	3798.24	7.194
380	0.0533	3396.15	3822.22	7.220
390	0.0539	3415.10	3846.21	7.245
400	0.0545	3434.08	3870.21	7.270
410	0.0551	3453.10	3894.23	7.294
420	0.0558	3472.15	3918.27	7.319
430	0.0564	3491.24	3942.34	7.343
440	0.0570	3510.36	3966.44	7.367

t (°C)	v $\left(\frac{m^3}{kg}\right)$	u $\left(\frac{kJ}{kg}\right)$	h $\left(\frac{kJ}{kg}\right)$	s $\left(\frac{kJ}{kg\,K}\right)$
Sat Vapour (g)	0.0235	2570.48	2758.67	5.745
450	0.0576	3529.53	3990.56	7.391
460	0.0582	3548.74	4014.71	7.415
470	0.0589	3567.99	4038.90	7.438
480	0.0595	3587.29	4063.12	7.461
490	0.0601	3606.64	4087.38	7.484
500	0.0607	3626.03	4111.67	7.507
510	0.0613	3645.47	4136.01	7.530
520	0.0619	3664.95	4160.38	7.552
530	0.0625	3684.49	4184.80	7.575
540	0.0631	3704.08	4209.26	7.597
550	0.0638	3723.72	4233.76	7.619
560	0.0644	3743.41	4258.31	7.641
570	0.0650	3763.16	4282.90	7.662
580	0.0656	3782.95	4307.54	7.684
590	0.0662	3802.80	4332.22	7.705
600	0.0668	3822.71	4356.95	7.727
610	0.0674	3842.66	4381.74	7.748
620	0.0680	3862.68	4406.56	7.769
630	0.0686	3882.74	4431.44	7.790
640	0.0692	3902.86	4456.37	7.810
650	0.0698	3923.04	4481.35	7.831
660	0.0704	3943.27	4506.38	7.851
670	0.0710	3963.56	4531.46	7.872
680	0.0716	3983.90	4556.59	7.892
690	0.0722	4004.30	4581.77	7.912
700	0.0728	4024.75	4607.00	7.932
710	0.0734	4045.26	4632.28	7.952
720	0.0740	4065.82	4657.61	7.972
730	0.0746	4086.44	4683.00	7.991
740	0.0752	4107.11	4708.44	8.011
750	0.0758	4127.84	4733.93	8.030
760	0.0764	4148.63	4759.47	8.049
770	0.0769	4169.46	4785.06	8.069
780	0.0775	4190.36	4810.70	8.088
790	0.0781	4211.31	4836.40	8.107
800	0.0787	4232.31	4862.14	8.126
810	0.0793	4253.37	4887.94	8.144
820	0.0799	4274.48	4913.79	8.163
830	0.0805	4295.65	4939.69	8.182
840	0.0811	4316.87	4965.65	8.200
850	0.0817	4338.14	4991.65	8.219
860	0.0823	4359.47	5017.70	8.237
870	0.0829	4380.85	5043.81	8.255

t (°C)	v $\left(\frac{m^3}{kg}\right)$	u $\left(\frac{kJ}{kg}\right)$	h $\left(\frac{kJ}{kg}\right)$	s $\left(\frac{kJ}{kg\,K}\right)$
Sat Vapour (g)	0.0235	2570.48	2758.67	5.745
880	0.0835	4402.28	5069.96	8.273
890	0.0841	4423.76	5096.17	8.291
900	0.0846	4445.30	5122.42	8.309
910	0.0852	4466.89	5148.73	8.327
920	0.0858	4488.53	5175.08	8.345
930	0.0864	4510.22	5201.49	8.362
940	0.0870	4531.97	5227.94	8.380
950	0.0876	4553.76	5254.44	8.398
960	0.0882	4575.61	5280.99	8.415
970	0.0888	4597.50	5307.59	8.432
980	0.0893	4619.45	5334.24	8.450
990	0.0899	4641.44	5360.93	8.467
1000	0.0905	4663.49	5387.68	8.484
1010	0.0911	4685.58	5414.47	8.501
1020	0.0917	4707.73	5441.31	8.518
1030	0.0923	4729.92	5468.19	8.535
1040	0.0929	4752.15	5495.12	8.552
1050	0.0935	4774.44	5522.10	8.568
1060	0.0940	4796.78	5549.12	8.585
1070	0.0946	4819.16	5576.19	8.602
1080	0.0952	4841.58	5603.30	8.618
1090	0.0958	4864.06	5630.46	8.634
1100	0.0964	4886.58	5657.67	8.651
1110	0.0970	4909.14	5684.91	8.667
1120	0.0976	4931.76	5712.21	8.683
1130	0.0981	4954.41	5739.54	8.699
1140	0.0987	4977.11	5766.92	8.716
1150	0.0993	4999.86	5794.35	8.732
1160	0.0999	5022.65	5821.81	8.748
1170	0.1005	5045.48	5849.32	8.763
1180	0.1011	5068.36	5876.87	8.779
1190	0.1016	5091.28	5904.46	8.795
1200	0.1022	5114.24	5932.10	8.811
1210	0.1028	5137.24	5959.77	8.826
1220	0.1034	5160.29	5987.49	8.842
1230	0.1040	5183.38	6015.25	8.857
1240	0.1046	5206.51	6043.05	8.873
1250	0.1052	5229.68	6070.88	8.888
1260	0.1057	5252.89	6098.76	8.903
1270	0.1063	5276.14	6126.68	8.919
1280	0.1069	5299.44	6154.64	8.934
1290	0.1075	5322.77	6182.63	8.949
1300	0.1081	5346.14	6210.67	8.964

Table 3(45) Superheated Steam Properties (Pressure Base) Pressure = 85.0 bar and saturation temperature, Tsat = 299.271°C

t (°C)	$v\left(\frac{m^3}{kg}\right)$	$u\left(\frac{kJ}{kg}\right)$	$h\left(\frac{kJ}{kg}\right)$	$s\left(\frac{kJ}{kg\,K}\right)$	t (°C)	$v\left(\frac{m^3}{kg}\right)$	$u\left(\frac{kJ}{kg}\right)$	$h\left(\frac{kJ}{kg}\right)$	$s\left(\frac{kJ}{kg\,K}\right)$	t (°C)	$v\left(\frac{m^3}{kg}\right)$	$u\left(\frac{kJ}{kg}\right)$	$h\left(\frac{kJ}{kg}\right)$	$s\left(\frac{kJ}{kg\,K}\right)$
Sat Vapour (g)	0.0219	2565.00	2751.00	5.712	Sat Vapour (g)	0.0219	2565.00	2751.00	5.712	Sat Vapour (g)	0.0219	2565.00	2751.00	5.712
10	0.0233	2608.08	2806.52	5.808	450	0.0544	3535.61	3998.20	7.371	880	0.0788	4410.48	5080.13	8.252
20	0.0246	2644.76	2853.73	5.888	460	0.0550	3554.89	4022.43	7.395	890	0.0793	4432.00	5106.38	8.270
30	0.0257	2677.37	2895.86	5.959	470	0.0556	3574.20	4046.70	7.418	900	0.0799	4453.57	5132.68	8.288
40	0.0267	2707.18	2934.46	6.022	480	0.0562	3593.57	4071.00	7.441	910	0.0805	4475.19	5159.02	8.306
50	0.0277	2734.92	2970.46	6.081	490	0.0567	3612.98	4095.34	7.464	920	0.0810	4496.87	5185.42	8.324
60	0.0286	2761.07	3004.44	6.135	500	0.0573	3632.43	4119.71	7.487	930	0.0816	4518.60	5211.86	8.341
70	0.0295	2785.97	3036.82	6.186	510	0.0579	3651.93	4144.12	7.510	940	0.0821	4540.38	5238.36	8.359
80	0.0304	2809.88	3067.91	6.234	520	0.0585	3671.48	4168.57	7.532	950	0.0827	4562.21	5264.90	8.376
90	0.0312	2832.96	3097.95	6.279	530	0.0591	3691.08	4193.06	7.555	960	0.0832	4584.08	5291.49	8.394
100	0.0320	2855.37	3127.10	6.323	540	0.0596	3710.72	4217.59	7.577	970	0.0838	4606.01	5318.12	8.411
110	0.0327	2877.23	3155.52	6.365	550	0.0602	3730.42	4242.16	7.599	980	0.0843	4627.99	5344.81	8.428
120	0.0335	2898.61	3183.32	6.405	560	0.0608	3750.17	4266.77	7.621	990	0.0849	4650.02	5371.54	8.446
130	0.0342	2919.60	3210.59	6.444	570	0.0613	3769.97	4291.43	7.642	1000	0.0854	4672.10	5398.32	8.463
140	0.0350	2940.25	3237.40	6.482	580	0.0619	3789.82	4316.14	7.664	1010	0.0860	4694.22	5425.15	8.480
150	0.0357	2960.61	3263.82	6.519	590	0.0625	3809.72	4340.89	7.685	1020	0.0865	4716.40	5452.02	8.497
160	0.0364	2980.74	3289.90	6.555	600	0.0631	3829.68	4365.68	7.706	1030	0.0871	4738.62	5478.94	8.514
170	0.0371	3000.65	3315.69	6.590	610	0.0636	3849.69	4390.53	7.727	1040	0.0876	4760.89	5505.91	8.530
180	0.0377	3020.39	3341.23	6.624	620	0.0642	3869.75	4415.42	7.748	1050	0.0882	4783.20	5532.92	8.547
190	0.0384	3039.97	3366.54	6.658	630	0.0648	3889.87	4440.36	7.769	1060	0.0888	4805.57	5559.97	8.564
200	0.0391	3059.42	3391.66	6.690	640	0.0653	3910.04	4465.35	7.790	1070	0.0893	4827.98	5587.08	8.580
210	0.0397	3078.77	3416.61	6.722	650	0.0659	3930.26	4490.38	7.810	1080	0.0899	4850.44	5614.22	8.597
220	0.0404	3098.02	3441.42	6.754	660	0.0665	3950.54	4515.47	7.831	1090	0.0904	4872.94	5641.41	8.613
230	0.0410	3117.19	3466.10	6.785	670	0.0670	3970.88	4540.60	7.851	1100	0.0910	4895.49	5668.65	8.629
240	0.0417	3136.30	3490.66	6.815	680	0.0676	3991.27	4565.79	7.871	1110	0.0915	4918.08	5695.93	8.646
250	0.0423	3155.36	3515.14	6.845	690	0.0682	4011.71	4591.02	7.892	1120	0.0921	4940.72	5723.26	8.662
260	0.0430	3174.37	3539.53	6.875	700	0.0687	4032.21	4616.31	7.911	1130	0.0926	4963.41	5750.62	8.678
270	0.0436	3193.35	3563.85	6.904	710	0.0693	4052.76	4641.65	7.931	1140	0.0932	4986.13	5778.03	8.694
280	0.0442	3212.30	3588.11	6.932	720	0.0698	4073.37	4667.03	7.951	1150	0.0937	5008.91	5805.49	8.710
290	0.0448	3231.23	3612.33	6.961	730	0.0704	4094.03	4692.47	7.971	1160	0.0943	5031.72	5832.98	8.726
300	0.0455	3250.15	3636.50	6.989	740	0.0710	4114.75	4717.96	7.990	1170	0.0948	5054.58	5860.52	8.742
310	0.0461	3269.07	3660.64	7.016	750	0.0715	4135.52	4743.50	8.010	1180	0.0954	5077.49	5888.10	8.758
320	0.0467	3287.98	3684.75	7.043	760	0.0721	4156.35	4769.09	8.029	1190	0.0959	5100.43	5915.72	8.773
330	0.0473	3306.90	3708.84	7.070	770	0.0726	4177.23	4794.73	8.048	1200	0.0965	5123.42	5943.39	8.789
340	0.0479	3325.83	3732.93	7.097	780	0.0732	4198.17	4820.42	8.067	1210	0.0970	5146.45	5971.09	8.805
350	0.0485	3344.77	3757.00	7.123	790	0.0738	4219.16	4846.17	8.086	1220	0.0976	5169.52	5998.84	8.820
360	0.0491	3363.73	3781.07	7.149	800	0.0743	4240.20	4871.96	8.105	1230	0.0981	5192.64	6026.62	8.835
370	0.0497	3382.71	3805.14	7.174	810	0.0749	4261.30	4897.81	8.124	1240	0.0987	5215.79	6054.45	8.851
380	0.0503	3401.71	3829.22	7.200	820	0.0754	4282.45	4923.70	8.142	1250	0.0992	5238.99	6082.31	8.866
390	0.0509	3420.74	3853.30	7.225	830	0.0760	4303.65	4949.65	8.161	1260	0.0998	5262.22	6110.22	8.881
400	0.0515	3439.80	3877.40	7.250	840	0.0766	4324.91	4975.65	8.179	1270	0.1003	5285.50	6138.16	8.897
410	0.0521	3458.89	3901.52	7.275	850	0.0771	4346.22	5001.69	8.198	1280	0.1009	5308.82	6166.15	8.912
420	0.0527	3478.02	3925.65	7.299	860	0.0777	4367.59	5027.79	8.216	1290	0.1014	5332.17	6194.17	8.927
430	0.0533	3497.18	3949.81	7.323	870	0.0782	4389.01	5053.94	8.234	1300	0.1020	5355.57	6222.23	8.942
440	0.0538	3516.37	3973.99	7.347										

Table 3(46) Superheated Steam Properties (Pressure Base) Pressure = 90.0 bar and saturation temperature, Tsat = 303.345°C

t (°C)	v $\left(\frac{m^3}{kg}\right)$	u $\left(\frac{kJ}{kg}\right)$	h $\left(\frac{kJ}{kg}\right)$	s $\left(\frac{kJ}{kg\,K}\right)$
Sat Vapour (g)	*0.0205*	*2558.52*	*2742.92*	*5.679*
10	0.0219	2603.81	2800.82	5.779
20	0.0231	2641.76	2849.63	5.861
30	0.0242	2675.32	2892.98	5.933
40	0.0252	2705.88	2932.55	5.998
50	0.0261	2734.23	2969.34	6.057
60	0.0270	2760.91	3004.00	6.112
70	0.0279	2786.26	3036.96	6.164
80	0.0287	2810.56	3068.57	6.212
90	0.0295	2833.99	3099.06	6.258
100	0.0302	2856.72	3128.62	6.302
110	0.0310	2878.85	3157.42	6.344
120	0.0317	2900.50	3185.55	6.385
130	0.0324	2921.72	3213.13	6.424
140	0.0331	2942.59	3240.23	6.463
150	0.0338	2963.16	3266.92	6.500
160	0.0344	2983.47	3293.25	6.536
170	0.0351	3003.57	3319.27	6.571
180	0.0357	3023.47	3345.03	6.605
190	0.0364	3043.21	3370.55	6.638
200	0.0370	3062.82	3395.86	6.671
210	0.0376	3082.31	3421.00	6.703
220	0.0383	3101.69	3445.99	6.735
230	0.0389	3121.00	3470.83	6.766
240	0.0395	3140.23	3495.56	6.796
250	0.0401	3159.40	3520.19	6.826
260	0.0407	3178.53	3544.73	6.856
270	0.0413	3197.61	3569.19	6.885
280	0.0419	3216.67	3593.59	6.914
290	0.0425	3235.70	3617.94	6.942
300	0.0431	3254.72	3642.24	6.970
310	0.0436	3273.74	3666.50	6.997
320	0.0442	3292.74	3690.73	7.024
330	0.0448	3311.75	3714.94	7.051
340	0.0454	3330.77	3739.13	7.078
350	0.0459	3349.79	3763.31	7.104
360	0.0465	3368.84	3787.48	7.130
370	0.0471	3387.90	3811.66	7.156
380	0.0477	3406.98	3835.83	7.181
390	0.0482	3426.08	3860.02	7.206
400	0.0488	3445.22	3884.21	7.231
410	0.0493	3464.38	3908.42	7.256
420	0.0499	3483.57	3932.64	7.280
430	0.0505	3502.80	3956.88	7.305
440	0.0510	3522.07	3981.15	7.328

t (°C)	v $\left(\frac{m^3}{kg}\right)$	u $\left(\frac{kJ}{kg}\right)$	h $\left(\frac{kJ}{kg}\right)$	s $\left(\frac{kJ}{kg\,K}\right)$
Sat Vapour (g)	*0.0205*	*2558.52*	*2742.92*	*5.679*
450	0.0516	3541.37	4005.44	7.352
460	0.0521	3560.71	4029.76	7.376
470	0.0527	3580.10	4054.11	7.399
480	0.0532	3599.52	4078.49	7.422
490	0.0538	3618.99	4102.90	7.445
500	0.0543	3638.50	4127.34	7.468
510	0.0549	3658.06	4151.83	7.491
520	0.0554	3677.67	4176.35	7.513
530	0.0560	3697.33	4200.90	7.536
540	0.0565	3717.03	4225.50	7.558
550	0.0570	3736.78	4250.14	7.580
560	0.0576	3756.58	4274.82	7.602
570	0.0581	3776.44	4299.55	7.623
580	0.0587	3796.34	4324.32	7.645
590	0.0592	3816.30	4349.13	7.666
600	0.0597	3836.31	4373.99	7.687
610	0.0603	3856.37	4398.89	7.708
620	0.0608	3876.48	4423.85	7.729
630	0.0614	3896.64	4448.84	7.750
640	0.0619	3916.86	4473.89	7.771
650	0.0624	3937.14	4498.98	7.791
660	0.0630	3957.46	4524.12	7.812
670	0.0635	3977.84	4549.32	7.832
680	0.0640	3998.28	4574.56	7.852
690	0.0646	4018.77	4599.84	7.872
700	0.0651	4039.31	4625.18	7.892
710	0.0656	4059.91	4650.57	7.912
720	0.0662	4080.56	4676.01	7.932
730	0.0667	4101.26	4701.50	7.951
740	0.0672	4122.02	4727.04	7.971
750	0.0678	4142.84	4752.62	7.990
760	0.0683	4163.71	4778.26	8.009
770	0.0688	4184.63	4803.95	8.029
780	0.0693	4205.60	4829.69	8.048
790	0.0699	4226.63	4855.48	8.067
800	0.0704	4247.72	4881.32	8.085
810	0.0709	4268.85	4907.21	8.104
820	0.0715	4290.04	4933.15	8.123
830	0.0720	4311.29	4959.14	8.141
840	0.0725	4332.58	4985.18	8.160
850	0.0730	4353.93	5011.27	8.178
860	0.0736	4375.33	5037.41	8.196
870	0.0741	4396.78	5063.60	8.215

t (°C)	v $\left(\frac{m^3}{kg}\right)$	u $\left(\frac{kJ}{kg}\right)$	h $\left(\frac{kJ}{kg}\right)$	s $\left(\frac{kJ}{kg\,K}\right)$
Sat Vapour (g)	*0.0205*	*2558.52*	*2742.92*	*5.679*
880	0.0746	4418.29	5089.84	8.233
890	0.0751	4439.84	5116.13	8.251
900	0.0757	4461.45	5142.46	8.268
910	0.0762	4483.11	5168.85	8.286
920	0.0767	4504.82	5195.28	8.304
930	0.0772	4526.58	5221.77	8.322
940	0.0778	4548.39	5248.30	8.339
950	0.0783	4570.26	5274.87	8.357
960	0.0788	4592.17	5301.50	8.374
970	0.0793	4614.13	5328.18	8.391
980	0.0799	4636.14	5354.90	8.409
990	0.0804	4658.20	5381.66	8.426
1000	0.0809	4680.30	5408.48	8.443
1010	0.0814	4702.46	5435.34	8.460
1020	0.0820	4724.66	5462.25	8.477
1030	0.0825	4746.92	5489.20	8.494
1040	0.0830	4769.22	5516.20	8.510
1050	0.0835	4791.56	5543.25	8.527
1060	0.0840	4813.95	5570.34	8.544
1070	0.0846	4836.39	5597.47	8.560
1080	0.0851	4858.88	5624.65	8.577
1090	0.0856	4881.41	5651.87	8.593
1100	0.0861	4903.99	5679.14	8.609
1110	0.0866	4926.61	5706.45	8.625
1120	0.0872	4949.28	5733.81	8.642
1130	0.0877	4971.99	5761.20	8.658
1140	0.0882	4994.74	5788.65	8.674
1150	0.0887	5017.54	5816.13	8.690
1160	0.0893	5040.38	5843.65	8.706
1170	0.0898	5063.27	5871.22	8.721
1180	0.0903	5086.20	5898.83	8.737
1190	0.0908	5109.17	5926.48	8.753
1200	0.0913	5132.18	5954.17	8.769
1210	0.0919	5155.24	5981.90	8.784
1220	0.0924	5178.33	6009.68	8.800
1230	0.0929	5201.47	6037.49	8.815
1240	0.0934	5224.65	6065.34	8.830
1250	0.0939	5247.87	6093.23	8.846
1260	0.0944	5271.13	6121.16	8.861
1270	0.0950	5294.43	6149.13	8.876
1280	0.0955	5317.77	6177.14	8.891
1290	0.0960	5341.14	6205.19	8.906
1300	0.0965	5364.56	6233.28	8.921

Table 3(47) Superheated Steam Properties (Pressure Base) Pressure = 95.0 bar and saturation temperature, Tsat = 307.249°C

t (°C)	v ($\frac{m^3}{kg}$)	u ($\frac{kJ}{kg}$)	h ($\frac{kJ}{kg}$)	s ($\frac{kJ}{kg\,K}$)
Sat Vapour (g)	0.0192	2552.00	2734.00	5.647
10	0.0206	2599.30	2794.80	5.750
20	0.0218	2638.55	2845.28	5.835
30	0.0228	2673.07	2889.85	5.909
40	0.0238	2704.39	2930.40	5.975
50	0.0247	2733.36	2967.98	6.035
60	0.0256	2760.55	3003.31	6.090
70	0.0264	2786.35	3036.86	6.142
80	0.0272	2811.04	3068.97	6.191
90	0.0279	2834.82	3099.91	6.238
100	0.0286	2857.86	3129.89	6.282
110	0.0293	2880.27	3159.04	6.325
120	0.0300	2902.17	3187.51	6.366
130	0.0307	2923.63	3215.40	6.405
140	0.0314	2944.72	3242.78	6.444
150	0.0320	2965.49	3269.74	6.481
160	0.0327	2985.99	3296.31	6.517
170	0.0333	3006.26	3322.57	6.552
180	0.0339	3026.33	3348.54	6.587
190	0.0345	3046.23	3374.26	6.620
200	0.0351	3065.98	3399.77	6.653
210	0.0357	3085.60	3425.10	6.685
220	0.0363	3105.13	3450.25	6.717
230	0.0369	3124.56	3475.27	6.748
240	0.0375	3143.91	3500.15	6.779
250	0.0381	3163.20	3524.93	6.809
260	0.0387	3182.43	3549.62	6.838
270	0.0392	3201.63	3574.22	6.867
280	0.0398	3220.79	3598.75	6.896
290	0.0403	3239.92	3623.23	6.924
300	0.0409	3259.04	3647.65	6.952
310	0.0415	3278.14	3672.03	6.980
320	0.0420	3297.24	3696.38	7.007
330	0.0426	3316.34	3720.70	7.034
340	0.0431	3335.44	3745.00	7.060
350	0.0437	3354.55	3769.29	7.086
360	0.0442	3373.67	3793.56	7.112
370	0.0447	3392.81	3817.84	7.138
380	0.0453	3411.96	3842.11	7.163
390	0.0458	3431.14	3866.38	7.189
400	0.0463	3450.35	3890.67	7.214
410	0.0469	3469.58	3914.97	7.238
420	0.0474	3488.85	3939.28	7.263
430	0.0479	3508.14	3963.61	7.287
440	0.0485	3527.48	3987.95	7.311

t (°C)	v ($\frac{m^3}{kg}$)	u ($\frac{kJ}{kg}$)	h ($\frac{kJ}{kg}$)	s ($\frac{kJ}{kg\,K}$)
Sat Vapour (g)	0.0192	2552.00	2734.00	5.647
450	0.0490	3546.84	4012.33	7.335
460	0.0495	3566.25	4036.72	7.358
470	0.0500	3585.70	4061.15	7.381
480	0.0506	3605.18	4085.60	7.405
490	0.0511	3624.71	4110.09	7.428
500	0.0516	3644.28	4134.61	7.450
510	0.0521	3663.90	4159.16	7.473
520	0.0527	3683.57	4183.75	7.495
530	0.0532	3703.28	4208.38	7.518
540	0.0537	3723.03	4233.05	7.540
550	0.0542	3742.84	4257.75	7.562
560	0.0547	3762.70	4282.50	7.584
570	0.0552	3782.60	4307.29	7.605
580	0.0557	3802.56	4332.12	7.627
590	0.0563	3822.56	4356.99	7.648
600	0.0568	3842.62	4381.91	7.669
610	0.0573	3862.73	4406.87	7.690
620	0.0578	3882.89	4431.88	7.711
630	0.0583	3903.11	4456.94	7.732
640	0.0588	3923.37	4482.04	7.753
650	0.0593	3943.69	4507.19	7.773
660	0.0598	3964.06	4532.39	7.794
670	0.0603	3984.49	4557.63	7.814
680	0.0608	4004.97	4582.92	7.834
690	0.0613	4025.50	4608.27	7.854
700	0.0618	4046.09	4633.66	7.874
710	0.0624	4066.73	4659.09	7.894
720	0.0629	4087.42	4684.58	7.913
730	0.0634	4108.17	4710.12	7.933
740	0.0639	4128.97	4735.71	7.952
750	0.0644	4149.82	4761.34	7.972
760	0.0649	4170.73	4787.03	7.991
770	0.0654	4191.69	4812.76	8.010
780	0.0659	4212.71	4838.55	8.029
790	0.0664	4233.77	4864.38	8.048
800	0.0669	4254.89	4890.27	8.067
810	0.0674	4276.07	4916.20	8.086
820	0.0679	4297.29	4942.19	8.104
830	0.0684	4318.57	4968.22	8.123
840	0.0689	4339.91	4994.30	8.141
850	0.0694	4361.29	5020.43	8.160
860	0.0699	4382.73	5046.61	8.178
870	0.0704	4404.21	5072.84	8.196

t (°C)	v ($\frac{m^3}{kg}$)	u ($\frac{kJ}{kg}$)	h ($\frac{kJ}{kg}$)	s ($\frac{kJ}{kg\,K}$)
Sat Vapour (g)	0.0192	2552.00	2734.00	5.647
880	0.0709	4425.75	5099.12	8.214
890	0.0714	4447.34	5125.45	8.232
900	0.0719	4468.98	5151.82	8.250
910	0.0724	4490.68	5178.25	8.268
920	0.0729	4512.42	5204.72	8.285
930	0.0734	4534.21	5231.24	8.303
940	0.0739	4556.06	5257.81	8.321
950	0.0744	4577.95	5284.42	8.338
960	0.0749	4599.89	5311.08	8.355
970	0.0754	4621.88	5337.79	8.373
980	0.0759	4643.93	5364.55	8.390
990	0.0764	4666.01	5391.35	8.407
1000	0.0768	4688.15	5418.20	8.424
1010	0.0773	4710.34	5445.10	8.441
1020	0.0778	4732.57	5472.04	8.458
1030	0.0783	4754.85	5499.02	8.475
1040	0.0788	4777.18	5526.06	8.491
1050	0.0793	4799.55	5553.13	8.508
1060	0.0798	4821.97	5580.25	8.525
1070	0.0803	4844.44	5607.42	8.541
1080	0.0808	4866.95	5634.63	8.558
1090	0.0813	4889.51	5661.89	8.574
1100	0.0818	4912.12	5689.18	8.590
1110	0.0823	4934.76	5716.52	8.606
1120	0.0828	4957.46	5743.91	8.623
1130	0.0833	4980.19	5771.34	8.639
1140	0.0838	5002.97	5798.81	8.655
1150	0.0843	5025.80	5826.32	8.671
1160	0.0848	5048.67	5853.87	8.687
1170	0.0853	5071.58	5881.47	8.702
1180	0.0857	5094.53	5909.10	8.718
1190	0.0862	5117.53	5936.78	8.734
1200	0.0867	5140.56	5964.50	8.749
1210	0.0872	5163.64	5992.26	8.765
1220	0.0877	5186.76	6020.06	8.780
1230	0.0882	5209.92	6047.90	8.796
1240	0.0887	5233.12	6075.77	8.811
1250	0.0892	5256.37	6103.69	8.826
1260	0.0897	5279.65	6131.65	8.842
1270	0.0902	5302.97	6159.64	8.857
1280	0.0907	5326.33	6187.68	8.872
1290	0.0912	5349.73	6216.75	8.887
1300	0.0917	5373.17	6243.86	8.902

Table 3(48) Superheated Steam Properties (Pressure Base) Pressure = 100.0 bar and saturation temperature, Tsat = 310.998°C

t (°C)	$v\left(\frac{m^3}{kg}\right)$	$u\left(\frac{kJ}{kg}\right)$	$h\left(\frac{kJ}{kg}\right)$	$s\left(\frac{kJ}{kg\,K}\right)$	t (°C)	$v\left(\frac{m^3}{kg}\right)$	$u\left(\frac{kJ}{kg}\right)$	$h\left(\frac{kJ}{kg}\right)$	$s\left(\frac{kJ}{kg\,K}\right)$	t (°C)	$v\left(\frac{m^3}{kg}\right)$	$u\left(\frac{kJ}{kg}\right)$	$h\left(\frac{kJ}{kg}\right)$	$s\left(\frac{kJ}{kg\,K}\right)$
Sat Vapour (g)	*0.0180*	*2545.17*	*2725.46*	*5.616*	Sat Vapour (g)	*0.0180*	*2545.17*	*2725.46*	*5.616*	Sat Vapour (g)	*0.0180*	*2545.17*	*2725.46*	*5.616*
10	0.0194	2594.55	2788.50	5.723	450	0.0467	3552.06	4018.89	7.318	880	0.0675	4432.90	5108.02	8.196
20	0.0206	2635.14	2840.68	5.810	460	0.0472	3571.52	4043.36	7.341	890	0.0680	4454.52	5134.38	8.214
30	0.0216	2670.64	2886.50	5.885	470	0.0477	3591.03	4067.86	7.365	900	0.0685	4476.20	5160.80	8.232
40	0.0225	2702.71	2928.02	5.952	480	0.0482	3610.58	4092.39	7.388	910	0.0689	4497.92	5187.26	8.250
50	0.0234	2732.31	2966.41	6.013	490	0.0487	3630.16	4116.95	7.411	920	0.0694	4519.70	5213.76	8.268
60	0.0242	2760.02	3002.41	6.070	500	0.0492	3649.80	4141.54	7.434	930	0.0699	4541.52	5240.32	8.285
70	0.0250	2786.26	3036.53	6.122	510	0.0497	3669.47	4166.17	7.456	940	0.0704	4563.40	5266.92	8.303
80	0.0258	2811.34	3069.15	6.172	520	0.0502	3689.19	4190.82	7.479	950	0.0708	4585.32	5293.57	8.320
90	0.0265	2835.46	3100.54	6.219	530	0.0507	3708.95	4215.52	7.501	960	0.0713	4607.29	5320.27	8.338
100	0.0272	2858.81	3130.91	6.263	540	0.0511	3728.76	4240.25	7.523	970	0.0718	4629.32	5347.02	8.355
110	0.0279	2881.50	3160.44	6.306	550	0.0516	3748.62	4265.02	7.545	980	0.0722	4651.39	5373.81	8.372
120	0.0286	2903.65	3189.24	6.347	560	0.0521	3768.53	4289.83	7.567	990	0.0727	4673.50	5400.64	8.389
130	0.0292	2925.34	3217.43	6.387	570	0.0526	3788.49	4314.68	7.588	1000	0.0732	4695.67	5427.53	8.406
140	0.0298	2946.64	3245.09	6.426	580	0.0531	3808.49	4339.57	7.610	1010	0.0737	4717.89	5454.45	8.423
150	0.0305	2967.61	3272.30	6.463	590	0.0536	3828.55	4364.51	7.631	1020	0.0741	4740.15	5481.43	8.440
160	0.0311	2988.30	3299.12	6.499	600	0.0541	3848.65	4389.48	7.652	1030	0.0746	4762.46	5508.45	8.457
170	0.0317	3008.74	3325.61	6.535	610	0.0546	3868.81	4414.50	7.673	1040	0.0751	4784.81	5535.51	8.473
180	0.0323	3028.97	3351.79	6.569	620	0.0551	3889.02	4439.57	7.694	1050	0.0755	4807.21	5562.62	8.490
190	0.0329	3049.03	3377.72	6.603	630	0.0555	3909.28	4464.68	7.715	1060	0.0760	4829.66	5589.77	8.507
200	0.0334	3068.92	3403.42	6.636	640	0.0560	3929.59	4489.84	7.736	1070	0.0765	4852.15	5616.97	8.523
210	0.0340	3088.69	3428.92	6.668	650	0.0565	3949.96	4515.04	7.756	1080	0.0770	4874.69	5644.21	8.540
220	0.0346	3108.34	3454.25	6.700	660	0.0570	3970.37	4540.29	7.777	1090	0.0774	4897.28	5671.49	8.556
230	0.0352	3127.89	3479.42	6.731	670	0.0575	3990.84	4565.59	7.797	1100	0.0779	4919.91	5698.82	8.572
240	0.0357	3147.37	3504.46	6.762	680	0.0580	4011.36	4590.93	7.817	1110	0.0784	4942.58	5726.19	8.588
250	0.0363	3166.77	3529.39	6.792	690	0.0584	4031.94	4616.33	7.837	1120	0.0788	4965.30	5753.60	8.605
260	0.0368	3186.11	3554.22	6.821	700	0.0589	4052.57	4641.77	7.857	1130	0.0793	4988.06	5781.06	8.621
270	0.0374	3205.41	3578.96	6.850	710	0.0594	4073.25	4667.25	7.877	1140	0.0798	5010.87	5808.56	8.637
280	0.0379	3224.68	3603.62	6.879	720	0.0599	4093.98	4692.79	7.896	1150	0.0802	5033.72	5836.09	8.653
290	0.0384	3243.91	3628.22	6.907	730	0.0604	4114.77	4718.38	7.916	1160	0.0807	5056.61	5863.68	8.668
300	0.0390	3263.12	3652.77	6.935	740	0.0608	4135.61	4744.01	7.935	1170	0.0812	5079.54	5891.30	8.684
310	0.0395	3282.31	3677.27	6.963	750	0.0613	4156.50	4769.69	7.955	1180	0.0816	5102.52	5918.96	8.700
320	0.0400	3301.50	3701.73	6.990	760	0.0618	4177.45	4795.42	7.974	1190	0.0821	5125.54	5946.67	8.716
330	0.0405	3320.68	3726.16	7.017	770	0.0623	4198.45	4821.20	7.993	1200	0.0826	5148.60	5974.41	8.731
340	0.0411	3339.87	3750.57	7.043	780	0.0628	4219.50	4847.03	8.012	1210	0.0831	5171.70	6002.19	8.747
350	0.0416	3359.06	3774.96	7.070	790	0.0632	4240.61	4872.91	8.031	1220	0.0835	5194.84	6030.02	8.762
360	0.0421	3378.26	3799.33	7.096	800	0.0637	4261.76	4898.84	8.050	1230	0.0840	5218.03	6057.88	8.778
370	0.0426	3397.47	3823.70	7.121	810	0.0642	4282.97	4924.82	8.068	1240	0.0845	5241.25	6085.79	8.793
380	0.0431	3416.70	3848.07	7.147	820	0.0647	4304.24	4950.84	8.087	1250	0.0849	5264.51	6113.73	8.808
390	0.0436	3435.96	3872.44	7.172	830	0.0651	4325.55	4976.92	8.105	1260	0.0854	5287.82	6141.71	8.823
400	0.0442	3455.23	3896.81	7.197	840	0.0656	4346.92	5003.04	8.124	1270	0.0859	5311.16	6169.73	8.839
410	0.0447	3474.53	3921.20	7.221	850	0.0661	4368.34	5029.21	8.142	1280	0.0863	5334.54	6197.79	8.854
420	0.0452	3493.87	3945.59	7.246	860	0.0666	4389.81	5055.43	8.160	1290	0.0868	5357.96	6225.88	8.869
430	0.0457	3513.23	3970.01	7.270	870	0.0670	4411.33	5081.70	8.178	1300	0.0873	5381.42	6254.01	8.884
440	0.0462	3532.62	3994.44	7.294										

Table 3(49) Superheated Steam Properties (Pressure Base) Pressure = 110.0 bar and saturation temperature, Tsat = 318.079°C

t (°C)	v $\left(\frac{m^3}{kg}\right)$	u $\left(\frac{kJ}{kg}\right)$	h $\left(\frac{kJ}{kg}\right)$	s $\left(\frac{kJ}{kg\ K}\right)$
Sat Vapour (g)	0.0160	2530.44	2706.31	5.554
10	0.0173	2584.36	2775.06	5.670
20	0.0185	2627.76	2830.80	5.762
30	0.0194	2665.27	2879.18	5.840
40	0.0203	2698.89	2922.68	5.910
50	0.0212	2729.73	2962.67	5.973
60	0.0220	2758.48	3000.02	6.030
70	0.0227	2785.61	3035.29	6.084
80	0.0234	2811.46	3068.91	6.135
90	0.0241	2836.26	3101.19	6.182
100	0.0247	2860.21	3132.35	6.228
110	0.0254	2883.45	3162.59	6.271
120	0.0260	2906.10	3192.04	6.313
130	0.0266	2928.25	3220.83	6.353
140	0.0272	2949.97	3249.05	6.392
150	0.0278	2971.33	3276.77	6.430
160	0.0283	2992.38	3304.07	6.466
170	0.0289	3013.16	3330.99	6.502
180	0.0294	3033.71	3357.59	6.537
190	0.0300	3054.06	3383.91	6.570
200	0.0305	3074.24	3409.98	6.604
210	0.0311	3094.27	3435.84	6.636
220	0.0316	3114.18	3461.50	6.668
230	0.0321	3133.98	3486.99	6.699
240	0.0326	3153.68	3512.34	6.730
250	0.0331	3173.31	3537.56	6.760
260	0.0336	3192.87	3562.66	6.790
270	0.0341	3212.37	3587.67	6.819
280	0.0346	3231.83	3612.59	6.848
290	0.0351	3251.25	3637.44	6.876
300	0.0356	3270.65	3662.22	6.904
310	0.0361	3290.02	3686.95	6.932
320	0.0366	3309.38	3711.63	6.959
330	0.0371	3328.72	3736.27	6.986
340	0.0375	3348.07	3760.89	7.012
350	0.0380	3367.42	3785.48	7.039
360	0.0385	3386.77	3810.05	7.064
370	0.0390	3406.13	3834.61	7.090
380	0.0394	3425.51	3859.16	7.116
390	0.0399	3444.90	3883.71	7.141
400	0.0404	3464.32	3908.26	7.166
410	0.0408	3483.75	3932.81	7.190
420	0.0413	3503.21	3957.37	7.215
430	0.0417	3522.71	3981.95	7.239
440	0.0422	3542.23	4006.53	7.263

t (°C)	v $\left(\frac{m^3}{kg}\right)$	u $\left(\frac{kJ}{kg}\right)$	h $\left(\frac{kJ}{kg}\right)$	s $\left(\frac{kJ}{kg\ K}\right)$
Sat Vapour (g)	0.0160	2530.44	2706.31	5.554
450	0.0427	3561.78	4031.14	7.287
460	0.0431	3581.37	4055.76	7.310
470	0.0436	3600.99	4080.41	7.333
480	0.0440	3620.65	4105.09	7.357
490	0.0445	3640.35	4129.78	7.380
500	0.0449	3660.10	4154.51	7.402
510	0.0454	3679.88	4179.27	7.425
520	0.0459	3699.70	4204.06	7.447
530	0.0463	3719.57	4228.88	7.470
540	0.0468	3739.49	4253.74	7.492
550	0.0472	3759.45	4278.64	7.513
560	0.0476	3779.46	4303.57	7.535
570	0.0481	3799.51	4328.54	7.557
580	0.0485	3819.61	4353.55	7.578
590	0.0490	3839.76	4378.60	7.600
600	0.0494	3859.96	4403.69	7.621
610	0.0499	3880.21	4428.82	7.642
620	0.0503	3900.51	4454.00	7.663
630	0.0508	3920.86	4479.22	7.683
640	0.0512	3941.26	4504.48	7.704
650	0.0516	3961.71	4529.79	7.724
660	0.0521	3982.21	4555.14	7.745
670	0.0525	4002.76	4580.54	7.765
680	0.0530	4023.37	4605.98	7.785
690	0.0534	4044.02	4631.47	7.805
700	0.0538	4064.73	4657.01	7.825
710	0.0543	4085.49	4682.59	7.845
720	0.0547	4106.31	4708.22	7.864
730	0.0552	4127.17	4733.90	7.884
740	0.0556	4148.09	4759.63	7.903
750	0.0560	4169.06	4785.40	7.922
760	0.0565	4190.08	4811.22	7.942
770	0.0569	4211.15	4837.08	7.961
780	0.0573	4232.27	4863.00	7.980
790	0.0578	4253.45	4888.96	7.999
800	0.0582	4274.68	4914.97	8.017
810	0.0586	4295.96	4941.03	8.036
820	0.0591	4317.29	4967.14	8.055
830	0.0595	4338.67	4993.29	8.073
840	0.0599	4360.10	5019.49	8.091
850	0.0604	4381.59	5045.74	8.110
860	0.0608	4403.12	5072.04	8.128
870	0.0612	4424.71	5098.38	8.146

t (°C)	v $\left(\frac{m^3}{kg}\right)$	u $\left(\frac{kJ}{kg}\right)$	h $\left(\frac{kJ}{kg}\right)$	s $\left(\frac{kJ}{kg\ K}\right)$
Sat Vapour (g)	0.0160	2530.44	2706.31	5.554
880	0.0617	4446.35	5124.78	8.164
890	0.0621	4468.03	5151.21	8.182
900	0.0625	4489.77	5177.70	8.200
910	0.0630	4511.56	5204.23	8.217
920	0.0634	4533.39	5230.81	8.235
930	0.0638	4555.28	5257.44	8.253
940	0.0643	4577.21	5284.11	8.270
950	0.0647	4599.19	5310.83	8.287
960	0.0651	4621.22	5337.59	8.305
970	0.0656	4643.30	5364.40	8.322
980	0.0660	4665.43	5391.26	8.339
990	0.0664	4687.60	5418.16	8.356
1000	0.0668	4709.83	5445.11	8.373
1010	0.0673	4732.10	5472.10	8.390
1020	0.0677	4754.41	5499.13	8.407
1030	0.0681	4776.77	5526.21	8.424
1040	0.0686	4799.18	5553.34	8.440
1050	0.0690	4821.64	5580.51	8.457
1060	0.0694	4844.14	5607.72	8.474
1070	0.0698	4866.68	5634.97	8.490
1080	0.0703	4889.27	5662.27	8.506
1090	0.0707	4911.90	5689.61	8.523
1100	0.0711	4934.58	5717.00	8.539
1110	0.0716	4957.31	5744.42	8.555
1120	0.0720	4980.07	5771.89	8.571
1130	0.0724	5002.88	5799.40	8.587
1140	0.0728	5025.73	5826.95	8.603
1150	0.0733	5048.63	5854.54	8.619
1160	0.0737	5071.57	5882.18	8.635
1170	0.0741	5094.55	5909.85	8.651
1180	0.0745	5117.57	5937.57	8.666
1190	0.0750	5140.63	5965.32	8.682
1200	0.0754	5163.74	5993.12	8.698
1210	0.0758	5186.88	6020.95	8.713
1220	0.0763	5210.07	6048.82	8.728
1230	0.0767	5233.30	6076.74	8.744
1240	0.0771	5256.56	6104.69	8.759
1250	0.0775	5279.87	6132.68	8.774
1260	0.0780	5303.21	6160.71	8.790
1270	0.0784	5326.60	6188.77	8.805
1280	0.0788	5350.02	6216.87	8.820
1290	0.0792	5373.48	6245.01	8.835
1300	0.0797	5396.98	6273.19	8.850

Table 3(50) Superheated Steam Properties (Pressure Base) Pressure = 120.0 bar and saturation temperature, Tsat = 324.675°C

t (°C)	v $\left(\frac{m^3}{kg}\right)$	u $\left(\frac{kJ}{kg}\right)$	h $\left(\frac{kJ}{kg}\right)$	s $\left(\frac{kJ}{kg\,K}\right)$	t (°C)	v $\left(\frac{m^3}{kg}\right)$	u $\left(\frac{kJ}{kg}\right)$	h $\left(\frac{kJ}{kg}\right)$	s $\left(\frac{kJ}{kg\,K}\right)$	t (°C)	v $\left(\frac{m^3}{kg}\right)$	u $\left(\frac{kJ}{kg}\right)$	h $\left(\frac{kJ}{kg}\right)$	s $\left(\frac{kJ}{kg\,K}\right)$
Sat Vapour (g)	0.0143	2514.27	2685.43	5.494	Sat Vapour (g)	0.0143	2514.27	2685.43	5.494	Sat Vapour (g)	0.0143	2514.27	2685.43	5.494
10	0.0156	2573.32	2760.59	5.619	450	0.0393	3570.69	4042.38	7.258	880	0.0568	4458.81	5140.33	8.134
20	0.0167	2619.71	2820.11	5.716	460	0.0397	3590.39	4067.15	7.282	890	0.0572	4480.55	5166.84	8.152
30	0.0177	2659.30	2871.14	5.798	470	0.0402	3610.13	4091.94	7.305	900	0.0576	4502.35	5193.40	8.170
40	0.0185	2694.50	2916.66	5.870	480	0.0406	3629.90	4116.75	7.328	910	0.0580	4524.19	5220.00	8.188
50	0.0193	2726.60	2958.27	5.934	490	0.0410	3649.71	4141.59	7.351	920	0.0584	4546.08	5246.64	8.205
60	0.0200	2756.39	2996.95	5.994	500	0.0414	3669.56	4166.45	7.374	930	0.0588	4568.03	5273.34	8.223
70	0.0207	2784.40	3033.37	6.049	510	0.0418	3689.44	4191.33	7.396	940	0.0592	4590.02	5300.07	8.240
80	0.0214	2811.01	3067.98	6.100	520	0.0422	3709.37	4216.25	7.419	950	0.0596	4612.05	5326.86	8.258
90	0.0221	2836.49	3101.13	6.149	530	0.0427	3729.34	4241.20	7.441	960	0.0600	4634.14	5353.69	8.275
100	0.0227	2861.04	3133.08	6.195	540	0.0431	3749.36	4266.18	7.463	970	0.0604	4656.27	5380.56	8.292
110	0.0233	2884.82	3164.02	6.239	550	0.0435	3769.41	4291.19	7.485	980	0.0608	4678.45	5407.48	8.309
120	0.0238	2907.96	3194.12	6.281	560	0.0439	3789.51	4316.24	7.507	990	0.0611	4700.68	5434.44	8.326
130	0.0244	2930.55	3223.49	6.322	570	0.0443	3809.66	4341.33	7.528	1000	0.0615	4722.96	5461.45	8.343
140	0.0250	2952.68	3252.24	6.361	580	0.0447	3829.86	4366.45	7.550	1010	0.0619	4745.28	5488.50	8.360
150	0.0255	2974.42	3280.46	6.399	590	0.0451	3850.10	4391.61	7.571	1020	0.0623	4767.64	5515.59	8.377
160	0.0260	2995.82	3308.22	6.436	600	0.0455	3870.38	4416.81	7.592	1030	0.0627	4790.06	5542.73	8.394
170	0.0266	3016.93	3335.58	6.472	610	0.0459	3890.72	4442.05	7.613	1040	0.0631	4812.51	5569.91	8.410
180	0.0271	3037.79	3362.59	6.507	620	0.0464	3911.11	4467.33	7.634	1050	0.0635	4835.02	5597.14	8.427
190	0.0276	3058.44	3389.29	6.541	630	0.0468	3931.54	4492.66	7.655	1060	0.0639	4857.57	5624.41	8.443
200	0.0281	3078.89	3415.73	6.574	640	0.0472	3952.02	4518.02	7.675	1070	0.0643	4880.16	5651.72	8.460
210	0.0286	3099.19	3441.92	6.607	650	0.0476	3972.56	4543.43	7.696	1080	0.0647	4902.80	5679.07	8.476
220	0.0290	3119.34	3467.91	6.639	660	0.0480	3993.14	4568.88	7.716	1090	0.0651	4925.48	5706.46	8.492
230	0.0295	3139.38	3493.71	6.670	670	0.0484	4013.77	4594.37	7.736	1100	0.0655	4948.20	5733.90	8.509
240	0.0300	3159.31	3519.35	6.701	680	0.0488	4034.46	4619.91	7.756	1110	0.0659	4970.97	5761.38	8.525
250	0.0305	3179.15	3544.85	6.731	690	0.0492	4055.19	4645.50	7.776	1120	0.0663	4993.79	5788.90	8.541
260	0.0309	3198.91	3570.23	6.761	700	0.0496	4075.98	4671.13	7.796	1130	0.0667	5016.64	5816.46	8.557
270	0.0314	3218.62	3595.49	6.790	710	0.0500	4096.81	4696.80	7.816	1140	0.0670	5039.54	5844.07	8.573
280	0.0319	3238.27	3620.66	6.819	720	0.0504	4117.70	4722.52	7.835	1150	0.0674	5062.48	5871.71	8.589
290	0.0323	3257.87	3645.74	6.847	730	0.0508	4138.64	4748.28	7.855	1160	0.0678	5085.46	5899.39	8.604
300	0.0328	3277.44	3670.76	6.875	740	0.0512	4159.63	4774.09	7.874	1170	0.0682	5108.48	5927.12	8.620
310	0.0332	3296.99	3695.71	6.903	750	0.0516	4180.67	4799.95	7.893	1180	0.0686	5131.55	5954.88	8.636
320	0.0337	3316.51	3720.60	6.930	760	0.0520	4201.76	4825.86	7.912	1190	0.0690	5154.65	5982.68	8.651
330	0.0341	3336.02	3745.46	6.957	770	0.0524	4222.90	4851.81	7.931	1200	0.0694	5177.80	6010.53	8.667
340	0.0346	3355.52	3770.27	6.984	780	0.0528	4244.09	4877.80	7.950	1210	0.0698	5200.99	6038.41	8.682
350	0.0350	3375.02	3795.05	7.010	790	0.0532	4265.34	4903.85	7.969	1220	0.0702	5224.21	6066.33	8.698
360	0.0354	3394.52	3819.81	7.036	800	0.0536	4286.63	4929.94	7.988	1230	0.0706	5247.48	6094.29	8.713
370	0.0359	3414.02	3844.56	7.062	810	0.0540	4307.98	4956.07	8.007	1240	0.0710	5270.79	6122.28	8.728
380	0.0363	3433.54	3869.29	7.087	820	0.0544	4329.37	4982.26	8.025	1250	0.0713	5294.13	6150.32	8.744
390	0.0367	3453.07	3894.01	7.112	830	0.0548	4350.82	5008.49	8.044	1260	0.0717	5317.52	6178.39	8.759
400	0.0372	3472.61	3918.72	7.137	840	0.0552	4372.32	5034.76	8.062	1270	0.0721	5340.94	6206.50	8.774
410	0.0376	3492.18	3943.44	7.162	850	0.0556	4393.86	5061.08	8.080	1280	0.0725	5364.40	6234.64	8.789
420	0.0380	3511.76	3968.16	7.186	860	0.0560	4415.46	5087.45	8.098	1290	0.0729	5387.90	6262.83	8.804
430	0.0385	3531.38	3992.89	7.210	870	0.0564	4437.11	5113.87	8.116	1300	0.0733	5411.44	6291.05	8.819
440	0.0389	3551.02	4017.63	7.234										

Table 3(51) Superheated Steam Properties (Pressure Base) Pressure = 130.0 bar and saturation temperature, Tsat = 330.854°C

t (°C)	v $\left(\frac{m^3}{kg}\right)$	u $\left(\frac{kJ}{kg}\right)$	h $\left(\frac{kJ}{kg}\right)$	s $\left(\frac{kJ}{kg\,K}\right)$
Sat Vapour (g)	0.0128	2496.55	2662.70	5.434
10	0.0141	2561.45	2745.14	5.569
20	0.0152	2611.05	2808.70	5.672
30	0.0161	2652.81	2862.47	5.757
40	0.0170	2689.61	2910.04	5.832
50	0.0177	2722.98	2953.28	5.898
60	0.0184	2753.82	2993.31	5.959
70	0.0191	2782.72	3030.87	6.015
80	0.0197	2810.09	3066.47	6.068
90	0.0203	2836.23	3100.49	6.117
100	0.0209	2861.37	3133.20	6.164
110	0.0215	2885.68	3164.84	6.208
120	0.0220	2909.30	3195.56	6.251
130	0.0226	2932.33	3225.51	6.292
140	0.0231	2954.87	3254.79	6.332
150	0.0236	2976.98	3283.50	6.370
160	0.0241	2998.73	3311.71	6.407
170	0.0246	3020.16	3339.50	6.444
180	0.0250	3041.33	3366.90	6.479
190	0.0255	3062.25	3393.98	6.513
200	0.0260	3082.98	3420.76	6.547
210	0.0264	3103.53	3447.29	6.579
220	0.0269	3123.92	3473.59	6.611
230	0.0273	3144.19	3499.70	6.643
240	0.0278	3164.34	3525.63	6.674
250	0.0282	3184.39	3551.40	6.704
260	0.0287	3204.35	3577.04	6.734
270	0.0291	3224.25	3602.55	6.763
280	0.0295	3244.08	3627.96	6.792
290	0.0300	3263.87	3653.28	6.821
300	0.0304	3283.61	3678.51	6.849
310	0.0308	3303.32	3703.68	6.877
320	0.0312	3323.00	3728.78	6.904
330	0.0316	3342.67	3753.84	6.931
340	0.0320	3362.32	3778.84	6.958
350	0.0325	3381.96	3803.82	6.984
360	0.0329	3401.60	3828.76	7.010
370	0.0333	3421.25	3853.68	7.036
380	0.0337	3440.90	3878.58	7.061
390	0.0341	3460.55	3903.47	7.086
400	0.0345	3480.23	3928.34	7.111
410	0.0349	3499.92	3953.22	7.136
420	0.0353	3519.63	3978.09	7.160
430	0.0357	3539.36	4002.97	7.184
440	0.0361	3559.12	4027.86	7.208

t (°C)	v $\left(\frac{m^3}{kg}\right)$	u $\left(\frac{kJ}{kg}\right)$	h $\left(\frac{kJ}{kg}\right)$	s $\left(\frac{kJ}{kg\,K}\right)$
Sat Vapour (g)	0.0128	2496.55	2662.70	5.434
450	0.0365	3578.90	4052.76	7.232
460	0.0368	3598.71	4077.67	7.255
470	0.0372	3618.56	4102.59	7.279
480	0.0376	3638.44	4127.54	7.302
490	0.0380	3658.35	4152.50	7.325
500	0.0384	3678.30	4177.49	7.348
510	0.0388	3698.28	4202.50	7.370
520	0.0392	3718.31	4227.54	7.393
530	0.0396	3738.38	4252.61	7.415
540	0.0399	3758.48	4277.71	7.437
550	0.0403	3778.63	4302.83	7.459
560	0.0407	3798.83	4328.00	7.480
570	0.0411	3819.06	4353.19	7.502
580	0.0415	3839.35	4378.42	7.523
590	0.0418	3859.67	4403.69	7.545
600	0.0422	3880.05	4429.00	7.566
610	0.0426	3900.47	4454.34	7.587
620	0.0430	3920.93	4479.72	7.607
630	0.0434	3941.45	4505.14	7.628
640	0.0437	3962.01	4530.61	7.649
650	0.0441	3982.62	4556.11	7.669
660	0.0445	4003.29	4581.66	7.689
670	0.0449	4024.00	4607.24	7.710
680	0.0452	4044.75	4632.87	7.730
690	0.0456	4065.56	4658.55	7.750
700	0.0460	4086.42	4684.26	7.769
710	0.0464	4107.33	4710.02	7.789
720	0.0467	4128.29	4735.83	7.809
730	0.0471	4149.29	4761.68	7.828
740	0.0475	4170.35	4787.57	7.847
750	0.0479	4191.46	4813.51	7.867
760	0.0482	4212.62	4839.50	7.886
770	0.0486	4233.83	4865.53	7.905
780	0.0490	4255.09	4891.60	7.924
790	0.0493	4276.40	4917.72	7.942
800	0.0497	4297.75	4943.89	7.961
810	0.0501	4319.16	4970.10	7.980
820	0.0504	4340.62	4996.35	7.998
830	0.0508	4362.13	5022.66	8.017
840	0.0512	4383.69	5049.01	8.035
850	0.0515	4405.29	5075.40	8.053
860	0.0519	4426.95	5101.84	8.071
870	0.0523	4448.66	5128.32	8.089

t (°C)	v $\left(\frac{m^3}{kg}\right)$	u $\left(\frac{kJ}{kg}\right)$	h $\left(\frac{kJ}{kg}\right)$	s $\left(\frac{kJ}{kg\,K}\right)$
Sat Vapour (g)	0.0128	2496.55	2662.70	5.434
880	0.0527	4470.41	5154.85	8.107
890	0.0530	4492.21	5181.43	8.125
900	0.0534	4514.07	5208.05	8.143
910	0.0538	4535.97	5234.72	8.161
920	0.0541	4557.91	5261.43	8.178
930	0.0545	4579.91	5288.18	8.196
940	0.0548	4601.95	5314.98	8.213
950	0.0552	4624.04	5341.83	8.230
960	0.0556	4646.18	5368.72	8.248
970	0.0559	4668.37	5395.65	8.265
980	0.0563	4690.60	5422.63	8.282
990	0.0567	4712.88	5449.65	8.299
1000	0.0570	4735.20	5476.71	8.316
1010	0.0574	4757.57	5503.82	8.333
1020	0.0578	4779.98	5530.97	8.349
1030	0.0581	4802.45	5558.17	8.366
1040	0.0585	4824.95	5585.40	8.383
1050	0.0589	4847.50	5612.68	8.399
1060	0.0592	4870.10	5640.01	8.416
1070	0.0596	4892.74	5667.37	8.432
1080	0.0600	4915.42	5694.78	8.448
1090	0.0603	4938.15	5722.22	8.465
1100	0.0607	4960.92	5749.71	8.481
1110	0.0610	4983.73	5777.24	8.497
1120	0.0614	5006.58	5804.81	8.513
1130	0.0618	5029.48	5832.42	8.529
1140	0.0621	5052.42	5860.07	8.545
1150	0.0625	5075.40	5887.77	8.561
1160	0.0629	5098.43	5915.50	8.576
1170	0.0632	5121.49	5943.27	8.592
1180	0.0636	5144.60	5971.08	8.608
1190	0.0639	5167.74	5998.93	8.623
1200	0.0643	5190.93	6026.81	8.639
1210	0.0647	5214.15	6054.74	8.654
1220	0.0650	5237.42	6082.70	8.670
1230	0.0654	5260.73	6110.71	8.685
1240	0.0657	5284.07	6138.75	8.700
1250	0.0661	5307.45	6166.82	8.715
1260	0.0665	5330.88	6194.94	8.731
1270	0.0668	5354.34	6223.09	8.746
1280	0.0672	5377.83	6251.28	8.761
1290	0.0675	5401.37	6279.50	8.776
1300	0.0679	5424.94	6307.76	8.790

Table 3(52) Superheated Steam Properties (Pressure Base) Pressure = 140.0 bar and saturation temperature, Tsat = 336.666°C

t (°C)	v $\left(\frac{m^3}{kg}\right)$	u $\left(\frac{kJ}{kg}\right)$	h $\left(\frac{kJ}{kg}\right)$	s $\left(\frac{kJ}{kg\,K}\right)$
Sat Vapour (g)	0.0115	2477.12	2637.93	5.373
10	0.0129	2548.79	2728.76	5.521
20	0.0139	2601.83	2796.63	5.629
30	0.0148	2645.84	2853.24	5.719
40	0.0156	2684.29	2902.91	5.796
50	0.0163	2718.95	2947.80	5.864
60	0.0170	2750.84	2989.18	5.926
70	0.0177	2780.62	3027.87	5.984
80	0.0183	2808.75	3064.45	6.037
90	0.0188	2835.55	3099.33	6.087
100	0.0194	2861.28	3132.80	6.135
110	0.0199	2886.11	3165.12	6.180
120	0.0204	2910.20	3196.46	6.223
130	0.0210	2933.67	3226.97	6.265
140	0.0214	2956.60	3256.77	6.305
150	0.0219	2979.08	3285.96	6.343
160	0.0224	3001.16	3314.62	6.381
170	0.0229	3022.91	3342.82	6.417
180	0.0233	3044.37	3370.62	6.453
190	0.0237	3065.58	3398.06	6.487
200	0.0242	3086.57	3425.19	6.521
210	0.0246	3107.36	3452.04	6.554
220	0.0250	3128.00	3478.65	6.586
230	0.0255	3148.48	3505.05	6.618
240	0.0259	3168.85	3531.26	6.649
250	0.0263	3189.10	3557.30	6.679
260	0.0267	3209.26	3583.19	6.709
270	0.0271	3229.35	3608.95	6.739
280	0.0275	3249.36	3634.59	6.768
290	0.0279	3269.32	3660.14	6.796
300	0.0283	3289.23	3685.59	6.824
310	0.0287	3309.10	3710.96	6.852
320	0.0291	3328.94	3736.27	6.879
330	0.0295	3348.75	3761.52	6.906
340	0.0299	3368.55	3786.71	6.933
350	0.0303	3388.33	3811.87	6.959
360	0.0306	3408.11	3836.99	6.986
370	0.0310	3427.89	3862.08	7.011
380	0.0314	3447.67	3887.15	7.037
390	0.0318	3467.45	3912.20	7.062
400	0.0321	3487.25	3937.23	7.087
410	0.0325	3507.06	3962.26	7.111
420	0.0329	3526.89	3987.29	7.136
430	0.0333	3546.73	4012.31	7.160
440	0.0336	3566.60	4037.34	7.184

t (°C)	v $\left(\frac{m^3}{kg}\right)$	u $\left(\frac{kJ}{kg}\right)$	h $\left(\frac{kJ}{kg}\right)$	s $\left(\frac{kJ}{kg\,K}\right)$
Sat Vapour (g)	0.0115	2477.12	2637.93	5.373
450	0.0340	3586.50	4062.37	7.208
460	0.0344	3606.42	4087.42	7.231
470	0.0347	3626.36	4112.47	7.255
480	0.0351	3646.35	4137.55	7.278
490	0.0354	3666.36	4162.64	7.301
500	0.0358	3686.41	4187.75	7.323
510	0.0362	3706.49	4212.88	7.346
520	0.0365	3726.61	4238.04	7.368
530	0.0369	3746.77	4263.22	7.390
540	0.0372	3766.97	4288.44	7.412
550	0.0376	3787.21	4313.68	7.434
560	0.0380	3807.49	4338.95	7.456
570	0.0383	3827.81	4364.25	7.478
580	0.0387	3848.18	4389.59	7.499
590	0.0390	3868.59	4414.96	7.520
600	0.0394	3889.04	4440.36	7.541
610	0.0397	3909.54	4465.81	7.562
620	0.0401	3930.09	4491.28	7.583
630	0.0404	3950.68	4516.80	7.604
640	0.0408	3971.32	4542.36	7.624
650	0.0411	3992.01	4567.96	7.645
660	0.0415	4012.75	4593.59	7.665
670	0.0418	4033.53	4619.27	7.685
680	0.0422	4054.36	4644.99	7.705
690	0.0425	4075.24	4670.75	7.725
700	0.0429	4096.17	4696.55	7.745
710	0.0432	4117.15	4722.39	7.764
720	0.0436	4138.17	4748.28	7.784
730	0.0439	4159.25	4774.21	7.803
740	0.0443	4180.37	4800.18	7.823
750	0.0446	4201.55	4826.20	7.842
760	0.0450	4222.77	4852.26	7.861
770	0.0453	4244.04	4878.37	7.880
780	0.0457	4265.36	4904.52	7.899
790	0.0460	4286.73	4930.71	7.918
800	0.0463	4308.15	4956.95	7.936
810	0.0467	4329.62	4983.24	7.955
820	0.0470	4351.14	5009.56	7.973
830	0.0474	4372.71	5035.94	7.992
840	0.0477	4394.32	5062.35	8.010
850	0.0481	4415.99	5088.82	8.028
860	0.0484	4437.70	5115.32	8.046
870	0.0487	4459.46	5141.87	8.064

t (°C)	v $\left(\frac{m^3}{kg}\right)$	u $\left(\frac{kJ}{kg}\right)$	h $\left(\frac{kJ}{kg}\right)$	s $\left(\frac{kJ}{kg\,K}\right)$
Sat Vapour (g)	0.0115	2477.12	2637.93	5.373
880	0.0491	4481.27	5168.47	8.082
890	0.0494	4503.13	5195.11	8.100
900	0.0498	4525.03	5221.79	8.118
910	0.0501	4546.98	5248.52	8.135
920	0.0505	4568.98	5275.29	8.153
930	0.0508	4591.03	5302.11	8.170
940	0.0511	4613.13	5328.97	8.188
950	0.0515	4635.27	5355.88	8.205
960	0.0518	4657.46	5382.82	8.222
970	0.0522	4679.69	5409.81	8.239
980	0.0525	4701.97	5436.85	8.257
990	0.0528	4724.30	5463.93	8.274
1000	0.0532	4746.67	5491.05	8.290
1010	0.0535	4769.08	5518.21	8.307
1020	0.0538	4791.55	5545.41	8.324
1030	0.0542	4814.05	5572.66	8.341
1040	0.0545	4836.60	5599.95	8.357
1050	0.0549	4859.20	5627.28	8.374
1060	0.0552	4881.84	5654.66	8.390
1070	0.0555	4904.52	5682.07	8.407
1080	0.0559	4927.25	5709.53	8.423
1090	0.0562	4950.02	5737.02	8.439
1100	0.0566	4972.83	5764.56	8.455
1110	0.0569	4995.68	5792.14	8.471
1120	0.0572	5018.58	5819.76	8.487
1130	0.0576	5041.52	5847.42	8.503
1140	0.0579	5064.50	5875.11	8.519
1150	0.0582	5087.52	5902.85	8.535
1160	0.0586	5110.58	5930.63	8.551
1170	0.0589	5133.69	5958.44	8.566
1180	0.0592	5156.83	5986.30	8.582
1190	0.0596	5180.02	6014.19	8.598
1200	0.0599	5203.24	6042.12	8.613
1210	0.0603	5226.50	6070.09	8.628
1220	0.0606	5249.81	6098.10	8.644
1230	0.0609	5273.15	6126.14	8.659
1240	0.0613	5296.53	6154.22	8.674
1250	0.0616	5319.95	6182.34	8.689
1260	0.0619	5343.41	6210.50	8.705
1270	0.0623	5366.90	6238.69	8.720
1280	0.0626	5390.43	6266.92	8.735
1290	0.0629	5414.00	6295.18	8.749
1300	0.0633	5437.61	6323.48	8.764

Table 3(53) Superheated Steam Properties (Pressure Base) Pressure = 150.0 bar and saturation temperature, Tsat = 342.155°C

t (°C)	v $\left(\frac{m^3}{kg}\right)$	u $\left(\frac{kJ}{kg}\right)$	h $\left(\frac{kJ}{kg}\right)$	s $\left(\frac{kJ}{kg\,K}\right)$
Sat Vapour (g)	0.0103	2455.73	2610.84	5.311
10	0.0117	2535.34	2711.47	5.473
20	0.0128	2592.08	2783.95	5.588
30	0.0137	2638.44	2843.52	5.681
40	0.0145	2678.58	2895.33	5.761
50	0.0152	2714.55	2941.88	5.831
60	0.0158	2747.49	2984.61	5.895
70	0.0164	2778.15	3024.44	5.954
80	0.0170	2807.03	3061.99	6.008
90	0.0175	2834.49	3097.71	6.059
100	0.0181	2860.80	3131.94	6.107
110	0.0186	2886.15	3164.94	6.153
120	0.0191	2910.71	3196.89	6.197
130	0.0196	2934.60	3227.95	6.239
140	0.0200	2957.93	3258.27	6.279
150	0.0205	2980.76	3287.93	6.318
160	0.0209	3003.18	3317.02	6.356
170	0.0214	3025.24	3345.63	6.393
180	0.0218	3046.99	3373.81	6.428
190	0.0222	3068.47	3401.60	6.463
200	0.0226	3089.72	3429.07	6.497
210	0.0230	3110.76	3456.24	6.530
220	0.0234	3131.62	3483.16	6.563
230	0.0238	3152.33	3509.84	6.594
240	0.0242	3172.90	3536.32	6.625
250	0.0246	3193.35	3562.63	6.656
260	0.0250	3213.71	3588.77	6.686
270	0.0254	3233.97	3614.76	6.716
280	0.0258	3254.16	3640.64	6.745
290	0.0261	3274.29	3666.40	6.773
300	0.0265	3294.36	3692.06	6.801
310	0.0269	3314.39	3717.64	6.829
320	0.0273	3334.38	3743.15	6.857
330	0.0276	3354.34	3768.58	6.884
340	0.0280	3374.28	3793.96	6.910
350	0.0283	3394.20	3819.30	6.937
360	0.0287	3414.11	3844.59	6.963
370	0.0291	3434.02	3869.85	6.989
380	0.0294	3453.93	3895.08	7.014
390	0.0298	3473.83	3920.28	7.039
400	0.0301	3493.75	3945.47	7.064
410	0.0305	3513.68	3970.65	7.089
420	0.0308	3533.62	3995.82	7.113
430	0.0312	3553.57	4020.98	7.138
440	0.0315	3573.55	4046.15	7.161

t (°C)	v $\left(\frac{m^3}{kg}\right)$	u $\left(\frac{kJ}{kg}\right)$	h $\left(\frac{kJ}{kg}\right)$	s $\left(\frac{kJ}{kg\,K}\right)$
Sat Vapour (g)	0.0103	2455.73	2610.84	5.311
450	0.0319	3593.55	4071.32	7.185
460	0.0322	3613.57	4096.50	7.209
470	0.0325	3633.62	4121.68	7.232
480	0.0329	3653.70	4146.88	7.255
490	0.0332	3673.81	4172.09	7.278
500	0.0336	3693.95	4197.32	7.301
510	0.0339	3714.13	4222.57	7.323
520	0.0342	3734.34	4247.85	7.346
530	0.0346	3754.59	4273.14	7.368
540	0.0349	3774.88	4298.46	7.390
550	0.0352	3795.20	4323.81	7.412
560	0.0356	3815.57	4349.19	7.433
570	0.0359	3835.98	4374.59	7.455
580	0.0362	3856.42	4400.03	7.476
590	0.0366	3876.91	4425.50	7.498
600	0.0369	3897.45	4451.01	7.519
610	0.0372	3918.03	4476.54	7.540
620	0.0376	3938.65	4502.12	7.560
630	0.0379	3959.32	4527.73	7.581
640	0.0382	3980.03	4553.38	7.602
650	0.0386	4000.79	4579.06	7.622
660	0.0389	4021.60	4604.79	7.642
670	0.0392	4042.45	4630.55	7.662
680	0.0395	4063.36	4656.35	7.682
690	0.0399	4084.30	4682.19	7.702
700	0.0402	4105.30	4708.08	7.722
710	0.0405	4126.34	4734.00	7.742
720	0.0408	4147.43	4759.97	7.761
730	0.0412	4168.57	4785.98	7.780
740	0.0415	4189.76	4812.03	7.800
750	0.0418	4211.00	4838.13	7.819
760	0.0421	4232.28	4864.26	7.838
770	0.0425	4253.62	4890.44	7.857
780	0.0428	4275.00	4916.66	7.876
790	0.0431	4296.43	4942.93	7.895
800	0.0434	4317.91	4969.24	7.913
810	0.0437	4339.43	4995.59	7.932
820	0.0441	4361.01	5021.99	7.950
830	0.0444	4382.63	5048.43	7.969
840	0.0447	4404.30	5074.91	7.987
850	0.0450	4426.02	5101.44	8.005
860	0.0453	4447.79	5128.01	8.023
870	0.0457	4469.60	5154.62	8.041

t (°C)	v $\left(\frac{m^3}{kg}\right)$	u $\left(\frac{kJ}{kg}\right)$	h $\left(\frac{kJ}{kg}\right)$	s $\left(\frac{kJ}{kg\,K}\right)$
Sat Vapour (g)	0.0103	2455.73	2610.84	5.311
880	0.0460	4491.47	5181.28	8.059
890	0.0463	4513.38	5207.98	8.077
900	0.0466	4535.33	5234.73	8.094
910	0.0469	4557.34	5261.52	8.112
920	0.0473	4579.39	5288.35	8.130
930	0.0476	4601.48	5315.22	8.147
940	0.0479	4623.63	5342.14	8.164
950	0.0482	4645.82	5369.10	8.182
960	0.0485	4668.05	5396.11	8.199
970	0.0489	4690.33	5423.15	8.216
980	0.0492	4712.66	5450.24	8.233
990	0.0495	4735.03	5477.38	8.250
1000	0.0498	4757.45	5504.55	8.267
1010	0.0501	4779.91	5531.76	8.284
1020	0.0504	4802.42	5559.02	8.300
1030	0.0508	4824.97	5586.32	8.317
1040	0.0511	4847.56	5613.66	8.334
1050	0.0514	4870.20	5641.04	8.350
1060	0.0517	4892.88	5668.47	8.367
1070	0.0520	4915.61	5695.93	8.383
1080	0.0523	4938.37	5723.43	8.399
1090	0.0527	4961.19	5750.98	8.415
1100	0.0530	4984.04	5778.56	8.431
1110	0.0533	5006.93	5806.19	8.448
1120	0.0536	5029.87	5833.85	8.464
1130	0.0539	5052.85	5861.55	8.479
1140	0.0542	5075.87	5889.30	8.495
1150	0.0545	5098.93	5917.08	8.511
1160	0.0549	5122.03	5944.90	8.527
1170	0.0552	5145.17	5972.76	8.542
1180	0.0555	5168.35	6000.66	8.558
1190	0.0558	5191.57	6028.59	8.574
1200	0.0561	5214.83	6056.56	8.589
1210	0.0564	5238.13	6084.57	8.604
1220	0.0567	5261.47	6112.62	8.620
1230	0.0571	5284.84	6140.71	8.635
1240	0.0574	5308.26	6168.83	8.650
1250	0.0577	5331.71	6196.98	8.665
1260	0.0580	5355.20	6225.18	8.680
1270	0.0583	5378.73	6253.41	8.695
1280	0.0586	5402.30	6281.67	8.710
1290	0.0589	5425.90	6309.98	8.725
1300	0.0593	5449.54	6338.31	8.740

Table 3(54) Superheated Steam Properties (Pressure Base) Pressure = 160.0 bar and saturation temperature, Tsat = 347.356°C

t (°C)	v $\left(\frac{m^3}{kg}\right)$	u $\left(\frac{kJ}{kg}\right)$	h $\left(\frac{kJ}{kg}\right)$	s $\left(\frac{kJ}{kg\,K}\right)$	t (°C)	v $\left(\frac{m^3}{kg}\right)$	u $\left(\frac{kJ}{kg}\right)$	h $\left(\frac{kJ}{kg}\right)$	s $\left(\frac{kJ}{kg\,K}\right)$	t (°C)	v $\left(\frac{m^3}{kg}\right)$	u $\left(\frac{kJ}{kg}\right)$	h $\left(\frac{kJ}{kg}\right)$	s $\left(\frac{kJ}{kg\,K}\right)$
Sat Vapour (g)	*0.0093*	*2432.00*	*2580.99*	*5.247*	*Sat Vapour (g)*	*0.0093*	*2432.00*	*2580.99*	*5.247*	*Sat Vapour (g)*	*0.0093*	*2432.00*	*2580.99*	*5.247*
10	0.0108	2521.12	2693.29	5.426	450	0.0300	3600.12	4079.67	7.164	880	0.0433	4501.07	5193.38	8.037
20	0.0118	2581.85	2770.72	5.548	460	0.0303	3620.24	4104.97	7.188	890	0.0436	4523.03	5220.14	8.055
30	0.0127	2630.64	2833.35	5.645	470	0.0306	3640.39	4130.28	7.211	900	0.0439	4545.04	5246.95	8.073
40	0.0134	2672.52	2887.34	5.728	480	0.0309	3660.57	4155.60	7.234	910	0.0442	4567.09	5273.79	8.090
50	0.0141	2709.80	2935.57	5.800	490	0.0313	3680.77	4180.94	7.257	920	0.0445	4589.19	5300.68	8.108
60	0.0147	2743.81	2979.65	5.865	500	0.0316	3701.00	4206.28	7.280	930	0.0448	4611.33	5327.61	8.125
70	0.0153	2775.35	3020.61	5.925	510	0.0319	3721.27	4231.65	7.302	940	0.0451	4633.52	5354.59	8.143
80	0.0159	2804.99	3059.13	5.980	520	0.0322	3741.57	4257.03	7.325	950	0.0454	4655.76	5381.60	8.160
90	0.0164	2833.10	3095.70	6.032	530	0.0325	3761.91	4282.44	7.347	960	0.0457	4678.04	5408.66	8.177
100	0.0169	2859.98	3130.68	6.081	540	0.0328	3782.28	4307.86	7.369	970	0.0460	4700.37	5435.76	8.194
110	0.0174	2885.85	3164.34	6.128	550	0.0332	3802.69	4333.32	7.391	980	0.0463	4722.74	5462.90	8.211
120	0.0179	2910.87	3196.89	6.172	560	0.0335	3823.13	4358.79	7.412	990	0.0466	4745.16	5490.09	8.228
130	0.0183	2935.18	3228.51	6.214	570	0.0338	3843.62	4384.30	7.434	1000	0.0469	4767.62	5517.31	8.245
140	0.0188	2958.89	3259.32	6.255	580	0.0341	3864.15	4409.84	7.455	1010	0.0472	4790.12	5544.58	8.262
150	0.0192	2982.08	3289.45	6.294	590	0.0344	3884.72	4435.40	7.476	1020	0.0475	4812.67	5571.89	8.278
160	0.0196	3004.83	3318.97	6.333	600	0.0347	3905.33	4461.00	7.497	1030	0.0477	4835.26	5599.23	8.295
170	0.0200	3027.20	3347.98	6.369	610	0.0350	3925.98	4486.63	7.518	1040	0.0480	4857.90	5626.62	8.312
180	0.0205	3049.24	3376.53	6.405	620	0.0354	3946.68	4512.30	7.539	1050	0.0483	4880.58	5654.05	8.328
190	0.0209	3070.98	3404.68	6.440	630	0.0357	3967.42	4538.00	7.560	1060	0.0486	4903.30	5681.52	8.344
200	0.0213	3092.48	3432.48	6.474	640	0.0360	3988.21	4563.74	7.580	1070	0.0489	4926.07	5709.04	8.361
210	0.0216	3113.76	3459.96	6.508	650	0.0363	4009.04	4589.51	7.601	1080	0.0492	4948.88	5736.59	8.377
220	0.0220	3134.85	3487.17	6.540	660	0.0366	4029.91	4615.32	7.621	1090	0.0495	4971.72	5764.18	8.393
230	0.0224	3155.77	3514.14	6.572	670	0.0369	4050.84	4641.16	7.641	1100	0.0498	4994.62	5791.81	8.409
240	0.0228	3176.54	3540.89	6.603	680	0.0372	4071.80	4667.05	7.661	1110	0.0501	5017.55	5819.47	8.425
250	0.0231	3197.19	3567.44	6.634	690	0.0375	4092.82	4692.97	7.681	1120	0.0504	5040.52	5847.18	8.441
260	0.0235	3217.73	3593.83	6.664	700	0.0378	4113.88	4718.94	7.701	1130	0.0507	5063.54	5874.93	8.457
270	0.0239	3238.18	3620.06	6.694	710	0.0381	4134.99	4744.94	7.720	1140	0.0510	5086.60	5902.72	8.473
280	0.0242	3258.54	3646.16	6.723	720	0.0384	4156.14	4770.98	7.740	1150	0.0513	5109.69	5930.54	8.489
290	0.0246	3278.83	3672.13	6.752	730	0.0387	4177.34	4797.07	7.759	1160	0.0516	5132.83	5958.40	8.504
300	0.0249	3299.06	3698.01	6.780	740	0.0390	4198.59	4823.20	7.778	1170	0.0519	5156.01	5986.30	8.520
310	0.0253	3319.24	3723.78	6.808	750	0.0393	4219.89	4849.36	7.797	1180	0.0522	5179.22	6014.24	8.536
320	0.0256	3339.38	3749.48	6.835	760	0.0396	4241.23	4875.57	7.817	1190	0.0525	5202.48	6042.22	8.551
330	0.0260	3359.48	3775.10	6.862	770	0.0400	4262.63	4901.82	7.835	1200	0.0528	5225.77	6070.23	8.567
340	0.0263	3379.56	3800.66	6.889	780	0.0403	4284.07	4928.11	7.854	1210	0.0531	5249.11	6098.28	8.582
350	0.0267	3399.62	3826.17	6.916	790	0.0406	4305.55	4954.45	7.873	1220	0.0534	5272.48	6126.37	8.597
360	0.0270	3419.66	3851.63	6.942	800	0.0409	4327.09	4980.83	7.892	1230	0.0537	5295.89	6154.49	8.612
370	0.0273	3439.69	3877.05	6.967	810	0.0412	4348.67	5007.25	7.910	1240	0.0540	5319.34	6182.65	8.628
380	0.0277	3459.72	3902.44	6.993	820	0.0415	4370.30	5033.71	7.929	1250	0.0543	5342.82	6210.85	8.643
390	0.0280	3479.75	3927.80	7.018	830	0.0418	4391.98	5060.21	7.947	1260	0.0545	5366.35	6239.08	8.658
400	0.0283	3499.78	3953.13	7.043	840	0.0421	4413.70	5086.76	7.965	1270	0.0548	5389.91	6267.34	8.673
410	0.0287	3519.82	3978.46	7.068	850	0.0424	4435.47	5113.35	7.983	1280	0.0551	5413.51	6295.65	8.688
420	0.0290	3539.87	4003.77	7.092	860	0.0427	4457.29	5139.98	8.001	1290	0.0554	5437.14	6323.98	8.703
430	0.0293	3559.93	4029.07	7.116	870	0.0430	4479.16	5166.66	8.019	1300	0.0557	5460.81	6352.36	8.717
440	0.0296	3580.02	4054.37	7.140										

Table 3(55) Superheated Steam Properties (Pressure Base) Pressure = 170.0 bar and saturation temperature, Tsat = 352.294°C

t (°C)	v $\left(\frac{m^3}{kg}\right)$	u $\left(\frac{kJ}{kg}\right)$	h $\left(\frac{kJ}{kg}\right)$	s $\left(\frac{kJ}{kg\,K}\right)$	t (°C)	v $\left(\frac{m^3}{kg}\right)$	u $\left(\frac{kJ}{kg}\right)$	h $\left(\frac{kJ}{kg}\right)$	s $\left(\frac{kJ}{kg\,K}\right)$	t (°C)	v $\left(\frac{m^3}{kg}\right)$	u $\left(\frac{kJ}{kg}\right)$	h $\left(\frac{kJ}{kg}\right)$	s $\left(\frac{kJ}{kg\,K}\right)$
Sat Vapour (g)	0.0084	2405.38	2547.74	5.179	Sat Vapour (g)	0.0084	2405.38	2547.74	5.179	Sat Vapour (g)	0.0084	2405.38	2547.74	5.179
10	0.0099	2506.13	2674.23	5.380	450	0.0283	3606.25	4087.48	7.144	880	0.0409	4510.14	5204.83	8.017
20	0.0109	2571.14	2756.96	5.509	460	0.0286	3626.47	4112.91	7.168	890	0.0411	4532.15	5231.65	8.035
30	0.0118	2622.47	2822.76	5.610	470	0.0289	3646.71	4138.34	7.191	900	0.0414	4554.20	5258.51	8.052
40	0.0125	2666.12	2878.97	5.695	480	0.0292	3666.98	4163.78	7.214	910	0.0417	4576.30	5285.41	8.070
50	0.0132	2704.75	2928.90	5.770	490	0.0295	3687.28	4189.23	7.237	920	0.0420	4598.45	5312.35	8.087
60	0.0138	2739.82	2974.35	5.837	500	0.0298	3707.60	4214.69	7.260	930	0.0423	4620.64	5339.34	8.105
70	0.0144	2772.25	3016.44	5.898	510	0.0301	3727.95	4240.16	7.282	940	0.0426	4642.88	5366.37	8.122
80	0.0149	2802.64	3055.92	5.954	520	0.0304	3748.34	4265.65	7.305	950	0.0428	4665.16	5393.44	8.139
90	0.0154	2831.40	3093.33	6.007	530	0.0307	3768.76	4291.16	7.327	960	0.0431	4687.48	5420.55	8.157
100	0.0159	2858.86	3129.05	6.056	540	0.0310	3789.21	4316.69	7.349	970	0.0434	4709.85	5447.70	8.174
110	0.0164	2885.23	3163.37	6.103	550	0.0313	3809.70	4342.25	7.371	980	0.0437	4732.27	5474.89	8.191
120	0.0168	2910.71	3196.52	6.148	560	0.0316	3830.23	4367.83	7.392	990	0.0440	4754.73	5502.13	8.207
130	0.0172	2935.44	3228.67	6.191	570	0.0319	3850.80	4393.43	7.414	1000	0.0442	4777.23	5529.40	8.224
140	0.0177	2959.53	3259.98	6.232	580	0.0322	3871.40	4419.06	7.435	1010	0.0445	4799.78	5556.72	8.241
150	0.0181	2983.07	3290.57	6.272	590	0.0325	3892.05	4444.72	7.457	1020	0.0448	4822.37	5584.07	8.258
160	0.0185	3006.15	3320.52	6.310	600	0.0328	3912.73	4470.42	7.478	1030	0.0451	4845.00	5611.47	8.274
170	0.0189	3028.81	3349.92	6.347	610	0.0331	3933.46	4496.14	7.498	1040	0.0454	4867.68	5638.91	8.291
180	0.0193	3051.13	3378.83	6.383	620	0.0334	3954.23	4521.89	7.519	1050	0.0456	4890.39	5666.38	8.307
190	0.0197	3073.14	3407.33	6.419	630	0.0337	3975.04	4547.68	7.540	1060	0.0459	4913.16	5693.90	8.324
200	0.0200	3094.89	3435.45	6.453	640	0.0340	3995.89	4573.50	7.560	1070	0.0462	4935.96	5721.46	8.340
210	0.0204	3116.40	3463.24	6.486	650	0.0343	4016.79	4599.36	7.581	1080	0.0465	4958.81	5749.05	8.356
220	0.0208	3137.71	3490.75	6.519	660	0.0346	4037.73	4625.25	7.601	1090	0.0468	4981.69	5776.69	8.372
230	0.0211	3158.84	3517.99	6.551	670	0.0349	4058.72	4651.18	7.621	1100	0.0470	5004.62	5804.36	8.388
240	0.0215	3179.82	3545.00	6.583	680	0.0351	4079.75	4677.14	7.641	1110	0.0473	5027.59	5832.07	8.404
250	0.0218	3200.66	3571.80	6.613	690	0.0354	4100.83	4703.14	7.661	1120	0.0476	5050.60	5859.82	8.420
260	0.0222	3221.38	3598.43	6.644	700	0.0357	4121.96	4729.18	7.681	1130	0.0479	5073.65	5887.61	8.436
270	0.0225	3242.00	3624.89	6.673	710	0.0360	4143.13	4755.26	7.700	1140	0.0482	5096.75	5915.44	8.452
280	0.0229	3262.53	3651.20	6.702	720	0.0363	4164.34	4781.38	7.720	1150	0.0484	5119.88	5943.31	8.468
290	0.0232	3282.98	3677.39	6.731	730	0.0366	4185.60	4807.54	7.739	1160	0.0487	5143.05	5971.21	8.484
300	0.0235	3303.37	3703.46	6.760	740	0.0369	4206.91	4833.74	7.758	1170	0.0490	5166.26	5999.15	8.499
310	0.0239	3323.70	3729.44	6.788	750	0.0372	4228.27	4859.98	7.777	1180	0.0493	5189.51	6027.13	8.515
320	0.0242	3343.98	3755.32	6.815	760	0.0374	4249.67	4886.26	7.796	1190	0.0495	5212.80	6055.14	8.530
330	0.0245	3364.22	3781.12	6.842	770	0.0377	4271.12	4912.57	7.815	1200	0.0498	5236.13	6083.20	8.546
340	0.0248	3384.44	3806.86	6.869	780	0.0380	4292.62	4938.94	7.834	1210	0.0501	5259.50	6111.28	8.561
350	0.0252	3404.62	3832.53	6.895	790	0.0383	4314.16	4965.34	7.853	1220	0.0504	5282.90	6139.41	8.576
360	0.0255	3424.79	3858.16	6.922	800	0.0386	4335.75	4991.78	7.872	1230	0.0507	5306.34	6167.57	8.591
370	0.0258	3444.95	3883.74	6.947	810	0.0389	4357.38	5018.26	7.890	1240	0.0509	5329.83	6195.77	8.607
380	0.0261	3465.09	3909.28	6.973	820	0.0392	4379.07	5044.79	7.908	1250	0.0512	5353.34	6224.00	8.622
390	0.0264	3485.24	3934.79	6.998	830	0.0394	4400.80	5071.36	7.927	1260	0.0515	5376.90	6252.26	8.637
400	0.0268	3505.38	3960.27	7.023	840	0.0397	4422.57	5097.97	7.945	1270	0.0518	5400.49	6280.57	8.652
410	0.0271	3525.53	3985.74	7.048	850	0.0400	4444.40	5124.62	7.963	1280	0.0520	5424.12	6308.90	8.667
420	0.0274	3545.69	4011.19	7.072	860	0.0403	4466.26	5151.31	7.981	1290	0.0523	5447.78	6337.28	8.681
430	0.0277	3565.86	4036.62	7.097	870	0.0406	4488.18	5178.05	7.999	1300	0.0526	5471.48	6365.68	8.696
440	0.0280	3586.04	4062.05	7.121										

Table 3(56) Superheated Steam Properties (Pressure Base) Pressure = 180.0 bar and saturation temperature, Tsat = 356.993°C

t (°C)	$v\left(\frac{m^3}{kg}\right)$	$u\left(\frac{kJ}{kg}\right)$	$h\left(\frac{kJ}{kg}\right)$	$s\left(\frac{kJ}{kg\,K}\right)$
Sat Vapour (g)	0.0075	2374.95	2510.03	5.106
10	0.0091	2490.37	2654.32	5.334
20	0.0102	2559.99	2742.72	5.471
30	0.0110	2613.97	2811.79	5.576
40	0.0117	2659.42	2870.27	5.664
50	0.0124	2699.40	2921.91	5.741
60	0.0130	2735.56	2968.72	5.809
70	0.0135	2768.87	3011.94	5.871
80	0.0140	2800.01	3052.39	5.929
90	0.0145	2829.42	3090.63	5.982
100	0.0150	2857.45	3127.08	6.032
110	0.0154	2884.33	3162.06	6.080
120	0.0159	2910.27	3195.79	6.125
130	0.0163	2935.40	3228.49	6.169
140	0.0167	2959.87	3260.29	6.210
150	0.0171	2983.76	3291.32	6.250
160	0.0175	3007.15	3321.69	6.289
170	0.0179	3030.11	3351.47	6.326
180	0.0182	3052.71	3380.75	6.363
190	0.0186	3074.98	3409.59	6.398
200	0.0189	3096.97	3438.03	6.432
210	0.0193	3118.71	3466.13	6.466
220	0.0196	3140.24	3493.91	6.499
230	0.0200	3161.58	3521.42	6.531
240	0.0203	3182.75	3548.69	6.563
250	0.0207	3203.78	3575.74	6.594
260	0.0210	3224.68	3602.60	6.624
270	0.0213	3245.47	3629.29	6.654
280	0.0216	3266.16	3655.82	6.683
290	0.0220	3286.77	3682.21	6.712
300	0.0223	3307.31	3708.48	6.740
310	0.0226	3327.79	3734.64	6.768
320	0.0229	3348.21	3760.71	6.796
330	0.0232	3368.59	3786.69	6.823
340	0.0235	3388.94	3812.60	6.850
350	0.0238	3409.25	3838.44	6.876
360	0.0241	3429.55	3864.22	6.903
370	0.0245	3449.82	3889.96	6.928
380	0.0248	3470.09	3915.65	6.954
390	0.0251	3490.35	3941.31	6.979
400	0.0254	3510.60	3966.93	7.004
410	0.0256	3530.86	3992.54	7.029
420	0.0259	3551.12	4018.12	7.053
430	0.0262	3571.39	4043.69	7.078
440	0.0265	3591.68	4069.25	7.102

t (°C)	$v\left(\frac{m^3}{kg}\right)$	$u\left(\frac{kJ}{kg}\right)$	$h\left(\frac{kJ}{kg}\right)$	$s\left(\frac{kJ}{kg\,K}\right)$
Sat Vapour (g)	0.0075	2374.95	2510.03	5.106
450	0.0268	3611.97	4094.80	7.125
460	0.0271	3632.29	4120.35	7.149
470	0.0274	3652.63	4145.90	7.172
480	0.0277	3672.99	4171.45	7.195
490	0.0280	3693.37	4197.01	7.218
500	0.0283	3713.78	4222.58	7.241
510	0.0286	3734.22	4248.16	7.264
520	0.0288	3754.69	4273.76	7.286
530	0.0291	3775.19	4299.38	7.308
540	0.0294	3795.73	4325.01	7.330
550	0.0297	3816.30	4350.66	7.352
560	0.0300	3836.90	4376.34	7.374
570	0.0303	3857.54	4402.04	7.395
580	0.0305	3878.22	4427.76	7.417
590	0.0308	3898.94	4453.51	7.438
600	0.0311	3919.70	4479.29	7.459
610	0.0314	3940.49	4505.10	7.480
620	0.0316	3961.33	4530.94	7.500
630	0.0319	3982.21	4556.82	7.521
640	0.0322	4003.13	4582.72	7.542
650	0.0325	4024.10	4608.66	7.562
660	0.0328	4045.11	4634.63	7.582
670	0.0330	4066.16	4660.64	7.602
680	0.0333	4087.25	4686.68	7.622
690	0.0336	4108.39	4712.76	7.642
700	0.0339	4129.58	4738.87	7.662
710	0.0341	4150.81	4765.03	7.681
720	0.0344	4172.08	4791.22	7.701
730	0.0347	4193.40	4817.45	7.720
740	0.0349	4214.77	4843.72	7.739
750	0.0352	4236.18	4870.02	7.758
760	0.0355	4257.64	4896.37	7.777
770	0.0358	4279.14	4922.76	7.796
780	0.0360	4300.69	4949.18	7.815
790	0.0363	4322.29	4975.65	7.834
800	0.0366	4343.93	5002.16	7.852
810	0.0368	4365.62	5028.71	7.871
820	0.0371	4387.35	5055.29	7.889
830	0.0374	4409.13	5081.92	7.908
840	0.0376	4430.96	5108.59	7.926
850	0.0379	4452.83	5135.30	7.944
860	0.0382	4474.75	5162.06	7.962
870	0.0385	4496.71	5188.85	7.980

t (°C)	$v\left(\frac{m^3}{kg}\right)$	$u\left(\frac{kJ}{kg}\right)$	$h\left(\frac{kJ}{kg}\right)$	$s\left(\frac{kJ}{kg\,K}\right)$
Sat Vapour (g)	0.0075	2374.95	2510.03	5.106
880	0.0387	4518.72	5215.68	7.998
890	0.0390	4540.78	5242.56	8.015
900	0.0393	4562.88	5269.47	8.033
910	0.0395	4585.02	5296.43	8.051
920	0.0398	4607.21	5323.43	8.068
930	0.0401	4629.45	5350.47	8.086
940	0.0403	4651.73	5377.55	8.103
950	0.0406	4674.05	5404.67	8.120
960	0.0409	4696.42	5431.83	8.137
970	0.0411	4718.83	5459.03	8.154
980	0.0414	4741.29	5486.28	8.171
990	0.0417	4763.79	5513.56	8.188
1000	0.0419	4786.33	5540.88	8.205
1010	0.0422	4808.92	5568.25	8.222
1020	0.0425	4831.55	5595.65	8.238
1030	0.0427	4854.22	5623.09	8.255
1040	0.0430	4876.94	5650.57	8.271
1050	0.0432	4899.69	5678.10	8.288
1060	0.0435	4922.49	5705.66	8.304
1070	0.0438	4945.34	5733.26	8.320
1080	0.0440	4968.22	5760.90	8.337
1090	0.0443	4991.14	5788.57	8.353
1100	0.0446	5014.11	5816.29	8.369
1110	0.0448	5037.11	5844.04	8.385
1120	0.0451	5060.16	5871.84	8.401
1130	0.0454	5083.25	5899.67	8.417
1140	0.0456	5106.37	5927.53	8.432
1150	0.0459	5129.54	5955.44	8.448
1160	0.0461	5152.74	5983.38	8.464
1170	0.0464	5175.99	6011.36	8.479
1180	0.0467	5199.27	6039.38	8.495
1190	0.0469	5222.59	6067.43	8.510
1200	0.0472	5245.95	6095.52	8.526
1210	0.0475	5269.35	6123.64	8.541
1220	0.0477	5292.79	6151.81	8.556
1230	0.0480	5316.26	6180.00	8.572
1240	0.0482	5339.77	6208.24	8.587
1250	0.0485	5363.32	6236.50	8.602
1260	0.0488	5386.91	6264.80	8.617
1270	0.0490	5410.53	6293.14	8.632
1280	0.0493	5434.19	6321.51	8.647
1290	0.0496	5457.88	6349.92	8.661
1300	0.0498	5481.61	6378.36	8.676

Table 3(57) Superheated Steam Properties (Pressure Base) Pressure = 190.0 bar and saturation temperature, Tsat = 361.473°C

t (°C)	v $\left(\frac{m^3}{kg}\right)$	u $\left(\frac{kJ}{kg}\right)$	h $\left(\frac{kJ}{kg}\right)$	s $\left(\frac{kJ}{kg\,K}\right)$
Sat Vapour (g)	0.0067	2339.16	2466.03	5.026
10	0.0084	2473.84	2633.56	5.288
20	0.0095	2548.42	2728.02	5.433
30	0.0103	2605.13	2800.48	5.543
40	0.0110	2652.43	2861.26	5.634
50	0.0116	2693.79	2914.62	5.713
60	0.0122	2731.03	2962.80	5.782
70	0.0127	2765.23	3007.15	5.846
80	0.0132	2797.12	3048.56	5.904
90	0.0137	2827.18	3087.63	5.958
100	0.0142	2855.77	3124.81	6.009
110	0.0146	2883.16	3160.43	6.058
120	0.0150	2909.54	3194.75	6.103
130	0.0154	2935.09	3227.97	6.147
140	0.0158	2959.93	3260.25	6.189
150	0.0162	2984.16	3291.73	6.229
160	0.0166	3007.87	3322.51	6.268
170	0.0169	3031.13	3352.68	6.306
180	0.0173	3053.99	3382.32	6.343
190	0.0176	3076.52	3411.49	6.378
200	0.0180	3098.75	3440.25	6.413
210	0.0183	3120.72	3468.64	6.447
220	0.0186	3142.46	3496.70	6.480
230	0.0190	3164.01	3524.48	6.512
240	0.0193	3185.37	3552.00	6.544
250	0.0196	3206.58	3579.30	6.575
260	0.0199	3227.66	3606.38	6.605
270	0.0202	3248.62	3633.29	6.635
280	0.0206	3269.48	3660.03	6.665
290	0.0209	3290.24	3686.63	6.693
300	0.0212	3310.93	3713.09	6.722
310	0.0215	3331.55	3739.44	6.750
320	0.0218	3352.11	3765.69	6.778
330	0.0221	3372.63	3791.84	6.805
340	0.0224	3393.10	3817.92	6.832
350	0.0227	3413.54	3843.92	6.858
360	0.0229	3433.95	3869.86	6.885
370	0.0232	3454.35	3895.75	6.910
380	0.0235	3474.73	3921.59	6.936
390	0.0238	3495.10	3947.39	6.961
400	0.0241	3515.46	3973.16	6.986
410	0.0244	3535.82	3998.89	7.011
420	0.0247	3556.19	4024.61	7.036
430	0.0249	3576.56	4050.31	7.060
440	0.0252	3596.94	4075.99	7.084

t (°C)	v $\left(\frac{m^3}{kg}\right)$	u $\left(\frac{kJ}{kg}\right)$	h $\left(\frac{kJ}{kg}\right)$	s $\left(\frac{kJ}{kg\,K}\right)$
Sat Vapour (g)	0.0067	2339.16	2466.03	5.026
450	0.0255	3617.33	4101.66	7.108
460	0.0258	3637.74	4127.33	7.131
470	0.0260	3658.17	4152.99	7.155
480	0.0263	3678.62	4178.66	7.178
490	0.0266	3699.09	4204.33	7.201
500	0.0269	3719.59	4230.01	7.223
510	0.0271	3740.11	4255.70	7.246
520	0.0274	3760.66	4281.40	7.268
530	0.0277	3781.24	4307.11	7.290
540	0.0279	3801.85	4332.84	7.312
550	0.0282	3822.50	4358.59	7.334
560	0.0285	3843.18	4384.36	7.356
570	0.0288	3863.89	4410.16	7.377
580	0.0290	3884.64	4435.97	7.399
590	0.0293	3905.43	4461.81	7.420
600	0.0295	3926.26	4487.68	7.441
610	0.0298	3947.12	4513.58	7.462
620	0.0301	3968.03	4539.50	7.483
630	0.0303	3988.98	4565.46	7.503
640	0.0306	4009.96	4591.44	7.524
650	0.0309	4030.99	4617.46	7.544
660	0.0311	4052.06	4643.51	7.564
670	0.0314	4073.18	4669.59	7.584
680	0.0317	4094.33	4695.71	7.604
690	0.0319	4115.53	4721.86	7.624
700	0.0322	4136.78	4748.05	7.644
710	0.0324	4158.07	4774.28	7.663
720	0.0327	4179.40	4800.54	7.683
730	0.0330	4200.78	4826.84	7.702
740	0.0332	4222.20	4853.17	7.721
750	0.0335	4243.67	4879.55	7.740
760	0.0337	4265.18	4905.96	7.759
770	0.0340	4286.73	4932.41	7.778
780	0.0342	4308.34	4958.90	7.797
790	0.0345	4329.98	4985.43	7.816
800	0.0348	4351.68	5012.00	7.834
810	0.0350	4373.42	5038.61	7.853
820	0.0353	4395.20	5065.26	7.871
830	0.0355	4417.03	5091.95	7.890
840	0.0358	4438.90	5118.68	7.908
850	0.0360	4460.82	5145.45	7.926
860	0.0363	4482.79	5172.25	7.944
870	0.0365	4504.80	5199.10	7.962

t (°C)	v $\left(\frac{m^3}{kg}\right)$	u $\left(\frac{kJ}{kg}\right)$	h $\left(\frac{kJ}{kg}\right)$	s $\left(\frac{kJ}{kg\,K}\right)$
Sat Vapour (g)	0.0067	2339.16	2466.03	5.026
880	0.0368	4526.85	5225.99	7.980
890	0.0371	4548.95	5252.92	7.997
900	0.0373	4571.10	5279.89	8.015
910	0.0376	4593.29	5306.90	8.032
920	0.0378	4615.52	5333.95	8.050
930	0.0381	4637.80	5361.04	8.067
940	0.0383	4660.12	5388.18	8.085
950	0.0386	4682.49	5415.35	8.102
960	0.0388	4704.90	5442.56	8.119
970	0.0391	4727.35	5469.81	8.136
980	0.0393	4749.85	5497.10	8.153
990	0.0396	4772.39	5524.43	8.170
1000	0.0398	4794.97	5551.80	8.187
1010	0.0401	4817.60	5579.21	8.203
1020	0.0403	4840.26	5606.66	8.220
1030	0.0406	4862.97	5634.14	8.237
1040	0.0408	4885.73	5661.67	8.253
1050	0.0411	4908.52	5689.24	8.269
1060	0.0413	4931.36	5716.84	8.286
1070	0.0416	4954.23	5744.48	8.302
1080	0.0418	4977.15	5772.16	8.318
1090	0.0421	5000.11	5799.88	8.334
1100	0.0423	5023.11	5827.64	8.350
1110	0.0426	5046.15	5855.43	8.366
1120	0.0428	5069.23	5883.27	8.382
1130	0.0431	5092.35	5911.14	8.398
1140	0.0433	5115.51	5939.04	8.414
1150	0.0436	5138.71	5966.99	8.430
1160	0.0438	5161.95	5994.97	8.445
1170	0.0441	5185.22	6022.99	8.461
1180	0.0443	5208.54	6051.04	8.476
1190	0.0446	5231.89	6079.13	8.492
1200	0.0448	5255.28	6107.25	8.507
1210	0.0451	5278.71	6135.42	8.522
1220	0.0453	5302.18	6163.61	8.538
1230	0.0456	5325.68	6191.84	8.553
1240	0.0458	5349.23	6220.11	8.568
1250	0.0461	5372.80	6248.41	8.583
1260	0.0463	5396.42	6276.75	8.598
1270	0.0466	5420.07	6305.12	8.613
1280	0.0468	5443.75	6333.52	8.628
1290	0.0471	5467.47	6361.96	8.643
1300	0.0473	5491.23	6390.43	8.657

Table 3(58) Superheated Steam Properties (Pressure Base) Pressure = 200.0 bar and saturation temperature, Tsat = 107.109365.747°C

t (°C)	v ($\frac{m^3}{kg}$)	u ($\frac{kJ}{kg}$)	h ($\frac{kJ}{kg}$)	s ($\frac{kJ}{kg\,K}$)	t (°C)	v ($\frac{m^3}{kg}$)	u ($\frac{kJ}{kg}$)	h ($\frac{kJ}{kg}$)	s ($\frac{kJ}{kg\,K}$)	t (°C)	v ($\frac{m^3}{kg}$)	u ($\frac{kJ}{kg}$)	h ($\frac{kJ}{kg}$)	s ($\frac{kJ}{kg\,K}$)
Sat Vapour (g)	*0.0059*	*2294.81*	*2412.06*	*4.931*	*Sat Vapour (g)*	*0.0059*	*2294.81*	*2412.06*	*4.931*	*Sat Vapour (g)*	*0.0059*	*2294.81*	*2412.06*	*4.931*
10	0.0078	2456.54	2611.97	5.242	450	0.0243	3622.34	4108.10	7.091	880	0.0351	4534.56	5235.79	7.962
20	0.0088	2536.42	2712.87	5.396	460	0.0246	3642.84	4133.88	7.114	890	0.0353	4556.71	5262.78	7.980
30	0.0096	2595.97	2788.82	5.511	470	0.0248	3663.36	4159.66	7.138	900	0.0355	4578.89	5289.80	7.998
40	0.0103	2645.16	2851.94	5.604	480	0.0251	3683.89	4185.44	7.161	910	0.0358	4601.12	5316.86	8.015
50	0.0110	2687.91	2907.04	5.685	490	0.0253	3704.45	4211.22	7.184	920	0.0360	4623.40	5343.96	8.033
60	0.0115	2726.24	2956.60	5.756	500	0.0256	3725.03	4237.00	7.206	930	0.0363	4645.72	5371.10	8.050
70	0.0120	2761.34	3002.08	5.821	510	0.0259	3745.63	4262.79	7.229	940	0.0365	4668.08	5398.28	8.067
80	0.0125	2793.98	3044.44	5.880	520	0.0261	3766.26	4288.59	7.251	950	0.0368	4690.49	5425.50	8.084
90	0.0130	2824.69	3084.33	5.936	530	0.0264	3786.92	4314.40	7.274	960	0.0370	4712.94	5452.76	8.102
100	0.0134	2853.84	3122.24	5.987	540	0.0266	3807.61	4340.23	7.296	970	0.0372	4735.43	5480.06	8.119
110	0.0138	2881.73	3158.50	6.036	550	0.0269	3828.33	4366.08	7.317	980	0.0375	4757.97	5507.39	8.136
120	0.0142	2908.56	3193.40	6.082	560	0.0271	3849.08	4391.94	7.339	990	0.0377	4780.55	5534.77	8.152
130	0.0146	2934.52	3227.15	6.126	570	0.0274	3869.87	4417.82	7.361	1000	0.0380	4803.17	5562.19	8.169
140	0.0150	2959.72	3259.91	6.169	580	0.0277	3890.69	4443.73	7.382	1010	0.0382	4825.83	5589.64	8.186
150	0.0154	2984.29	3291.83	6.209	590	0.0279	3911.55	4469.65	7.403	1020	0.0384	4848.54	5617.13	8.203
160	0.0157	3008.31	3323.01	6.249	600	0.0282	3932.44	4495.61	7.424	1030	0.0387	4871.28	5644.66	8.219
170	0.0161	3031.86	3353.56	6.287	610	0.0284	3953.38	4521.59	7.445	1040	0.0389	4894.07	5672.23	8.236
180	0.0164	3055.00	3383.55	6.324	620	0.0287	3974.35	4547.59	7.466	1050	0.0391	4916.90	5699.84	8.252
190	0.0168	3077.78	3413.05	6.359	630	0.0289	3995.36	4573.63	7.486	1060	0.0394	4939.77	5727.49	8.268
200	0.0171	3100.25	3442.12	6.394	640	0.0292	4016.41	4599.69	7.507	1070	0.0396	4962.68	5755.17	8.284
210	0.0174	3122.44	3470.80	6.428	650	0.0294	4037.50	4625.79	7.527	1080	0.0399	4985.64	5782.89	8.301
220	0.0177	3144.40	3499.14	6.461	660	0.0297	4058.63	4651.91	7.547	1090	0.0401	5008.63	5810.65	8.317
230	0.0181	3166.14	3527.19	6.494	670	0.0299	4079.81	4678.07	7.567	1100	0.0403	5031.66	5838.45	8.333
240	0.0184	3187.70	3554.96	6.526	680	0.0302	4101.02	4704.26	7.587	1110	0.0406	5054.74	5866.28	8.349
250	0.0187	3209.09	3582.49	6.557	690	0.0304	4122.28	4730.49	7.607	1120	0.0408	5077.85	5894.16	8.365
260	0.0190	3230.34	3609.80	6.587	700	0.0307	4143.58	4756.75	7.627	1130	0.0411	5101.00	5922.06	8.380
270	0.0193	3251.47	3636.92	6.617	710	0.0309	4164.93	4783.05	7.646	1140	0.0413	5124.19	5950.01	8.396
280	0.0196	3272.48	3663.87	6.647	720	0.0312	4186.32	4809.38	7.666	1150	0.0415	5147.43	5977.99	8.412
290	0.0199	3293.40	3690.66	6.676	730	0.0314	4207.75	4835.74	7.685	1160	0.0418	5170.69	6006.01	8.428
300	0.0202	3314.23	3717.32	6.704	740	0.0316	4229.22	4862.15	7.704	1170	0.0420	5194.00	6034.06	8.443
310	0.0204	3334.99	3743.85	6.733	750	0.0319	4250.75	4888.58	7.723	1180	0.0422	5217.35	6062.15	8.459
320	0.0207	3355.69	3770.28	6.760	760	0.0321	4272.31	4915.06	7.742	1190	0.0425	5240.73	6090.28	8.474
330	0.0210	3376.34	3796.60	6.788	770	0.0324	4293.92	4941.58	7.761	1200	0.0427	5264.15	6118.44	8.489
340	0.0213	3396.94	3822.84	6.815	780	0.0326	4315.57	4968.13	7.780	1210	0.0430	5287.61	6146.63	8.505
350	0.0216	3417.50	3849.01	6.841	790	0.0329	4337.27	4994.72	7.799	1220	0.0432	5311.11	6174.86	8.520
360	0.0219	3438.03	3875.10	6.867	800	0.0331	4359.01	5021.35	7.817	1230	0.0434	5334.64	6203.13	8.535
370	0.0221	3458.54	3901.14	6.893	810	0.0334	4380.80	5048.02	7.836	1240	0.0437	5358.21	6231.43	8.550
380	0.0224	3479.03	3927.12	6.919	820	0.0336	4402.63	5074.73	7.854	1250	0.0439	5381.82	6259.76	8.565
390	0.0227	3499.51	3953.06	6.944	830	0.0338	4424.51	5101.47	7.872	1260	0.0441	5405.46	6288.13	8.580
400	0.0229	3519.98	3978.97	6.969	840	0.0341	4446.43	5128.26	7.891	1270	0.0444	5429.14	6316.54	8.595
410	0.0232	3540.44	4004.84	6.994	850	0.0343	4468.40	5155.08	7.909	1280	0.0446	5452.85	6344.97	8.610
420	0.0235	3560.91	4030.68	7.019	860	0.0346	4490.41	5181.95	7.927	1290	0.0448	5476.60	6373.44	8.625
430	0.0238	3581.38	4056.50	7.043	870	0.0348	4512.46	5208.85	7.945	1300	0.0451	5500.38	6401.95	8.639
440	0.0240	3601.86	4082.31	7.067										

Table 3(59) Superheated Steam Properties (Pressure Base) Pressure = 210.0 bar and saturation temperature, Tsat = 369.826°C

t (°C)	v $\left(\frac{m^3}{kg}\right)$	u $\left(\frac{kJ}{kg}\right)$	h $\left(\frac{kJ}{kg}\right)$	s $\left(\frac{kJ}{kg\,K}\right)$	t (°C)	v $\left(\frac{m^3}{kg}\right)$	u $\left(\frac{kJ}{kg}\right)$	h $\left(\frac{kJ}{kg}\right)$	s $\left(\frac{kJ}{kg\,K}\right)$	t (°C)	v $\left(\frac{m^3}{kg}\right)$	u $\left(\frac{kJ}{kg}\right)$	h $\left(\frac{kJ}{kg}\right)$	s $\left(\frac{kJ}{kg\,K}\right)$
Sat Vapour (g)	0.0050	2233.49	2338.36	4.808	Sat Vapour (g)	0.0050	2233.49	2338.36	4.808	Sat Vapour (g)	0.0050	2233.49	2338.36	4.808
10	0.0072	2438.46	2589.54	5.196	450	0.0232	3627.02	4114.13	7.074	880	0.0335	4541.86	5245.10	7.946
20	0.0083	2524.00	2697.28	5.360	460	0.0234	3647.61	4140.02	7.098	890	0.0337	4564.05	5272.14	7.964
30	0.0091	2586.48	2776.81	5.479	470	0.0237	3668.21	4165.91	7.121	900	0.0339	4586.28	5299.21	7.981
40	0.0097	2637.60	2842.30	5.576	480	0.0240	3688.83	4191.80	7.145	910	0.0342	4608.55	5326.32	7.999
50	0.0104	2681.76	2899.16	5.658	490	0.0242	3709.47	4217.68	7.168	920	0.0344	4630.87	5353.47	8.016
60	0.0109	2721.20	2950.10	5.731	500	0.0245	3730.13	4243.57	7.190	930	0.0346	4653.23	5380.66	8.034
70	0.0114	2757.19	2996.71	5.797	510	0.0247	3750.81	4269.46	7.213	940	0.0349	4675.63	5407.89	8.051
80	0.0119	2790.59	3040.03	5.857	520	0.0249	3771.52	4295.36	7.235	950	0.0351	4698.08	5435.16	8.068
90	0.0123	2821.94	3080.74	5.913	530	0.0252	3792.25	4321.27	7.257	960	0.0353	4720.57	5462.46	8.085
100	0.0127	2851.66	3119.37	5.966	540	0.0254	3813.02	4347.19	7.279	970	0.0356	4743.10	5489.81	8.102
110	0.0132	2880.04	3156.27	6.015	550	0.0257	3833.81	4373.13	7.301	980	0.0358	4765.67	5517.19	8.119
120	0.0135	2907.33	3191.75	6.062	560	0.0259	3854.63	4399.08	7.323	990	0.0360	4788.29	5544.61	8.136
130	0.0139	2933.68	3226.01	6.106	570	0.0262	3875.49	4425.05	7.344	1000	0.0362	4810.95	5572.07	8.153
140	0.0143	2959.26	3259.25	6.149	580	0.0264	3896.38	4451.04	7.366	1010	0.0365	4833.65	5599.57	8.169
150	0.0146	2984.16	3291.60	6.190	590	0.0267	3917.31	4477.06	7.387	1020	0.0367	4856.39	5627.10	8.186
160	0.0150	3008.49	3323.19	6.230	600	0.0269	3938.27	4503.09	7.408	1030	0.0369	4879.17	5654.68	8.202
170	0.0153	3032.33	3354.11	6.268	610	0.0271	3959.26	4529.15	7.429	1040	0.0372	4901.99	5682.29	8.219
180	0.0157	3055.74	3384.45	6.305	620	0.0274	3980.30	4555.24	7.450	1050	0.0374	4924.86	5709.94	8.235
190	0.0160	3078.77	3414.28	6.341	630	0.0276	4001.37	4581.35	7.470	1060	0.0376	4947.76	5737.62	8.252
200	0.0163	3101.47	3443.65	6.376	640	0.0279	4022.49	4607.49	7.491	1070	0.0378	4970.71	5765.35	8.268
210	0.0166	3123.89	3472.63	6.410	650	0.0281	4043.64	4633.67	7.511	1080	0.0381	4993.69	5793.11	8.284
220	0.0169	3146.05	3501.24	6.444	660	0.0283	4064.83	4659.87	7.531	1090	0.0383	5016.72	5820.91	8.300
230	0.0172	3167.99	3529.54	6.476	670	0.0286	4086.06	4686.10	7.551	1100	0.0385	5039.79	5848.74	8.316
240	0.0175	3189.73	3557.56	6.508	680	0.0288	4107.34	4712.36	7.571	1110	0.0387	5062.89	5876.62	8.332
250	0.0178	3211.31	3585.32	6.540	690	0.0290	4128.65	4738.66	7.591	1120	0.0390	5086.04	5904.53	8.348
260	0.0181	3232.73	3612.86	6.570	700	0.0293	4150.01	4764.99	7.611	1130	0.0392	5109.22	5932.47	8.364
270	0.0184	3254.02	3640.19	6.600	710	0.0295	4171.41	4791.35	7.630	1140	0.0394	5132.44	5960.45	8.379
280	0.0187	3275.19	3667.34	6.630	720	0.0298	4192.85	4817.75	7.650	1150	0.0397	5155.71	5988.47	8.395
290	0.0190	3296.25	3694.34	6.659	730	0.0300	4214.34	4844.18	7.669	1160	0.0399	5179.01	6016.52	8.411
300	0.0192	3317.23	3721.18	6.688	740	0.0302	4235.87	4870.65	7.688	1170	0.0401	5202.34	6044.61	8.426
310	0.0195	3338.13	3747.89	6.716	750	0.0305	4257.44	4897.15	7.707	1180	0.0403	5225.72	6072.74	8.442
320	0.0198	3358.96	3774.49	6.744	760	0.0307	4279.05	4923.69	7.726	1190	0.0406	5249.13	6100.90	8.457
330	0.0201	3379.74	3800.99	6.771	770	0.0309	4300.71	4950.27	7.745	1200	0.0408	5272.58	6129.09	8.472
340	0.0203	3400.46	3827.39	6.798	780	0.0312	4322.42	4976.88	7.764	1210	0.0410	5296.07	6157.32	8.488
350	0.0206	3421.15	3853.71	6.825	790	0.0314	4344.16	5003.53	7.782	1220	0.0412	5319.60	6185.59	8.503
360	0.0209	3441.80	3879.95	6.851	800	0.0316	4365.95	5030.22	7.801	1230	0.0415	5343.16	6213.88	8.518
370	0.0211	3462.42	3906.14	6.877	810	0.0319	4387.79	5056.95	7.820	1240	0.0417	5366.75	6242.22	8.533
380	0.0214	3483.02	3932.26	6.903	820	0.0321	4409.67	5083.71	7.838	1250	0.0419	5390.39	6270.58	8.548
390	0.0217	3503.60	3958.34	6.928	830	0.0323	4431.59	5110.51	7.856	1260	0.0421	5414.06	6298.98	8.563
400	0.0219	3524.18	3984.38	6.953	840	0.0326	4453.56	5137.36	7.874	1270	0.0424	5437.76	6327.42	8.578
410	0.0222	3544.74	4010.38	6.978	850	0.0328	4475.57	5164.23	7.892	1280	0.0426	5461.50	6355.89	8.593
420	0.0224	3565.31	4036.35	7.002	860	0.0330	4497.62	5191.15	7.910	1290	0.0428	5485.28	6384.39	8.608
430	0.0227	3585.87	4062.29	7.027	870	0.0333	4519.72	5218.11	7.928	1300	0.0430	5509.09	6412.92	8.622
440	0.0229	3606.44	4088.22	7.051										

Table 3(60) Superheated Steam Properties (Pressure Base) Pressure = 220.0 bar and saturation temperature, Tsat = 373.706°C

t (°C)	v (m³/kg)	u (kJ/kg)	h (kJ/kg)	s (kJ/kg K)
Sat Vapour (g)	0.0036	2092.42	2172.60	4.544
10	0.0067	2419.46	2566.12	5.150
20	0.0077	2511.06	2681.12	5.324
30	0.0085	2576.59	2764.35	5.447
40	0.0092	2629.69	2832.28	5.547
50	0.0098	2675.29	2890.92	5.632
60	0.0103	2715.84	2943.25	5.706
70	0.0108	2752.75	2991.00	5.774
80	0.0113	2786.90	3035.28	5.835
90	0.0117	2818.90	3076.82	5.892
100	0.0121	2849.19	3116.16	5.945
110	0.0125	2878.07	3153.71	5.995
120	0.0129	2905.80	3189.75	6.042
130	0.0133	2932.56	3224.54	6.087
140	0.0136	2958.50	3258.25	6.130
150	0.0140	2983.74	3291.03	6.172
160	0.0143	3008.38	3323.02	6.211
170	0.0146	3032.51	3354.31	6.250
180	0.0149	3056.18	3385.00	6.287
190	0.0153	3079.46	3415.15	6.324
200	0.0156	3102.39	3444.83	6.359
210	0.0159	3125.03	3474.09	6.393
220	0.0162	3147.40	3502.97	6.427
230	0.0165	3169.54	3531.53	6.459
240	0.0167	3191.47	3559.79	6.491
250	0.0170	3213.22	3587.78	6.523
260	0.0173	3234.81	3615.54	6.554
270	0.0176	3256.26	3643.08	6.584
280	0.0179	3277.58	3670.44	6.614
290	0.0181	3298.80	3697.62	6.643
300	0.0184	3319.92	3724.65	6.671
310	0.0187	3340.95	3751.54	6.700
320	0.0189	3361.92	3778.31	6.727
330	0.0192	3382.82	3804.97	6.755
340	0.0194	3403.67	3831.53	6.782
350	0.0197	3424.47	3858.01	6.809
360	0.0200	3445.23	3884.40	6.835
370	0.0202	3465.97	3910.73	6.861
380	0.0205	3486.67	3937.00	6.887
390	0.0207	3507.36	3963.21	6.912
400	0.0210	3528.04	3989.38	6.937
410	0.0212	3548.70	4015.51	6.962
420	0.0215	3569.36	4041.60	6.987
430	0.0217	3590.02	4067.66	7.011
440	0.0220	3610.68	4093.71	7.035

t (°C)	v (m³/kg)	u (kJ/kg)	h (kJ/kg)	s (kJ/kg K)
Sat Vapour (g)	0.0036	2092.42	2172.60	4.544
450	0.0222	3631.35	4119.73	7.059
460	0.0224	3652.02	4145.74	7.082
470	0.0227	3672.71	4171.74	7.106
480	0.0229	3693.41	4197.73	7.129
490	0.0232	3714.13	4223.72	7.152
500	0.0234	3734.87	4249.70	7.175
510	0.0236	3755.63	4275.69	7.197
520	0.0239	3776.41	4301.69	7.220
530	0.0241	3797.22	4327.69	7.242
540	0.0243	3818.06	4353.71	7.264
550	0.0246	3838.92	4379.73	7.286
560	0.0248	3859.82	4405.78	7.307
570	0.0251	3880.74	4431.83	7.329
580	0.0253	3901.70	4457.91	7.350
590	0.0255	3922.69	4484.01	7.371
600	0.0257	3943.72	4510.12	7.392
610	0.0260	3964.78	4536.26	7.413
620	0.0262	3985.88	4562.43	7.434
630	0.0264	4007.01	4588.62	7.455
640	0.0267	4028.18	4614.84	7.475
650	0.0269	4049.40	4641.08	7.495
660	0.0271	4070.65	4667.35	7.516
670	0.0274	4091.94	4693.66	7.536
680	0.0276	4113.27	4719.99	7.556
690	0.0278	4134.64	4746.36	7.575
700	0.0280	4156.05	4772.75	7.595
710	0.0283	4177.50	4799.18	7.615
720	0.0285	4199.00	4825.65	7.634
730	0.0287	4220.54	4852.14	7.653
740	0.0289	4242.12	4878.67	7.673
750	0.0292	4263.74	4905.24	7.692
760	0.0294	4285.40	4931.84	7.711
770	0.0296	4307.11	4958.48	7.729
780	0.0298	4328.86	4985.15	7.748
790	0.0301	4350.66	5011.86	7.767
800	0.0303	4372.49	5038.61	7.785
810	0.0305	4394.38	5065.39	7.804
820	0.0307	4416.30	5092.21	7.822
830	0.0309	4438.27	5119.07	7.840
840	0.0312	4460.28	5145.96	7.859
850	0.0314	4482.33	5172.89	7.877
860	0.0316	4504.43	5199.86	7.895
870	0.0318	4526.57	5226.87	7.913

t (°C)	v (m³/kg)	u (kJ/kg)	h (kJ/kg)	s (kJ/kg K)
Sat Vapour (g)	0.0036	2092.42	2172.60	4.544
880	0.0321	4548.76	5253.92	7.930
890	0.0323	4570.98	5281.00	7.948
900	0.0325	4593.25	5308.12	7.966
910	0.0327	4615.57	5335.28	7.983
920	0.0329	4637.92	5362.48	8.000
930	0.0332	4660.32	5389.72	8.018
940	0.0334	4682.76	5416.99	8.035
950	0.0336	4705.25	5444.30	8.052
960	0.0338	4727.77	5471.66	8.069
970	0.0340	4750.34	5499.04	8.086
980	0.0343	4772.95	5526.47	8.103
990	0.0345	4795.60	5553.94	8.120
1000	0.0347	4818.30	5581.44	8.137
1010	0.0349	4841.03	5608.98	8.153
1020	0.0351	4863.81	5636.55	8.170
1030	0.0353	4886.62	5664.17	8.187
1040	0.0356	4909.48	5691.82	8.203
1050	0.0358	4932.38	5719.51	8.219
1060	0.0360	4955.32	5747.24	8.236
1070	0.0362	4978.30	5775.00	8.252
1080	0.0364	5001.32	5802.80	8.268
1090	0.0366	5024.37	5830.64	8.284
1100	0.0369	5047.47	5858.51	8.300
1110	0.0371	5070.61	5886.42	8.316
1120	0.0373	5093.78	5914.37	8.332
1130	0.0375	5117.00	5942.35	8.348
1140	0.0377	5140.25	5970.37	8.363
1150	0.0379	5163.54	5998.42	8.379
1160	0.0382	5186.87	6026.51	8.395
1170	0.0384	5210.24	6054.63	8.410
1180	0.0386	5233.64	6082.79	8.426
1190	0.0388	5257.09	6110.98	8.441
1200	0.0390	5280.56	6139.21	8.456
1210	0.0392	5304.08	6167.47	8.472
1220	0.0395	5327.63	6195.77	8.487
1230	0.0397	5351.22	6224.10	8.502
1240	0.0399	5374.84	6252.46	8.517
1250	0.0401	5398.50	6280.86	8.532
1260	0.0403	5422.20	6309.29	8.547
1270	0.0405	5445.93	6337.76	8.562
1280	0.0408	5469.70	6366.26	8.577
1290	0.0410	5493.50	6394.79	8.591
1300	0.0412	5517.33	6423.35	8.606

STEAM TABLES
[With SI Units]

B.B. Arora
Department of Mechanical Engineering
Delhi Technological University (Formerly DCE)
Delhi

Akhilesh Arora
Department of Mechanical Engineering
Delhi Technological University (Formerly DCE)
Delhi

Vipin
Department of Production Engineering
Delhi Technological University (Formerly DCE)
Delhi

Part of the Main Book
Not to be Sold Separately

I.K. International Publishing House Pvt. Ltd.

NEW DELHI

Published by
I.K. International Publishing House Pvt. Ltd.
S-25, Green Park Extension
Uphaar Cinema Market
New Delhi–110 016 (India)
E-mail:info@ikinternational.com

ISBN: 978-93-84588-53-3

Published by Krishan Makhijani for I.K. International Publishing House Pvt. Ltd., S-25, Green Park Extension, Uphaar Cinema Market, New Delhi–110 016 and Printed by Rekha Printers Pvt. Ltd., Okhla Industrial Area, Phase II, New Delhi–110 020.

Tables

Table 3(A1) Superheated Steam Properties (Pressure Base) Pressure = 0.01 bar and saturation temperature, Tsat = 6.97°C

t (°C)	v $\left(\frac{m^3}{kg}\right)$	u $\left(\frac{kJ}{kg}\right)$	h $\left(\frac{kJ}{kg}\right)$	s $\left(\frac{kJ}{kg\,K}\right)$
Sat Vapour (g)	129.19	2384.50	2513.69	8.975
10	133.82	2398.68	2532.50	9.041
20	138.44	2412.80	2551.25	9.104
30	143.07	2426.92	2569.99	9.166
40	147.69	2441.04	2588.74	9.225
50	152.32	2455.19	2607.50	9.283
60	156.94	2469.36	2626.29	9.339
70	161.56	2483.55	2645.11	9.394
80	166.17	2497.78	2663.96	9.447
90	170.79	2512.05	2682.84	9.499
100	175.41	2526.35	2701.76	9.549
110	180.03	2540.70	2720.72	9.598
120	184.64	2555.09	2739.73	9.646
130	189.26	2569.52	2758.78	9.693
140	193.88	2584.00	2777.88	9.739
150	198.49	2598.54	2797.03	9.784
160	203.11	2613.12	2816.23	9.829
170	207.73	2627.75	2835.48	9.872
180	212.34	2642.44	2854.78	9.914
190	216.96	2657.18	2874.14	9.956
200	221.57	2671.98	2893.56	9.997
210	226.19	2686.84	2913.03	10.037
220	230.81	2701.75	2932.56	10.076
230	235.42	2716.72	2952.14	10.115
240	240.04	2731.75	2971.79	10.153
250	244.65	2746.84	2991.49	10.191
260	249.27	2761.99	3011.25	10.228
270	253.88	2777.19	3031.08	10.264
280	258.50	2792.46	3050.96	10.300
290	263.12	2807.80	3070.91	10.335
300	267.73	2823.19	3090.92	10.370
310	272.35	2838.64	3110.99	10.404
320	276.96	2854.16	3131.12	10.438
330	281.58	2869.74	3151.32	10.471
340	286.19	2885.39	3171.58	10.504
350	290.81	2901.10	3191.91	10.537
360	295.42	2916.87	3212.29	10.569
370	300.04	2932.71	3232.75	10.601
380	304.65	2948.61	3253.27	10.632
390	309.27	2964.58	3273.85	10.663
400	313.89	2980.61	3294.50	10.694
410	318.50	2996.71	3315.21	10.724
420	323.12	3012.88	3336.00	10.754
430	327.73	3029.11	3356.84	10.783
440	332.35	3045.41	3377.76	10.812

t (°C)	v $\left(\frac{m^3}{kg}\right)$	u $\left(\frac{kJ}{kg}\right)$	h $\left(\frac{kJ}{kg}\right)$	s $\left(\frac{kJ}{kg\,K}\right)$
Sat Vapour (g)	129.19	2384.50	2513.69	8.975
450	336.96	3061.78	3398.74	10.841
460	341.58	3078.21	3419.79	10.870
470	346.19	3094.71	3440.91	10.898
480	350.81	3111.28	3462.09	10.926
490	355.42	3127.92	3483.35	10.954
500	360.04	3144.63	3504.67	10.982
510	364.65	3161.41	3526.06	11.009
520	369.27	3178.25	3547.52	11.036
530	373.88	3195.16	3569.05	11.063
540	378.50	3212.15	3590.65	11.089
550	383.11	3229.20	3612.32	11.115
560	387.73	3246.33	3634.06	11.141
570	392.35	3263.52	3655.86	11.167
580	396.96	3280.78	3677.74	11.193
590	401.58	3298.12	3699.69	11.218
600	406.19	3315.52	3721.71	11.243
610	410.81	3333.00	3743.80	11.268
620	415.42	3350.54	3765.97	11.293
630	420.04	3368.16	3788.20	11.318
640	424.65	3385.85	3810.50	11.342
650	429.27	3403.61	3832.88	11.366
660	433.88	3421.44	3855.32	11.390
670	438.50	3439.34	3877.84	11.414
680	443.11	3457.31	3900.43	11.438
690	447.73	3475.36	3923.09	11.461
700	452.34	3493.47	3945.82	11.485
710	456.96	3511.66	3968.62	11.508
720	461.57	3529.92	3991.49	11.531
730	466.19	3548.25	4014.43	11.553
740	470.80	3566.64	4037.45	11.576
750	475.42	3585.11	4060.53	11.599
760	480.04	3603.65	4083.69	11.621
770	484.65	3622.27	4106.92	11.643
780	489.27	3640.95	4130.21	11.665
790	493.88	3659.70	4153.58	11.687
800	498.50	3678.52	4177.02	11.709
810	503.11	3697.41	4200.53	11.731
820	507.73	3716.38	4224.10	11.752
830	512.34	3735.41	4247.75	11.774
840	516.96	3754.51	4271.47	11.795
850	521.57	3773.68	4295.25	11.816
860	526.19	3792.92	4319.10	11.837
870	530.80	3812.23	4343.03	11.858

t (°C)	v $\left(\frac{m^3}{kg}\right)$	u $\left(\frac{kJ}{kg}\right)$	h $\left(\frac{kJ}{kg}\right)$	s $\left(\frac{kJ}{kg\,K}\right)$
Sat Vapour (g)	129.19	2384.50	2513.69	8.975
880	535.42	3831.60	4367.02	11.879
890	540.03	3851.05	4391.08	11.899
900	544.65	3870.56	4415.21	11.920
910	549.26	3890.14	4439.40	11.940
920	553.88	3909.78	4463.66	11.961
930	558.49	3929.50	4487.99	11.981
940	563.11	3949.28	4512.39	12.001
950	567.72	3969.13	4536.85	12.021
960	572.34	3989.04	4561.38	12.041
970	576.95	4009.02	4585.97	12.060
980	581.57	4029.06	4610.63	12.080
990	586.19	4049.17	4635.36	12.100
1000	590.80	4069.35	4660.15	12.119
1010	595.42	4089.58	4685.00	12.138
1020	600.03	4109.89	4709.92	12.158
1030	604.65	4130.25	4734.90	12.177
1040	609.26	4150.68	4759.94	12.196
1050	613.88	4171.17	4785.05	12.215
1060	618.49	4191.72	4810.22	12.234
1070	623.11	4212.34	4835.45	12.252
1080	627.72	4233.02	4860.74	12.271
1090	632.34	4253.76	4886.09	12.290
1100	636.95	4274.56	4911.51	12.308
1110	641.57	4295.42	4936.98	12.327
1120	646.18	4316.33	4962.52	12.345
1130	650.80	4337.31	4988.11	12.363
1140	655.41	4358.35	5013.77	12.381
1150	660.03	4379.45	5039.48	12.399
1160	664.64	4400.61	5065.25	12.417
1170	669.26	4421.82	5091.08	12.435
1180	673.87	4443.09	5116.96	12.453
1190	678.49	4464.42	5142.91	12.471
1200	683.10	4485.80	5168.91	12.488
1210	687.72	4507.24	5194.96	12.506
1220	692.34	4528.74	5221.07	12.523
1230	696.95	4550.29	5247.24	12.541
1240	701.57	4571.90	5273.46	12.558
1250	706.18	4593.56	5299.74	12.575
1260	710.80	4615.28	5326.07	12.592
1270	715.41	4637.05	5352.46	12.609
1280	720.03	4658.87	5378.89	12.626
1290	724.64	4680.74	5405.39	12.643
1300	729.26	4702.67	5431.93	12.660

Table 3(A2) Superheated Steam Properties (Pressure Base) Pressure = 0.02 bar and saturation temperature, Tsat = 17.495°C

t (°C)	$v\left(\frac{m^3}{kg}\right)$	$u\left(\frac{kJ}{kg}\right)$	$h\left(\frac{kJ}{kg}\right)$	$s\left(\frac{kJ}{kg\,K}\right)$
Sat Vapour (g)	66.99	2398.91	2532.89	8.723
10	69.31	2413.19	2551.81	8.787
20	71.63	2427.38	2570.64	8.848
30	73.94	2441.56	2589.45	8.908
40	76.26	2455.74	2608.26	8.966
50	78.57	2469.94	2627.08	9.022
60	80.88	2484.16	2645.93	9.076
70	83.19	2498.41	2664.80	9.129
80	85.50	2512.70	2683.71	9.181
90	87.81	2527.02	2702.64	9.231
100	90.12	2541.37	2721.62	9.281
110	92.43	2555.77	2740.64	9.329
120	94.74	2570.22	2759.70	9.376
130	97.05	2584.71	2778.81	9.422
140	99.36	2599.25	2797.97	9.467
150	101.67	2613.84	2817.18	9.511
160	103.98	2628.48	2836.44	9.554
170	106.29	2643.18	2855.75	9.596
180	108.59	2657.93	2875.12	9.638
190	110.90	2672.73	2894.54	9.679
200	113.21	2687.59	2914.01	9.719
210	115.52	2702.51	2933.55	9.758
220	117.83	2717.48	2953.14	9.797
230	120.14	2732.52	2972.79	9.835
240	122.44	2747.61	2992.50	9.873
250	124.75	2762.76	3012.27	9.910
260	127.06	2777.97	3032.09	9.946
270	129.37	2793.25	3051.98	9.982
280	131.68	2808.58	3071.94	10.017
290	133.98	2823.98	3091.95	10.052
300	136.29	2839.44	3112.02	10.086
310	138.60	2854.96	3132.16	10.120
320	140.91	2870.55	3152.36	10.153
330	143.22	2886.20	3172.63	10.186
340	145.52	2901.91	3192.96	10.219
350	147.83	2917.69	3213.35	10.251
360	150.14	2933.53	3233.81	10.282
370	152.45	2949.44	3254.33	10.314
380	154.75	2965.41	3274.92	10.345
390	157.06	2981.45	3295.57	10.375
400	159.37	2997.55	3316.29	10.405
410	161.68	3013.72	3337.07	10.435
420	163.99	3029.96	3357.93	10.465
430	166.29	3046.26	3378.85	10.494
440	168.60	3062.63	3399.83	10.523

t (°C)	$v\left(\frac{m^3}{kg}\right)$	$u\left(\frac{kJ}{kg}\right)$	$h\left(\frac{kJ}{kg}\right)$	$s\left(\frac{kJ}{kg\,K}\right)$
Sat Vapour (g)	66.99	2398.91	2532.89	8.723
450	170.91	3079.07	3420.88	10.552
460	173.22	3095.57	3442.01	10.580
470	175.52	3112.15	3463.20	10.608
480	177.83	3128.79	3484.45	10.636
490	180.14	3145.50	3505.78	10.663
500	182.45	3162.28	3527.17	10.690
510	184.75	3179.13	3548.64	10.717
520	187.06	3196.05	3570.17	10.744
530	189.37	3213.04	3591.78	10.771
540	191.68	3230.09	3613.45	10.797
550	193.99	3247.22	3635.19	10.823
560	196.29	3264.42	3657.00	10.849
570	198.60	3281.69	3678.89	10.874
580	200.91	3299.02	3700.84	10.900
590	203.22	3316.43	3722.86	10.925
600	205.52	3333.91	3744.96	10.950
610	207.83	3351.46	3767.13	10.975
620	210.14	3369.08	3789.36	10.999
630	212.45	3386.78	3811.67	11.023
640	214.75	3404.54	3834.05	11.048
650	217.06	3422.37	3856.50	11.072
660	219.37	3440.28	3879.02	11.095
670	221.68	3458.26	3901.61	11.119
680	223.98	3476.30	3924.27	11.143
690	226.29	3494.42	3947.01	11.166
700	228.60	3512.61	3969.81	11.189
710	230.91	3530.87	3992.69	11.212
720	233.22	3549.21	4015.64	11.235
730	235.52	3567.61	4038.65	11.257
740	237.83	3586.08	4061.74	11.280
750	240.14	3604.63	4084.90	11.302
760	242.45	3623.24	4108.13	11.325
770	244.75	3641.93	4131.43	11.347
780	247.06	3660.68	4154.80	11.369
790	249.37	3679.51	4178.25	11.390
800	251.68	3698.40	4201.76	11.412
810	253.98	3717.37	4225.34	11.433
820	256.29	3736.41	4248.99	11.455
830	258.60	3755.51	4272.71	11.476
840	260.91	3774.68	4296.50	11.497
850	263.21	3793.93	4320.36	11.518
860	265.52	3813.24	4344.28	11.539
870	267.83	3832.62	4368.28	11.560

t (°C)	$v\left(\frac{m^3}{kg}\right)$	$u\left(\frac{kJ}{kg}\right)$	$h\left(\frac{kJ}{kg}\right)$	$s\left(\frac{kJ}{kg\,K}\right)$
Sat Vapour (g)	66.99	2398.91	2532.89	8.723
880	270.14	3852.07	4392.34	11.581
890	272.45	3871.58	4416.47	11.601
900	274.75	3891.16	4440.67	11.622
910	277.06	3910.82	4464.94	11.642
920	279.37	3930.53	4489.27	11.662
930	281.68	3950.32	4513.67	11.682
940	283.98	3970.17	4538.13	11.702
950	286.29	3990.08	4562.67	11.722
960	288.60	4010.07	4587.26	11.742
970	290.91	4030.11	4611.93	11.761
980	293.21	4050.23	4636.65	11.781
990	295.52	4070.40	4661.45	11.800
1000	297.83	4090.65	4686.30	11.820
1010	300.14	4110.95	4711.22	11.839
1020	302.44	4131.32	4736.21	11.858
1030	304.75	4151.75	4761.25	11.877
1040	307.06	4172.25	4786.36	11.896
1050	309.37	4192.80	4811.54	11.915
1060	311.67	4213.42	4836.77	11.934
1070	313.98	4234.10	4862.07	11.952
1080	316.29	4254.84	4887.42	11.971
1090	318.60	4275.65	4912.84	11.989
1100	320.90	4296.51	4938.32	12.008
1110	323.21	4317.43	4963.86	12.026
1120	325.52	4338.42	4989.46	12.044
1130	327.83	4359.46	5015.11	12.062
1140	330.14	4380.56	5040.83	12.080
1150	332.44	4401.72	5066.60	12.098
1160	334.75	4422.93	5092.43	12.116
1170	337.06	4444.21	5118.32	12.134
1180	339.37	4465.54	5144.27	12.152
1190	341.67	4486.93	5170.27	12.169
1200	343.98	4508.37	5196.33	12.187
1210	346.29	4529.87	5222.45	12.204
1220	348.60	4551.42	5248.62	12.222
1230	350.90	4573.03	5274.84	12.239
1240	353.21	4594.70	5301.12	12.256
1250	355.52	4616.42	5327.45	12.273
1260	357.83	4638.19	5353.84	12.290
1270	360.13	4660.01	5380.28	12.307
1280	362.44	4681.89	5406.78	12.324
1290	364.75	4703.82	5433.32	12.341
1300	367.06	4725.81	5459.92	12.358

Table 3(A3) Superheated Steam Properties (Pressure Base) Pressure = 0.04 bar and saturation temperature, Tsat = 28.96°C

t (°C)	$v\left(\frac{m^3}{kg}\right)$	$u\left(\frac{kJ}{kg}\right)$	$h\left(\frac{kJ}{kg}\right)$	$s\left(\frac{kJ}{kg\,K}\right)$	t (°C)	$v\left(\frac{m^3}{kg}\right)$	$u\left(\frac{kJ}{kg}\right)$	$h\left(\frac{kJ}{kg}\right)$	$s\left(\frac{kJ}{kg\,K}\right)$	t (°C)	$v\left(\frac{m^3}{kg}\right)$	$u\left(\frac{kJ}{kg}\right)$	$h\left(\frac{kJ}{kg}\right)$	$s\left(\frac{kJ}{kg\,K}\right)$
Sat Vapour (g)	34.79	2414.51	2553.68	8.473	Sat Vapour (g)	34.79	2414.51	2553.68	8.473	Sat Vapour (g)	34.79	2414.51	2553.68	8.473
10	35.96	2428.90	2572.72	8.535	450	86.78	3097.98	3445.08	10.264	880	136.39	3874.44	4420.01	11.284
20	37.12	2443.19	2591.66	8.595	460	87.93	3114.57	3466.28	10.292	890	137.55	3894.03	4444.21	11.305
30	38.28	2457.45	2610.56	8.653	470	89.08	3131.22	3487.55	10.320	900	138.70	3913.69	4468.49	11.325
40	39.44	2471.71	2629.46	8.709	480	90.24	3147.94	3508.89	10.347	910	139.85	3933.42	4492.83	11.345
50	40.59	2485.99	2648.37	8.763	490	91.39	3164.73	3530.29	10.375	920	141.01	3953.22	4517.24	11.365
60	41.75	2500.28	2667.29	8.816	500	92.55	3181.59	3551.77	10.401	930	142.16	3973.08	4541.72	11.385
70	42.91	2514.60	2686.23	8.868	510	93.70	3198.52	3573.31	10.428	940	143.31	3993.00	4566.26	11.405
80	44.06	2528.94	2705.20	8.918	520	94.85	3215.52	3594.93	10.455	950	144.47	4012.99	4590.87	11.425
90	45.22	2543.33	2724.21	8.967	530	96.01	3232.58	3616.61	10.481	960	145.62	4033.05	4615.54	11.444
100	46.38	2557.75	2743.26	9.015	540	97.16	3249.72	3638.37	10.507	970	146.78	4053.17	4640.28	11.464
110	47.53	2572.22	2762.35	9.062	550	98.31	3266.93	3660.19	10.533	980	147.93	4073.36	4665.08	11.483
120	48.69	2586.73	2781.48	9.108	560	99.47	3284.21	3682.08	10.558	990	149.08	4093.61	4689.94	11.502
130	49.84	2601.29	2800.65	9.153	570	100.62	3301.56	3704.05	10.583	1000	150.24	4113.93	4714.87	11.522
140	51.00	2615.89	2819.88	9.197	580	101.78	3318.98	3726.08	10.609	1010	151.39	4134.30	4739.87	11.541
150	52.15	2630.55	2839.16	9.240	590	102.93	3336.47	3748.19	10.633	1020	152.54	4154.75	4764.92	11.560
160	53.31	2645.26	2858.48	9.282	600	104.08	3354.03	3770.36	10.658	1030	153.70	4175.25	4790.04	11.579
170	54.46	2660.02	2877.87	9.324	610	105.24	3371.66	3792.61	10.683	1040	154.85	4195.82	4815.22	11.598
180	55.62	2674.84	2897.30	9.365	620	106.39	3389.36	3814.93	10.707	1050	156.01	4216.44	4840.47	11.616
190	56.77	2689.71	2916.79	9.405	630	107.55	3407.14	3837.32	10.731	1060	157.16	4237.13	4865.77	11.635
200	57.92	2704.64	2936.34	9.444	640	108.70	3424.98	3859.78	10.755	1070	158.31	4257.88	4891.14	11.654
210	59.08	2719.63	2955.94	9.483	650	109.85	3442.90	3882.31	10.779	1080	159.47	4278.70	4916.57	11.672
220	60.23	2734.67	2975.61	9.521	660	111.01	3460.89	3904.91	10.803	1090	160.62	4299.57	4942.05	11.690
230	61.39	2749.78	2995.33	9.558	670	112.16	3478.94	3927.59	10.826	1100	161.78	4320.50	4967.60	11.709
240	62.54	2764.94	3015.11	9.595	680	113.31	3497.07	3950.33	10.849	1110	162.93	4341.49	4993.21	11.727
250	63.70	2780.17	3034.95	9.631	690	114.47	3515.27	3973.15	10.872	1120	164.08	4362.54	5018.87	11.745
260	64.85	2795.45	3054.85	9.667	700	115.62	3533.55	3996.04	10.895	1130	165.24	4383.65	5044.60	11.763
270	66.00	2810.80	3074.81	9.702	710	116.78	3551.89	4018.99	10.918	1140	166.39	4404.82	5070.38	11.781
280	67.16	2826.21	3094.84	9.737	720	117.93	3570.30	4042.02	10.941	1150	167.54	4426.04	5096.22	11.799
290	68.31	2841.68	3114.93	9.771	730	119.08	3588.79	4065.12	10.963	1160	168.70	4447.32	5122.12	11.817
300	69.47	2857.21	3135.08	9.805	740	120.24	3607.34	4088.29	10.986	1170	169.85	4468.66	5148.07	11.834
310	70.62	2872.81	3155.29	9.838	750	121.39	3625.97	4111.53	11.008	1180	171.01	4490.06	5174.08	11.852
320	71.77	2888.47	3175.56	9.871	760	122.55	3644.66	4134.84	11.030	1190	172.16	4511.51	5200.15	11.869
330	72.93	2904.19	3195.90	9.903	770	123.70	3663.43	4158.23	11.052	1200	173.31	4533.02	5226.27	11.887
340	74.08	2919.98	3216.31	9.935	780	124.85	3682.27	4181.68	11.074	1210	174.47	4554.58	5252.45	11.904
350	75.24	2935.83	3236.78	9.967	790	126.01	3701.17	4205.20	11.095	1220	175.62	4576.20	5278.68	11.922
360	76.39	2951.75	3257.31	9.998	800	127.16	3720.15	4228.79	11.117	1230	176.77	4597.87	5304.97	11.939
370	77.54	2967.73	3277.91	10.029	810	128.31	3739.19	4252.45	11.138	1240	177.93	4619.60	5331.31	11.956
380	78.70	2983.78	3298.57	10.060	820	129.47	3758.31	4276.18	11.159	1250	179.08	4641.38	5357.71	11.973
390	79.85	2999.89	3319.30	10.090	830	130.62	3777.49	4299.98	11.180	1260	180.24	4663.21	5384.16	11.990
400	81.01	3016.07	3340.10	10.120	840	131.78	3796.74	4323.85	11.201	1270	181.39	4685.10	5410.66	12.007
410	82.16	3032.32	3360.96	10.149	850	132.93	3816.07	4347.79	11.222	1280	182.54	4707.04	5437.21	12.024
420	83.31	3048.64	3381.89	10.178	860	134.08	3835.46	4371.79	11.243	1290	183.70	4729.03	5463.82	12.040
430	84.47	3065.02	3402.89	10.207	870	135.24	3854.91	4395.86	11.264	1300	184.85	4751.07	5490.48	12.057
440	85.62	3081.46	3423.95	10.236										

Table 3(A4) Superheated Steam Properties (Pressure Base) Pressure = 0.06 bar and saturation temperature, Tsat = 36.159°C

t (°C)	v (m^3/kg)	u (kJ/kg)	h (kJ/kg)	s ($kJ/kg\ K$)	t (°C)	v (m^3/kg)	u (kJ/kg)	h (kJ/kg)	s ($kJ/kg\ K$)	t (°C)	v (m^3/kg)	u (kJ/kg)	h (kJ/kg)	s ($kJ/kg\ K$)
Sat Vapour (g)	23.73	2424.24	2566.64	8.329	Sat Vapour (g)	23.73	2424.24	2566.64	8.329	Sat Vapour (g)	23.73	2424.24	2566.64	8.329
10	24.51	2438.70	2585.77	8.390	450	58.40	3109.90	3460.31	10.097	880	91.48	3888.53	4437.42	11.112
20	25.29	2453.06	2604.79	8.449	460	59.17	3126.53	3481.56	10.125	890	92.25	3908.17	4461.67	11.132
30	26.06	2467.39	2623.76	8.505	470	59.94	3143.23	3502.88	10.153	900	93.02	3927.88	4486.00	11.152
40	26.84	2481.71	2642.72	8.560	480	60.71	3160.00	3524.27	10.180	910	93.79	3947.66	4510.39	11.172
50	27.61	2496.03	2661.69	8.614	490	61.48	3176.84	3545.73	10.207	920	94.56	3967.50	4534.85	11.192
60	28.38	2510.38	2680.66	8.666	500	62.25	3193.75	3567.25	10.234	930	95.33	3987.41	4559.37	11.212
70	29.15	2524.74	2699.66	8.717	510	63.02	3210.73	3588.85	10.260	940	96.10	4007.38	4583.96	11.232
80	29.92	2539.14	2718.68	8.766	520	63.79	3227.78	3610.51	10.286	950	96.87	4027.42	4608.61	11.252
90	30.70	2553.57	2737.74	8.815	530	64.56	3244.90	3632.25	10.312	960	97.63	4047.52	4633.33	11.271
100	31.47	2568.04	2756.83	8.862	540	65.33	3262.09	3654.05	10.338	970	98.40	4067.69	4658.12	11.291
110	32.24	2582.55	2775.97	8.908	550	66.10	3279.35	3675.92	10.364	980	99.17	4087.93	4682.96	11.310
120	33.01	2597.10	2795.14	8.953	560	66.87	3296.68	3697.87	10.389	990	99.94	4108.22	4707.88	11.329
130	33.78	2611.70	2814.37	8.998	570	67.63	3314.08	3719.89	10.414	1000	100.71	4128.58	4732.85	11.348
140	34.55	2626.35	2833.64	9.041	580	68.40	3331.55	3741.97	10.439	1010	101.48	4149.01	4757.89	11.367
150	35.32	2641.06	2852.96	9.083	590	69.17	3349.09	3764.13	10.464	1020	102.25	4169.49	4782.99	11.386
160	36.09	2655.81	2872.34	9.125	600	69.94	3366.70	3786.36	10.489	1030	103.02	4190.04	4808.16	11.405
170	36.86	2670.62	2891.76	9.166	610	70.71	3384.39	3808.66	10.513	1040	103.79	4210.65	4833.38	11.424
180	37.63	2685.48	2911.24	9.206	620	71.48	3402.14	3831.03	10.537	1050	104.56	4231.33	4858.67	11.443
190	38.40	2700.40	2930.78	9.246	630	72.25	3419.97	3853.47	10.561	1060	105.33	4252.06	4884.02	11.461
200	39.17	2715.37	2950.37	9.285	640	73.02	3437.86	3875.98	10.585	1070	106.10	4272.85	4909.43	11.480
210	39.94	2730.40	2970.02	9.323	650	73.79	3455.83	3898.56	10.609	1080	106.87	4293.71	4934.90	11.498
220	40.71	2745.49	2989.73	9.361	660	74.56	3473.87	3921.22	10.632	1090	107.63	4314.62	4960.43	11.516
230	41.48	2760.64	3009.50	9.398	670	75.33	3491.98	3943.94	10.656	1100	108.40	4335.60	4986.02	11.535
240	42.25	2775.85	3029.33	9.434	680	76.10	3510.16	3966.74	10.679	1110	109.17	4356.63	5011.67	11.553
250	43.02	2791.13	3049.22	9.470	690	76.87	3528.41	3989.60	10.702	1120	109.94	4377.72	5037.38	11.571
260	43.78	2806.46	3069.17	9.505	700	77.63	3546.73	4012.54	10.725	1130	110.71	4398.88	5063.14	11.589
270	44.55	2821.85	3089.18	9.540	710	78.40	3565.13	4035.55	10.747	1140	111.48	4420.08	5088.97	11.607
280	45.32	2837.30	3109.25	9.574	720	79.17	3583.59	4058.63	10.770	1150	112.25	4441.35	5114.85	11.624
290	46.09	2852.82	3129.38	9.608	730	79.94	3602.13	4081.78	10.792	1160	113.02	4462.67	5140.79	11.642
300	46.86	2868.40	3149.58	9.642	740	80.71	3620.73	4105.00	10.815	1170	113.79	4484.05	5166.78	11.660
310	47.63	2884.05	3169.84	9.675	750	81.48	3639.41	4128.30	10.837	1180	114.56	4505.49	5192.84	11.677
320	48.40	2899.75	3190.16	9.707	760	82.25	3658.16	4151.66	10.859	1190	115.33	4526.98	5218.94	11.695
330	49.17	2915.52	3210.55	9.739	770	83.02	3676.97	4175.09	10.880	1200	116.10	4548.53	5245.11	11.712
340	49.94	2931.36	3231.00	9.771	780	83.79	3695.86	4198.59	10.902	1210	116.87	4570.13	5271.32	11.730
350	50.71	2947.26	3251.52	9.802	790	84.56	3714.82	4222.16	10.924	1220	117.63	4591.79	5297.60	11.747
360	51.48	2963.22	3272.10	9.833	800	85.33	3733.84	4245.81	10.945	1230	118.40	4613.50	5323.92	11.764
370	52.25	2979.25	3292.74	9.864	810	86.10	3752.94	4269.52	10.966	1240	119.17	4635.27	5350.30	11.781
380	53.02	2995.35	3313.46	9.894	820	86.87	3772.10	4293.30	10.987	1250	119.94	4657.09	5376.74	11.798
390	53.79	3011.51	3334.24	9.924	830	87.63	3791.34	4317.15	11.008	1260	120.71	4678.96	5403.22	11.815
400	54.56	3027.74	3355.08	9.954	840	88.40	3810.64	4341.06	11.029	1270	121.48	4700.88	5429.76	11.832
410	55.33	3044.04	3375.99	9.983	850	89.17	3830.01	4365.05	11.050	1280	122.25	4722.86	5456.36	11.849
420	56.10	3060.40	3396.97	10.012	860	89.94	3849.45	4389.10	11.071	1290	123.02	4744.89	5483.00	11.865
430	56.86	3076.83	3418.02	10.041	870	90.71	3868.96	4413.23	11.091	1300	123.79	4766.96	5509.69	11.882
440	57.63	3093.33	3439.13	10.069										

Table 3(A5) Superheated Steam Properties (Pressure Base) Pressure = 0.08 bar and saturation temperature, Tsat = 41.508°C

t (°C)	v (m³/kg)	u (kJ/kg)	h (kJ/kg)	s (kJ/kg K)	t (°C)	v (m³/kg)	u (kJ/kg)	h (kJ/kg)	s (kJ/kg K)	t (°C)	v (m³/kg)	u (kJ/kg)	h (kJ/kg)	s (kJ/kg K)
Sat Vapour (g)	18.10	2431.42	2576.21	8.227	Sat Vapour (g)	18.10	2431.42	2576.21	8.227	Sat Vapour (g)	18.10	2431.42	2576.21	8.227
10	18.68	2445.95	2595.42	8.287	450	44.11	3118.77	3471.65	9.979	880	68.92	3899.02	4450.38	10.990
20	19.27	2460.37	2614.50	8.345	460	44.69	3135.44	3492.94	10.007	890	69.50	3918.70	4474.67	11.010
30	19.85	2474.75	2633.54	8.401	470	45.26	3152.18	3514.29	10.034	900	70.07	3938.45	4499.03	11.030
40	20.43	2489.11	2652.55	8.456	480	45.84	3168.99	3535.72	10.061	910	70.65	3958.26	4523.46	11.050
50	21.01	2503.48	2671.56	8.509	490	46.42	3185.87	3557.21	10.088	920	71.23	3978.14	4547.95	11.070
60	21.59	2517.87	2690.58	8.560	500	47.00	3202.81	3578.78	10.115	930	71.80	3998.08	4572.51	11.090
70	22.17	2532.27	2709.62	8.610	510	47.57	3219.83	3600.41	10.141	940	72.38	4018.09	4597.13	11.110
80	22.75	2546.70	2728.68	8.659	520	48.15	3236.92	3622.11	10.168	950	72.96	4038.16	4621.82	11.129
90	23.33	2561.17	2747.78	8.707	530	48.73	3254.07	3643.88	10.193	960	73.53	4058.30	4646.58	11.149
100	23.90	2575.68	2766.91	8.754	540	49.30	3271.30	3665.73	10.219	970	74.11	4078.50	4671.40	11.168
110	24.48	2590.22	2786.08	8.799	550	49.88	3288.60	3687.64	10.245	980	74.69	4098.77	4696.28	11.187
120	25.06	2604.81	2805.30	8.844	560	50.46	3305.97	3709.62	10.270	990	75.27	4119.10	4721.22	11.207
130	25.64	2619.45	2824.56	8.888	570	51.03	3323.40	3731.68	10.295	1000	75.84	4139.50	4746.23	11.226
140	26.22	2634.13	2843.86	8.931	580	51.61	3340.91	3753.80	10.320	1010	76.42	4159.95	4771.31	11.245
150	26.79	2648.87	2863.22	8.973	590	52.19	3358.49	3776.00	10.345	1020	77.00	4180.47	4796.44	11.264
160	27.37	2663.66	2882.63	9.014	600	52.76	3376.14	3798.26	10.369	1030	77.57	4201.06	4821.64	11.282
170	27.95	2678.50	2902.09	9.055	610	53.34	3393.86	3820.60	10.393	1040	78.15	4221.70	4846.90	11.301
180	28.53	2693.39	2921.61	9.095	620	53.92	3411.66	3843.01	10.417	1050	78.73	4242.41	4872.22	11.320
190	29.10	2708.35	2941.18	9.134	630	54.50	3429.52	3865.49	10.441	1060	79.30	4263.17	4897.60	11.338
200	29.68	2723.35	2960.81	9.172	640	55.07	3447.46	3888.04	10.465	1070	79.88	4284.00	4923.04	11.357
210	30.26	2738.42	2980.50	9.210	650	55.65	3465.46	3910.66	10.489	1080	80.46	4304.89	4948.55	11.375
220	30.84	2753.55	3000.24	9.248	660	56.23	3483.54	3933.35	10.512	1090	81.03	4325.83	4974.11	11.393
230	31.41	2768.73	3020.04	9.284	670	56.80	3501.69	3956.12	10.535	1100	81.61	4346.84	4999.73	11.412
240	31.99	2783.98	3039.91	9.320	680	57.38	3519.91	3978.95	10.558	1110	82.19	4367.90	5025.41	11.430
250	32.57	2799.28	3059.83	9.356	690	57.96	3538.20	4001.86	10.581	1120	82.77	4389.03	5051.15	11.448
260	33.15	2814.65	3079.81	9.391	700	58.53	3556.56	4024.83	10.604	1130	83.34	4410.21	5076.95	11.466
270	33.72	2830.08	3099.86	9.426	710	59.11	3574.99	4047.88	10.627	1140	83.92	4431.45	5102.80	11.483
280	34.30	2845.57	3119.97	9.460	720	59.69	3593.49	4071.00	10.649	1150	84.50	4452.75	5128.72	11.501
290	34.88	2861.12	3140.14	9.493	730	60.27	3612.06	4094.19	10.671	1160	85.07	4474.10	5154.68	11.519
300	35.45	2876.73	3160.37	9.527	740	60.84	3630.71	4117.45	10.694	1170	85.65	4495.51	5180.71	11.536
310	36.03	2892.41	3180.66	9.559	750	61.42	3649.42	4140.78	10.716	1180	86.23	4516.98	5206.79	11.554
320	36.61	2908.15	3201.02	9.592	760	62.00	3668.21	4164.18	10.737	1190	86.80	4538.50	5232.93	11.571
330	37.19	2923.96	3221.45	9.624	770	62.57	3687.06	4187.65	10.759	1200	87.38	4560.08	5259.12	11.589
340	37.76	2939.83	3241.93	9.655	780	63.15	3705.99	4211.18	10.781	1210	87.96	4581.71	5285.37	11.606
350	38.34	2955.77	3262.49	9.686	790	63.73	3724.98	4234.79	10.802	1220	88.53	4603.40	5311.67	11.623
360	38.92	2971.77	3283.10	9.717	800	64.30	3744.04	4258.47	10.824	1230	89.11	4625.14	5338.03	11.640
370	39.49	2987.84	3303.79	9.747	810	64.88	3763.18	4282.22	10.845	1240	89.69	4646.93	5364.44	11.657
380	40.07	3003.97	3324.54	9.778	820	65.46	3782.38	4306.04	10.866	1250	90.27	4668.78	5390.90	11.674
390	40.65	3020.17	3345.35	9.807	830	66.03	3801.65	4329.92	10.887	1260	90.84	4690.68	5417.41	11.691
400	41.23	3036.43	3366.23	9.837	840	66.61	3820.99	4353.88	10.908	1270	91.42	4712.63	5443.98	11.708
410	41.80	3052.76	3387.18	9.866	850	67.19	3840.39	4377.90	10.928	1280	92.00	4734.63	5470.60	11.725
420	42.38	3069.16	3408.20	9.895	860	67.77	3859.87	4401.99	10.949	1290	92.57	4756.69	5497.27	11.741
430	42.96	3085.63	3429.28	9.923	870	68.34	3879.41	4426.15	10.969	1300	93.15	4778.79	5523.99	11.758
440	43.53	3102.17	3450.43	9.951										

Table 3(A6) Superheated Steam Properties (Pressure Base) Pressure = 0.1 bar and saturation temperature, Tsat = 45.806°C

t (°C)	$v\left(\frac{m^3}{kg}\right)$	$u\left(\frac{kJ}{kg}\right)$	$h\left(\frac{kJ}{kg}\right)$	$s\left(\frac{kJ}{kg\,K}\right)$	t (°C)	$v\left(\frac{m^3}{kg}\right)$	$u\left(\frac{kJ}{kg}\right)$	$h\left(\frac{kJ}{kg}\right)$	$s\left(\frac{kJ}{kg\,K}\right)$	t (°C)	$v\left(\frac{m^3}{kg}\right)$	$u\left(\frac{kJ}{kg}\right)$	$h\left(\frac{kJ}{kg}\right)$	$s\left(\frac{kJ}{kg\,K}\right)$
Sat Vapour (g)	14.67	2437.16	2583.86	8.149	Sat Vapour (g)	14.67	2437.16	2583.86	8.149	Sat Vapour (g)	14.67	2437.16	2583.86	8.149
10	15.14	2451.75	2603.14	8.208	450	35.49	3125.91	3480.77	9.888	880	55.33	3907.47	4460.80	10.896
20	15.61	2466.22	2622.28	8.266	460	35.95	3142.61	3502.09	9.916	890	55.80	3927.17	4485.12	10.916
30	16.07	2480.64	2641.36	8.321	470	36.41	3159.38	3523.47	9.943	900	56.26	3946.95	4509.51	10.936
40	16.54	2495.05	2660.42	8.375	480	36.87	3176.22	3544.93	9.970	910	56.72	3966.79	4533.97	10.956
50	17.00	2509.46	2679.47	8.427	490	37.33	3193.13	3566.45	9.997	920	57.18	3986.69	4558.49	10.976
60	17.47	2523.87	2698.53	8.478	500	37.79	3210.10	3588.05	10.023	930	57.64	4006.66	4583.08	10.995
70	17.93	2538.31	2717.61	8.528	510	38.26	3227.15	3609.71	10.050	940	58.10	4026.70	4607.73	11.015
80	18.39	2552.78	2736.71	8.576	520	38.72	3244.27	3631.44	10.076	950	58.56	4046.80	4632.45	11.035
90	18.86	2567.27	2755.84	8.624	530	39.18	3261.46	3653.25	10.102	960	59.03	4066.97	4657.23	11.054
100	19.32	2581.81	2775.00	8.670	540	39.64	3278.71	3675.12	10.127	970	59.49	4087.20	4682.08	11.074
110	19.78	2596.38	2794.20	8.715	550	40.10	3296.04	3697.06	10.153	980	59.95	4107.49	4706.99	11.093
120	20.24	2611.00	2813.45	8.760	560	40.56	3313.44	3719.08	10.178	990	60.41	4127.85	4731.96	11.112
130	20.71	2625.67	2832.74	8.803	570	41.03	3330.91	3741.16	10.203	1000	60.87	4148.28	4757.00	11.131
140	21.17	2640.38	2852.08	8.846	580	41.49	3348.45	3763.32	10.228	1010	61.33	4168.76	4782.10	11.150
150	21.63	2655.14	2871.46	8.888	590	41.95	3366.06	3785.54	10.252	1020	61.80	4189.31	4807.26	11.169
160	22.09	2669.96	2890.90	8.929	600	42.41	3383.74	3807.84	10.276	1030	62.26	4209.92	4832.48	11.188
170	22.56	2684.83	2910.39	8.969	610	42.87	3401.49	3830.21	10.301	1040	62.72	4230.59	4857.77	11.206
180	23.02	2699.75	2929.94	9.008	620	43.33	3419.31	3852.65	10.325	1050	63.18	4251.32	4883.12	11.225
190	23.48	2714.74	2949.54	9.047	630	43.79	3437.21	3875.16	10.349	1060	63.64	4272.11	4908.52	11.243
200	23.94	2729.77	2969.20	9.086	640	44.26	3455.17	3897.74	10.372	1070	64.10	4292.96	4933.99	11.262
210	24.40	2744.87	2988.91	9.123	650	44.72	3473.21	3920.39	10.396	1080	64.56	4313.88	4959.52	11.280
220	24.87	2760.02	3008.68	9.160	660	45.18	3491.32	3943.11	10.419	1090	65.03	4334.85	4985.11	11.298
230	25.33	2775.23	3028.52	9.197	670	45.64	3509.50	3965.91	10.442	1100	65.49	4355.88	5010.76	11.316
240	25.79	2790.51	3048.41	9.233	680	46.10	3527.75	3988.77	10.465	1110	65.95	4376.97	5036.46	11.334
250	26.25	2805.84	3068.36	9.268	690	46.56	3546.07	4011.71	10.488	1120	66.41	4398.12	5062.23	11.352
260	26.71	2821.24	3088.37	9.303	700	47.03	3564.46	4034.72	10.511	1130	66.87	4419.33	5088.05	11.370
270	27.18	2836.69	3108.45	9.337	710	47.49	3582.92	4057.80	10.533	1140	67.33	4440.59	5113.93	11.388
280	27.64	2852.21	3128.58	9.371	720	47.95	3601.46	4080.94	10.556	1150	67.80	4461.91	5139.86	11.406
290	28.10	2867.79	3148.78	9.405	730	48.41	3620.06	4104.16	10.578	1160	68.26	4483.29	5165.86	11.423
300	28.56	2883.43	3169.04	9.438	740	48.87	3638.73	4127.45	10.600	1170	68.72	4504.73	5191.91	11.441
310	29.02	2899.14	3189.37	9.470	750	49.33	3657.48	4150.81	10.622	1180	69.18	4526.22	5218.01	11.458
320	29.48	2914.91	3209.76	9.502	760	49.80	3676.29	4174.24	10.644	1190	69.64	4547.76	5244.18	11.476
330	29.95	2930.75	3230.21	9.534	770	50.26	3695.18	4197.74	10.666	1200	70.10	4569.36	5270.39	11.493
340	30.41	2946.65	3250.73	9.566	780	50.72	3714.13	4221.31	10.687	1210	70.56	4591.02	5296.66	11.510
350	30.87	2962.61	3271.31	9.597	790	51.18	3733.16	4244.95	10.708	1220	71.03	4612.73	5322.99	11.528
360	31.33	2978.64	3291.95	9.627	800	51.64	3752.25	4268.66	10.730	1230	71.49	4634.49	5349.37	11.545
370	31.79	2994.74	3312.67	9.657	810	52.10	3771.41	4292.44	10.751	1240	71.95	4656.31	5375.80	11.562
380	32.25	3010.90	3333.45	9.687	820	52.56	3790.64	4316.29	10.772	1250	72.41	4678.18	5402.28	11.579
390	32.72	3027.13	3354.29	9.717	830	53.03	3809.94	4340.20	10.793	1260	72.87	4700.10	5428.82	11.595
400	33.18	3043.42	3375.20	9.746	840	53.49	3829.31	4364.19	10.814	1270	73.33	4722.07	5455.41	11.612
410	33.64	3059.79	3396.18	9.775	850	53.95	3848.75	4388.24	10.834	1280	73.80	4744.10	5482.05	11.629
420	34.10	3076.21	3417.23	9.804	860	54.41	3868.25	4412.36	10.855	1290	74.26	4766.18	5508.74	11.646
430	34.56	3092.71	3438.34	9.832	870	54.87	3887.83	4436.55	10.875	1300	74.72	4788.31	5535.49	11.662
440	35.02	3109.28	3459.52	9.860										

Table 3(A7) Superheated Steam Properties (Pressure Base) Pressure = 0.2 bar and saturation temperature, Tsat = 60.507°C

t (°C)	v $\left(\frac{m^3}{kg}\right)$	u $\left(\frac{kJ}{kg}\right)$	h $\left(\frac{kJ}{kg}\right)$	s $\left(\frac{kJ}{kg\ K}\right)$	t (°C)	v $\left(\frac{m^3}{kg}\right)$	u $\left(\frac{kJ}{kg}\right)$	h $\left(\frac{kJ}{kg}\right)$	s $\left(\frac{kJ}{kg\ K}\right)$	t (°C)	v $\left(\frac{m^3}{kg}\right)$	u $\left(\frac{kJ}{kg}\right)$	h $\left(\frac{kJ}{kg}\right)$	s $\left(\frac{kJ}{kg\ K}\right)$
Sat Vapour (g)	7.648	2455.98	2608.95	7.90725	Sat Vapour (g)	7.648	2455.98	2608.95	7.90725	Sat Vapour (g)	7.648	2455.98	2608.95	7.90725
10	7.884	2470.78	2628.46	7.965	450	18.071	3149.65	3511.07	9.607	880	27.995	3935.54	4495.45	10.604
20	8.119	2485.44	2647.82	8.021	460	18.302	3166.46	3532.49	9.635	890	28.226	3955.35	4519.87	10.624
30	8.353	2500.04	2667.1	8.074	470	18.532	3183.33	3553.97	9.661	900	28.457	3975.21	4544.35	10.644
40	8.587	2514.6	2686.34	8.127	480	18.763	3200.27	3575.53	9.688	910	28.688	3995.15	4568.9	10.664
50	8.820	2529.15	2705.55	8.177	490	18.994	3217.28	3597.16	9.715	920	28.919	4015.15	4593.52	10.684
60	9.053	2543.7	2724.76	8.227	500	19.225	3234.35	3618.85	9.741	930	29.149	4035.21	4618.2	10.704
70	9.286	2558.26	2743.97	8.275	510	19.456	3251.5	3640.62	9.767	940	29.380	4055.34	4642.95	10.723
80	9.518	2572.85	2763.2	8.322	520	19.687	3268.72	3662.45	9.792	950	29.611	4075.54	4667.76	10.742
90	9.750	2587.46	2782.46	8.368	530	19.917	3286.01	3684.36	9.818	960	29.842	4095.8	4692.63	10.762
100	9.982	2602.1	2801.74	8.413	540	20.148	3303.37	3706.34	9.843	970	30.073	4116.12	4717.57	10.781
110	10.214	2616.79	2821.06	8.457	550	20.379	3320.8	3728.38	9.868	980	30.303	4136.5	4742.57	10.8
120	10.445	2631.51	2840.42	8.501	560	20.610	3338.3	3750.5	9.893	990	30.534	4156.95	4767.63	10.819
130	10.677	2646.28	2859.82	8.543	570	20.841	3355.87	3772.69	9.918	1000	30.765	4177.46	4792.76	10.838
140	10.909	2661.09	2879.26	8.585	580	21.072	3373.51	3794.94	9.943	1010	30.996	4198.04	4817.95	10.857
150	11.140	2675.96	2898.76	8.625	590	21.302	3391.23	3817.27	9.967	1020	31.226	4218.67	4843.2	10.876
160	11.371	2690.87	2918.3	8.665	600	21.533	3409.01	3839.67	9.991	1030	31.457	4239.37	4868.51	10.894
170	11.603	2705.84	2937.9	8.705	610	21.764	3426.86	3862.14	10.015	1040	31.688	4260.13	4893.89	10.913
180	11.834	2720.86	2957.54	8.743	620	21.995	3444.79	3884.69	10.039	1050	31.919	4280.95	4919.32	10.931
190	12.065	2735.94	2977.25	8.781	630	22.226	3462.79	3907.3	10.062	1060	32.150	4301.83	4944.81	10.95
200	12.297	2751.08	2997.01	8.819	640	22.456	3480.86	3929.98	10.086	1070	32.380	4322.76	4970.37	10.968
210	12.528	2766.27	3016.82	8.856	650	22.687	3498.99	3952.74	10.109	1080	32.611	4343.76	4995.98	10.986
220	12.759	2781.52	3036.69	8.892	660	22.918	3517.2	3975.56	10.132	1090	32.842	4364.82	5021.66	11.004
230	12.990	2796.83	3056.63	8.928	670	23.149	3535.49	3998.46	10.155	1100	33.073	4385.94	5047.39	11.022
240	13.221	2812.2	3076.62	8.963	680	23.380	3553.84	4021.43	10.178	1110	33.303	4407.11	5073.18	11.04
250	13.452	2827.63	3096.67	8.997	690	23.610	3572.26	4044.47	10.2	1120	33.534	4428.34	5099.02	11.058
260	13.683	2843.12	3116.78	9.032	700	23.841	3590.75	4067.58	10.223	1130	33.765	4449.63	5124.93	11.076
270	13.914	2858.67	3136.95	9.065	710	24.072	3609.32	4090.76	10.245	1140	33.996	4470.98	5150.89	11.093
280	14.145	2874.29	3157.19	9.099	720	24.303	3627.95	4114.01	10.267	1150	34.226	4492.38	5176.91	11.111
290	14.376	2889.96	3177.49	9.131	730	24.534	3646.66	4137.33	10.289	1160	34.457	4513.84	5202.98	11.129
300	14.607	2905.7	3197.85	9.164	740	24.764	3665.43	4160.72	10.311	1170	34.688	4535.35	5229.11	11.146
310	14.838	2921.51	3218.27	9.196	750	24.995	3684.28	4184.18	10.333	1180	34.919	4556.92	5255.3	11.163
320	15.069	2937.38	3238.76	9.227	760	25.226	3703.19	4207.71	10.355	1190	35.150	4578.55	5281.54	11.181
330	15.300	2953.31	3259.31	9.259	770	25.457	3722.18	4231.31	10.376	1200	35.380	4600.22	5307.83	11.198
340	15.531	2969.31	3279.92	9.289	780	25.688	3741.23	4254.98	10.398	1210	35.611	4621.96	5334.18	11.215
350	15.762	2985.37	3300.6	9.32	790	25.918	3760.35	4278.72	10.419	1220	35.842	4643.74	5360.58	11.232
360	15.993	3001.5	3321.35	9.35	800	26.149	3779.55	4302.53	10.44	1230	36.073	4665.58	5387.04	11.249
370	16.224	3017.69	3342.16	9.38	810	26.380	3798.81	4326.41	10.461	1240	36.303	4687.48	5413.54	11.266
380	16.455	3033.95	3363.04	9.409	820	26.611	3818.14	4350.35	10.482	1250	36.534	4709.42	5440.1	11.283
390	16.685	3050.28	3383.99	9.439	830	26.842	3837.54	4374.37	10.503	1260	36.765	4731.42	5466.71	11.299
400	16.916	3066.67	3405	9.467	840	27.072	3857	4398.45	10.523	1270	36.996	4753.46	5493.38	11.316
410	17.147	3083.13	3426.08	9.496	850	27.303	3876.54	4422.6	10.544	1280	37.226	4775.56	5520.09	11.333
420	17.378	3099.66	3447.22	9.524	860	27.534	3896.14	4446.82	10.564	1290	37.457	4797.71	5546.86	11.349
430	17.609	3116.26	3468.44	9.552	870	27.765	3915.81	4471.1	10.584	1300	37.688	4819.91	5573.67	11.366
440	17.840	3132.92	3489.72	9.58										

Table 3(A8) Superheated Steam Properties (Pressure Base) Pressure = 0.3 bar and saturation temperature, Tsat = 69.094°C

t (°C)	v $\left(\frac{m^3}{kg}\right)$	u $\left(\frac{kJ}{kg}\right)$	h $\left(\frac{kJ}{kg}\right)$	s $\left(\frac{kJ}{kg\,K}\right)$
Sat Vapour (g)	5.229	2467.70	2624.56	7.76751
10	5.387	2482.66	2644.26	7.824
20	5.544	2497.47	2663.79	7.879
30	5.701	2512.19	2683.22	7.932
40	5.857	2526.87	2702.59	7.983
50	6.013	2541.53	2721.93	8.033
60	6.169	2556.18	2741.25	8.082
70	6.324	2570.83	2760.57	8.129
80	6.480	2585.5	2779.89	8.176
90	6.635	2600.2	2799.24	8.221
100	6.790	2614.92	2818.61	8.265
110	6.944	2629.68	2838.01	8.308
120	7.099	2644.48	2857.45	8.351
130	7.254	2659.32	2876.93	8.393
140	7.408	2674.2	2896.45	8.434
150	7.563	2689.14	2916.02	8.474
160	7.717	2704.12	2935.64	8.513
170	7.872	2719.15	2955.3	8.552
180	8.026	2734.24	2975.02	8.59
190	8.180	2749.39	2994.79	8.628
200	8.335	2764.59	3014.62	8.665
210	8.489	2779.84	3034.5	8.701
220	8.643	2795.16	3054.45	8.737
230	8.797	2810.53	3074.44	8.772
240	8.951	2825.97	3094.5	8.807
250	9.105	2841.46	3114.62	8.841
260	9.259	2857.01	3134.8	8.875
270	9.414	2872.63	3155.04	8.908
280	9.568	2888.31	3175.34	8.941
290	9.722	2904.05	3195.7	8.973
300	9.876	2919.85	3216.12	9.005
310	10.030	2935.72	3236.61	9.037
320	10.184	2951.65	3257.17	9.068
330	10.338	2967.65	3277.78	9.099
340	10.492	2983.71	3298.46	9.13
350	10.646	2999.84	3319.21	9.16
360	10.800	3016.03	3340.02	9.19
370	10.954	3032.28	3360.9	9.219
380	11.108	3048.61	3381.84	9.249
390	11.262	3065	3402.85	9.277
400	11.416	3081.46	3423.93	9.306
410	11.570	3097.98	3445.07	9.334
420	11.724	3114.57	3466.28	9.362
430	11.878	3131.23	3487.56	9.39
440	12.031	3147.96	3508.91	9.418

t (°C)	v $\left(\frac{m^3}{kg}\right)$	u $\left(\frac{kJ}{kg}\right)$	h $\left(\frac{kJ}{kg}\right)$	s $\left(\frac{kJ}{kg\,K}\right)$
Sat Vapour (g)	5.229	2467.70	2624.56	7.76751
450	12.185	3164.76	3530.32	9.445
460	12.339	3181.63	3551.81	9.472
470	12.493	3198.56	3573.36	9.498
480	12.647	3215.57	3594.98	9.525
490	12.801	3232.64	3616.67	9.551
500	12.955	3249.79	3638.44	9.577
510	13.109	3267	3660.27	9.603
520	13.263	3284.28	3682.17	9.628
530	13.417	3301.64	3704.14	9.654
540	13.571	3319.06	3726.18	9.679
550	13.725	3336.56	3748.29	9.704
560	13.878	3354.12	3770.48	9.728
570	14.032	3371.76	3792.73	9.753
580	14.186	3389.47	3815.05	9.777
590	14.340	3407.24	3837.45	9.801
600	14.494	3425.09	3859.91	9.825
610	14.648	3443.01	3882.45	9.849
620	14.802	3461	3905.06	9.873
630	14.956	3479.07	3927.74	9.896
640	15.110	3497.2	3950.49	9.92
650	15.264	3515.4	3973.31	9.943
660	15.417	3533.68	3996.2	9.966
670	15.571	3552.03	4019.16	9.988
680	15.725	3570.44	4042.2	10.011
690	15.879	3588.93	4065.3	10.034
700	16.033	3607.49	4088.47	10.056
710	16.187	3626.12	4111.72	10.078
720	16.341	3644.82	4135.03	10.1
730	16.495	3663.58	4158.42	10.122
740	16.648	3682.42	4181.87	10.144
750	16.802	3701.33	4205.4	10.166
760	16.956	3720.31	4228.99	10.187
770	17.110	3739.36	4252.66	10.208
780	17.264	3758.48	4276.39	10.23
790	17.418	3777.66	4300.2	10.251
800	17.572	3796.92	4324.07	10.272
810	17.726	3816.24	4348.01	10.293
820	17.879	3835.64	4372.02	10.313
830	18.033	3855.1	4396.09	10.334
840	18.187	3874.62	4420.24	10.354
850	18.341	3894.22	4444.45	10.375
860	18.495	3913.88	4468.73	10.395
870	18.649	3933.61	4493.07	10.415

t (°C)	v $\left(\frac{m^3}{kg}\right)$	u $\left(\frac{kJ}{kg}\right)$	h $\left(\frac{kJ}{kg}\right)$	s $\left(\frac{kJ}{kg\,K}\right)$
Sat Vapour (g)	5.229	2467.70	2624.56	7.76751
880	18.803	3953.41	4517.48	10.435
890	18.956	3973.27	4541.96	10.455
900	19.110	3993.2	4566.51	10.475
910	19.264	4013.19	4591.12	10.495
920	19.418	4033.25	4615.79	10.515
930	19.572	4053.38	4640.53	10.534
940	19.726	4073.57	4665.34	10.553
950	19.880	4093.82	4690.2	10.573
960	20.033	4114.13	4715.14	10.592
970	20.187	4134.52	4740.13	10.611
980	20.341	4154.96	4765.19	10.63
990	20.495	4175.46	4790.31	10.649
1000	20.649	4196.03	4815.5	10.668
1010	20.803	4216.66	4840.74	10.687
1020	20.957	4237.35	4866.05	10.705
1030	21.110	4258.11	4891.42	10.724
1040	21.264	4278.92	4916.85	10.742
1050	21.418	4299.79	4942.34	10.761
1060	21.572	4320.73	4967.88	10.779
1070	21.726	4341.72	4993.49	10.797
1080	21.880	4362.77	5019.16	10.815
1090	22.034	4383.88	5044.89	10.833
1100	22.187	4405.05	5070.67	10.851
1110	22.341	4426.28	5096.51	10.869
1120	22.495	4447.56	5122.41	10.887
1130	22.649	4468.9	5148.37	10.905
1140	22.803	4490.3	5174.38	10.922
1150	22.957	4511.75	5200.45	10.94
1160	23.111	4533.26	5226.58	10.957
1170	23.264	4554.83	5252.76	10.975
1180	23.418	4576.44	5278.99	10.992
1190	23.572	4598.12	5305.28	11.009
1200	23.726	4619.85	5331.62	11.026
1210	23.880	4641.63	5358.02	11.043
1220	24.034	4663.46	5384.47	11.06
1230	24.187	4685.35	5410.97	11.077
1240	24.341	4707.29	5437.53	11.094
1250	24.495	4729.28	5464.14	11.111
1260	24.649	4751.32	5490.79	11.127
1270	24.803	4773.42	5517.5	11.144
1280	24.957	4795.56	5544.26	11.161
1290	25.111	4817.76	5571.07	11.177
1300	25.264	4840	5597.93	11.193

Table 3(A9) Superheated Steam Properties (Pressure Base) Pressure = 0.4 bar and saturation temperature, Tsat = 75.856°C

t (°C)	v $\left(\frac{m^3}{kg}\right)$	u $\left(\frac{kJ}{kg}\right)$	h $\left(\frac{kJ}{kg}\right)$	s $\left(\frac{kJ}{kg\,K}\right)$	t (°C)	v $\left(\frac{m^3}{kg}\right)$	u $\left(\frac{kJ}{kg}\right)$	h $\left(\frac{kJ}{kg}\right)$	s $\left(\frac{kJ}{kg\,K}\right)$	t (°C)	v $\left(\frac{m^3}{kg}\right)$	u $\left(\frac{kJ}{kg}\right)$	h $\left(\frac{kJ}{kg}\right)$	s $\left(\frac{kJ}{kg\,K}\right)$
Sat Vapour (g)	*3.993*	*2476.34*	*2636.07*	*7.66905*	*Sat Vapour (g)*	*3.993*	*2476.34*	*2636.07*	*7.66905*	*Sat Vapour (g)*	*3.993*	*2476.34*	*2636.07*	*7.66905*
10	4.112	2491.43	2655.92	7.725	450	9.217	3176.09	3544.74	9.33	880	14.180	3966.8	4534	10.316
20	4.231	2506.36	2675.6	7.779	460	9.332	3193	3566.28	9.357	890	14.295	3986.71	4558.52	10.336
30	4.349	2521.2	2695.15	7.831	470	9.447	3209.98	3587.88	9.383	900	14.411	4006.69	4583.11	10.356
40	4.467	2535.97	2714.64	7.882	480	9.563	3227.04	3609.55	9.41	910	14.526	4026.72	4607.76	10.375
50	4.584	2550.72	2734.08	7.932	490	9.678	3244.16	3631.29	9.436	920	14.641	4046.83	4632.48	10.395
60	4.701	2565.45	2753.49	7.98	500	9.794	3261.35	3653.11	9.462	930	14.757	4067	4657.27	10.414
70	4.818	2580.18	2772.89	8.026	510	9.909	3278.62	3674.99	9.487	940	14.872	4087.23	4682.12	10.434
80	4.935	2594.92	2792.3	8.072	520	10.025	3295.95	3696.94	9.513	950	14.988	4107.52	4707.03	10.453
90	5.051	2609.68	2811.72	8.117	530	10.140	3313.35	3718.96	9.538	960	15.103	4127.88	4732	10.472
100	5.167	2624.47	2831.17	8.161	540	10.256	3330.83	3741.05	9.563	970	15.218	4148.31	4757.04	10.491
110	5.284	2639.29	2850.64	8.204	550	10.371	3348.37	3763.21	9.588	980	15.334	4168.79	4782.15	10.51
120	5.400	2654.15	2870.14	8.246	560	10.487	3365.99	3785.45	9.612	990	15.449	4189.34	4807.31	10.529
130	5.516	2669.05	2889.69	8.287	570	10.602	3383.67	3807.75	9.637	1000	15.565	4209.95	4832.54	10.548
140	5.632	2683.99	2909.27	8.327	580	10.717	3401.43	3830.12	9.661	1010	15.680	4230.62	4857.82	10.566
150	5.748	2698.98	2928.9	8.367	590	10.833	3419.26	3852.57	9.685	1020	15.795	4251.36	4883.17	10.585
160	5.864	2714.01	2948.57	8.406	600	10.948	3437.15	3875.08	9.709	1030	15.911	4272.15	4908.58	10.604
170	5.980	2729.1	2968.3	8.445	610	11.064	3455.12	3897.67	9.732	1040	16.026	4293.01	4934.05	10.622
180	6.096	2744.24	2988.07	8.482	620	11.179	3473.16	3920.33	9.756	1050	16.142	4313.92	4959.58	10.64
190	6.212	2759.44	3007.9	8.519	630	11.295	3491.28	3943.06	9.779	1060	16.257	4334.89	4985.17	10.659
200	6.327	2774.69	3027.78	8.556	640	11.410	3509.46	3965.85	9.802	1070	16.372	4355.93	5010.82	10.677
210	6.443	2790	3047.72	8.592	650	11.525	3527.71	3988.72	9.825	1080	16.488	4377.02	5036.53	10.695
220	6.559	2805.36	3067.71	8.627	660	11.641	3546.03	4011.67	9.848	1090	16.603	4398.17	5062.29	10.713
230	6.674	2820.79	3087.76	8.662	670	11.756	3564.43	4034.68	9.871	1100	16.719	4419.38	5088.12	10.731
240	6.790	2836.27	3107.87	8.697	680	11.872	3582.9	4057.76	9.894	1110	16.834	4440.64	5114	10.748
250	6.906	2851.81	3128.04	8.731	690	11.987	3601.43	4080.91	9.916	1120	16.949	4461.97	5139.94	10.766
260	7.021	2867.42	3148.27	8.764	700	12.102	3620.04	4104.14	9.938	1130	17.065	4483.34	5165.93	10.784
270	7.137	2883.08	3168.56	8.797	710	12.218	3638.71	4127.43	9.96	1140	17.180	4504.78	5191.98	10.801
280	7.253	2898.81	3188.91	8.83	720	12.333	3657.46	4150.79	9.982	1150	17.296	4526.27	5218.09	10.819
290	7.368	2914.6	3209.33	8.862	730	12.449	3676.28	4174.23	10.004	1160	17.411	4547.82	5244.25	10.836
300	7.484	2930.45	3229.8	8.894	740	12.564	3695.17	4197.73	10.026	1170	17.526	4569.42	5270.47	10.853
310	7.599	2946.37	3250.34	8.925	750	12.680	3714.12	4221.3	10.047	1180	17.642	4591.08	5296.74	10.871
320	7.715	2962.35	3270.94	8.956	760	12.795	3733.15	4244.95	10.069	1190	17.757	4612.79	5323.07	10.888
330	7.830	2978.39	3291.61	8.987	770	12.910	3752.25	4268.66	10.09	1200	17.873	4634.55	5349.45	10.905
340	7.946	2994.5	3312.34	9.017	780	13.026	3771.41	4292.44	10.111	1210	17.988	4656.37	5375.88	10.922
350	8.062	3010.67	3333.13	9.047	790	13.141	3790.64	4316.29	10.132	1220	18.103	4678.24	5402.37	10.939
360	8.177	3026.91	3354	9.077	800	13.257	3809.95	4340.21	10.153	1230	18.219	4700.16	5428.91	10.956
370	8.293	3043.22	3374.92	9.106	810	13.372	3829.32	4364.2	10.174	1240	18.334	4722.14	5455.5	10.973
380	8.408	3059.59	3395.91	9.135	820	13.487	3848.76	4388.25	10.195	1250	18.449	4744.16	5482.14	10.989
390	8.524	3076.03	3416.97	9.164	830	13.603	3868.26	4412.37	10.215	1260	18.565	4766.24	5508.83	11.006
400	8.639	3092.54	3438.1	9.192	840	13.718	3887.84	4436.57	10.235	1270	18.680	4788.37	5535.58	11.022
410	8.755	3109.11	3459.29	9.22	850	13.834	3907.48	4460.82	10.256	1280	18.796	4810.55	5562.37	11.039
420	8.870	3125.75	3480.55	9.248	860	13.949	3927.19	4485.15	10.276	1290	18.911	4832.78	5589.22	11.055
430	8.986	3142.46	3501.88	9.276	870	14.064	3946.96	4509.54	10.296	1300	19.026	4855.06	5616.11	11.072
440	9.101	3159.24	3523.28	9.303										

Table 3(A10) Superheated Steam Properties (Pressure Base) Pressure = 0.5 bar and saturation temperature, Tsat = 81.316°C

t (°C)	$v\left(\frac{m^3}{kg}\right)$	$u\left(\frac{kJ}{kg}\right)$	$h\left(\frac{kJ}{kg}\right)$	$s\left(\frac{kJ}{kg\,K}\right)$	*t (°C)*	$v\left(\frac{m^3}{kg}\right)$	$u\left(\frac{kJ}{kg}\right)$	$h\left(\frac{kJ}{kg}\right)$	$s\left(\frac{kJ}{kg\,K}\right)$	*t (°C)*	$v\left(\frac{m^3}{kg}\right)$	$u\left(\frac{kJ}{kg}\right)$	$h\left(\frac{kJ}{kg}\right)$	$s\left(\frac{kJ}{kg\,K}\right)$
Sat Vapour (g)	*3.240*	*2483.22*	*2645.24*	*7.59306*	*Sat Vapour (g)*	*3.240*	*2483.22*	*2645.24*	*7.59306*	*Sat Vapour (g)*	*3.240*	*2483.22*	*2645.24*	*7.59306*
10	3.336	2498.43	2665.23	7.649	450	7.423	3185.24	3556.4	9.242	880	11.394	3977.64	4547.35	10.224
20	3.431	2513.47	2685.03	7.702	460	7.516	3202.2	3577.97	9.268	890	11.487	3997.58	4571.91	10.244
30	3.526	2528.4	2704.7	7.754	470	7.608	3219.22	3599.62	9.295	900	11.579	4017.59	4596.53	10.263
40	3.620	2543.26	2724.28	7.804	480	7.700	3236.31	3621.33	9.321	910	11.671	4037.67	4621.22	10.283
50	3.715	2558.09	2743.81	7.853	490	7.793	3253.48	3643.11	9.347	920	11.764	4057.81	4645.98	10.303
60	3.808	2572.89	2763.31	7.901	500	7.885	3270.71	3664.96	9.373	930	11.856	4078.01	4670.8	10.322
70	3.902	2587.68	2782.79	7.947	510	7.977	3288.01	3686.88	9.398	940	11.948	4098.28	4695.68	10.341
80	3.996	2602.49	2802.27	7.993	520	8.070	3305.39	3708.88	9.423	950	12.040	4118.61	4720.63	10.36
90	4.089	2617.31	2821.76	8.037	530	8.162	3322.83	3730.94	9.448	960	12.133	4139	4745.64	10.38
100	4.182	2632.15	2841.26	8.08	540	8.255	3340.34	3753.07	9.473	970	12.225	4159.46	4770.72	10.399
110	4.275	2647.02	2860.79	8.123	550	8.347	3357.93	3775.27	9.498	980	12.317	4179.98	4795.85	10.417
120	4.368	2661.93	2880.35	8.165	560	8.439	3375.58	3797.55	9.523	990	12.410	4200.57	4821.05	10.436
130	4.461	2676.88	2899.95	8.206	570	8.532	3393.31	3819.89	9.547	1000	12.502	4221.21	4846.31	10.455
140	4.554	2691.87	2919.59	8.246	580	8.624	3411.1	3842.3	9.571	1010	12.594	4241.92	4871.63	10.474
150	4.647	2706.91	2939.27	8.285	590	8.716	3428.97	3864.79	9.595	1020	12.687	4262.68	4897.02	10.492
160	4.740	2721.99	2958.99	8.324	600	8.809	3446.91	3887.35	9.619	1030	12.779	4283.51	4922.46	10.511
170	4.833	2737.12	2978.76	8.362	610	8.901	3464.92	3909.97	9.642	1040	12.871	4304.4	4947.96	10.529
180	4.926	2752.31	2998.59	8.399	620	8.993	3483	3932.67	9.666	1050	12.964	4325.34	4973.53	10.547
190	5.018	2767.55	3018.46	8.436	630	9.086	3501.15	3955.44	9.689	1060	13.056	4346.35	4999.15	10.565
200	5.111	2782.84	3038.39	8.472	640	9.178	3519.37	3978.28	9.712	1070	13.148	4367.42	5024.83	10.584
210	5.204	2798.19	3058.37	8.508	650	9.271	3537.66	4001.19	9.735	1080	13.241	4388.54	5050.57	10.602
220	5.296	2813.6	3078.41	8.543	660	9.363	3556.03	4024.17	9.758	1090	13.333	4409.72	5076.37	10.619
230	5.389	2829.06	3098.51	8.578	670	9.455	3574.46	4047.22	9.78	1100	13.425	4430.96	5102.22	10.637
240	5.481	2844.59	3118.66	8.612	680	9.548	3592.97	4070.34	9.803	1110	13.518	4452.26	5128.14	10.655
250	5.574	2860.17	3138.87	8.646	690	9.640	3611.54	4093.53	9.825	1120	13.610	4473.61	5154.11	10.673
260	5.667	2875.82	3159.14	8.679	700	9.732	3630.19	4116.8	9.847	1130	13.702	4495.02	5180.13	10.69
270	5.759	2891.52	3179.48	8.712	710	9.825	3648.91	4140.13	9.869	1140	13.795	4516.49	5206.21	10.708
280	5.852	2907.29	3199.87	8.744	720	9.917	3667.69	4163.53	9.891	1150	13.887	4538.01	5232.35	10.725
290	5.944	2923.12	3220.32	8.776	730	10.009	3686.55	4187.01	9.913	1160	13.979	4559.59	5258.55	10.743
300	6.037	2939.01	3240.84	8.808	740	10.102	3705.47	4210.55	9.934	1170	14.071	4581.22	5284.79	10.76
310	6.129	2954.97	3261.42	8.839	750	10.194	3724.47	4234.16	9.956	1180	14.164	4602.91	5311.1	10.777
320	6.222	2970.98	3282.06	8.87	760	10.286	3743.53	4257.84	9.977	1190	14.256	4624.65	5337.45	10.794
330	6.314	2987.07	3302.77	8.9	770	10.379	3762.67	4281.6	9.999	1200	14.348	4646.44	5363.86	10.811
340	6.407	3003.22	3323.54	8.931	780	10.471	3781.87	4305.42	10.02	1210	14.441	4668.29	5390.32	10.828
350	6.499	3019.43	3344.38	8.96	790	10.563	3801.14	4329.3	10.041	1220	14.533	4690.19	5416.84	10.845
360	6.591	3035.71	3365.28	8.99	800	10.656	3820.48	4353.26	10.061	1230	14.625	4712.14	5443.41	10.862
370	6.684	3052.06	3386.25	9.019	810	10.748	3839.89	4377.29	10.082	1240	14.718	4734.14	5470.02	10.879
380	6.776	3068.47	3407.28	9.048	820	10.840	3859.37	4401.38	10.103	1250	14.810	4756.2	5496.7	10.895
390	6.869	3084.95	3428.38	9.076	830	10.933	3878.91	4425.54	10.123	1260	14.902	4778.3	5523.42	10.912
400	6.961	3101.49	3449.55	9.105	840	11.025	3898.53	4449.77	10.144	1270	14.995	4800.46	5550.19	10.928
410	7.054	3118.11	3470.78	9.133	850	11.117	3918.2	4474.06	10.164	1280	15.087	4822.67	5577.01	10.945
420	7.146	3134.79	3492.09	9.16	860	11.210	3937.95	4498.42	10.184	1290	15.179	4844.92	5603.88	10.961
430	7.238	3151.54	3513.45	9.188	870	11.302	3957.76	4522.85	10.204	1300	15.272	4867.23	5630.8	10.978
440	7.331	3168.35	3534.89	9.215										

Table 3(A11) Superheated Steam Properties (Pressure Base) Pressure = 0.6 bar and saturation temperature, Tsat = 85.926°C

t (°C)	v $\left(\frac{m^3}{kg}\right)$	u $\left(\frac{kJ}{kg}\right)$	h $\left(\frac{kJ}{kg}\right)$	s $\left(\frac{kJ}{kg\,K}\right)$
Sat Vapour (g)	2.732	2488.96	2652.88	7.53121
10	2.812	2504.28	2673	7.586
20	2.892	2519.42	2692.92	7.64
30	2.971	2534.43	2712.69	7.691
40	3.050	2549.37	2732.36	7.741
50	3.128	2564.27	2751.97	7.79
60	3.207	2579.13	2771.54	7.837
70	3.285	2593.99	2791.09	7.883
80	3.363	2608.85	2810.63	7.928
90	3.441	2623.72	2830.18	7.972
100	3.519	2638.61	2849.74	8.015
110	3.597	2653.53	2869.33	8.057
120	3.674	2668.48	2888.94	8.099
130	3.752	2683.48	2908.59	8.139
140	3.829	2698.51	2928.27	8.179
150	3.907	2713.58	2948	8.218
160	3.984	2728.71	2967.76	8.257
170	4.062	2743.88	2987.58	8.295
180	4.139	2759.11	3007.45	8.332
190	4.216	2774.38	3027.36	8.368
200	4.294	2789.71	3047.33	8.404
210	4.371	2805.1	3067.35	8.44
220	4.448	2820.54	3087.43	8.475
230	4.525	2836.04	3107.56	8.509
240	4.603	2851.6	3127.75	8.543
250	4.680	2867.22	3148	8.577
260	4.757	2882.9	3168.31	8.61
270	4.834	2898.64	3188.68	8.643
280	4.911	2914.44	3209.11	8.675
290	4.988	2930.31	3229.6	8.707
300	5.065	2946.24	3250.16	8.738
310	5.142	2962.23	3270.77	8.769
320	5.220	2978.28	3291.45	8.8
330	5.297	2994.4	3312.19	8.83
340	5.374	3010.58	3333	8.86
350	5.451	3026.83	3353.87	8.89
360	5.528	3043.14	3374.81	8.919
370	5.605	3059.52	3395.81	8.948
380	5.682	3075.97	3416.88	8.977
390	5.759	3092.48	3438.02	9.005
400	5.836	3109.06	3459.22	9.033
410	5.913	3125.71	3480.49	9.061
420	5.990	3142.42	3501.82	9.089
430	6.067	3159.2	3523.23	9.116
440	6.144	3176.06	3544.7	9.143

t (°C)	v $\left(\frac{m^3}{kg}\right)$	u $\left(\frac{kJ}{kg}\right)$	h $\left(\frac{kJ}{kg}\right)$	s $\left(\frac{kJ}{kg\,K}\right)$
Sat Vapour (g)	2.732	2488.96	2652.88	7.53121
450	6.221	3192.98	3566.24	9.17
460	6.298	3209.97	3587.85	9.196
470	6.375	3227.02	3609.53	9.223
480	6.452	3244.15	3631.27	9.249
490	6.529	3261.35	3653.09	9.275
500	6.606	3278.61	3674.98	9.3
510	6.683	3295.95	3696.93	9.326
520	6.760	3313.36	3718.96	9.351
530	6.837	3330.84	3741.06	9.376
540	6.914	3348.38	3763.22	9.401
550	6.991	3366	3785.46	9.425
560	7.068	3383.69	3807.77	9.45
570	7.145	3401.45	3830.15	9.474
580	7.222	3419.28	3852.59	9.498
590	7.299	3437.18	3875.11	9.522
600	7.376	3455.15	3897.7	9.545
610	7.453	3473.2	3920.37	9.569
620	7.530	3491.31	3943.1	9.592
630	7.607	3509.5	3965.9	9.615
640	7.684	3527.75	3988.77	9.638
650	7.761	3546.08	4011.72	9.661
660	7.838	3564.48	4034.73	9.684
670	7.915	3582.94	4057.82	9.706
680	7.992	3601.48	4080.97	9.729
690	8.069	3620.09	4104.2	9.751
700	8.146	3638.77	4127.5	9.773
710	8.222	3657.52	4150.86	9.795
720	8.299	3676.34	4174.3	9.817
730	8.376	3695.23	4197.81	9.839
740	8.453	3714.19	4221.38	9.86
750	8.530	3733.21	4245.03	9.882
760	8.607	3752.31	4268.74	9.903
770	8.684	3771.48	4292.53	9.924
780	8.761	3790.71	4316.38	9.945
790	8.838	3810.02	4340.3	9.966
800	8.915	3829.39	4364.29	9.987
810	8.992	3848.83	4388.35	10.007
820	9.069	3868.34	4412.47	10.028
830	9.146	3887.92	4436.66	10.048
840	9.223	3907.56	4460.92	10.069
850	9.300	3927.27	4485.25	10.089
860	9.377	3947.05	4509.64	10.109
870	9.454	3966.89	4534.1	10.129

t (°C)	v $\left(\frac{m^3}{kg}\right)$	u $\left(\frac{kJ}{kg}\right)$	h $\left(\frac{kJ}{kg}\right)$	s $\left(\frac{kJ}{kg\,K}\right)$
Sat Vapour (g)	2.732	2488.96	2652.88	7.53121
880	9.531	3986.8	4558.63	10.149
890	9.608	4006.77	4583.22	10.169
900	9.684	4026.81	4607.88	10.188
910	9.761	4046.92	4632.6	10.208
920	9.838	4067.09	4657.38	10.227
930	9.915	4087.32	4682.23	10.247
940	9.992	4107.62	4707.15	10.266
950	10.069	4127.98	4732.12	10.285
960	10.146	4148.4	4757.17	10.304
970	10.223	4168.89	4782.27	10.323
980	10.300	4189.44	4807.43	10.342
990	10.377	4210.05	4832.66	10.361
1000	10.454	4230.72	4857.95	10.379
1010	10.531	4251.46	4883.3	10.398
1020	10.608	4272.25	4908.71	10.417
1030	10.685	4293.11	4934.18	10.435
1040	10.762	4314.02	4959.72	10.453
1050	10.839	4335	4985.31	10.471
1060	10.915	4356.03	5010.96	10.49
1070	10.992	4377.13	5036.67	10.508
1080	11.069	4398.28	5062.43	10.526
1090	11.146	4419.49	5088.26	10.544
1100	11.223	4440.75	5114.14	10.561
1110	11.300	4462.08	5140.08	10.579
1120	11.377	4483.46	5166.08	10.597
1130	11.454	4504.89	5192.13	10.614
1140	11.531	4526.39	5218.24	10.632
1150	11.608	4547.93	5244.4	10.649
1160	11.685	4569.54	5270.62	10.666
1170	11.762	4591.19	5296.89	10.684
1180	11.839	4612.91	5323.22	10.701
1190	11.916	4634.67	5349.6	10.718
1200	11.992	4656.49	5376.04	10.735
1210	12.069	4678.36	5402.52	10.752
1220	12.146	4700.28	5429.06	10.769
1230	12.223	4722.26	5455.65	10.785
1240	12.300	4744.29	5482.3	10.802
1250	12.377	4766.37	5508.99	10.819
1260	12.454	4788.5	5535.74	10.835
1270	12.531	4810.68	5562.53	10.852
1280	12.608	4832.9	5589.38	10.868
1290	12.685	4855.18	5616.27	10.885
1300	12.762	4877.51	5643.22	10.901

Table 3(A12) Superheated Steam Properties (Pressure Base) Pressure = 0.7 bar and saturation temperature, Tsat = 89.931°C

t (°C)	$v\left(\frac{m^3}{kg}\right)$	$u\left(\frac{kJ}{kg}\right)$	$h\left(\frac{kJ}{kg}\right)$	$s\left(\frac{kJ}{kg\,K}\right)$	t (°C)	$v\left(\frac{m^3}{kg}\right)$	$u\left(\frac{kJ}{kg}\right)$	$h\left(\frac{kJ}{kg}\right)$	$s\left(\frac{kJ}{kg\,K}\right)$	t (°C)	$v\left(\frac{m^3}{kg}\right)$	$u\left(\frac{kJ}{kg}\right)$	$h\left(\frac{kJ}{kg}\right)$	$s\left(\frac{kJ}{kg\,K}\right)$
Sat Vapour (g)	2.365	2493.89	2659.45	7.47907	Sat Vapour (g)	2.365	2493.89	2659.45	7.47907	Sat Vapour (g)	2.365	2493.89	2659.45	7.47907
10	2.434	2509.32	2679.69	7.534	450	5.358	3199.7	3574.79	9.109	880	8.195	3994.76	4568.44	10.086
20	2.502	2524.54	2699.71	7.587	460	5.425	3216.72	3596.43	9.136	890	8.261	4014.77	4593.06	10.105
30	2.571	2539.63	2719.57	7.638	470	5.491	3233.81	3618.14	9.162	900	8.327	4034.83	4617.74	10.125
40	2.638	2554.65	2739.33	7.688	480	5.557	3250.97	3639.92	9.188	910	8.393	4054.96	4642.49	10.145
50	2.706	2569.6	2759.02	7.736	490	5.623	3268.19	3661.77	9.214	920	8.459	4075.16	4667.3	10.164
60	2.773	2584.53	2778.66	7.783	500	5.689	3285.49	3683.68	9.239	930	8.525	4095.42	4692.18	10.183
70	2.840	2599.44	2798.27	7.829	510	5.755	3302.86	3705.67	9.264	940	8.591	4115.74	4717.12	10.202
80	2.908	2614.35	2817.87	7.874	520	5.821	3320.29	3727.73	9.29	950	8.657	4136.13	4742.12	10.222
90	2.974	2629.27	2837.47	7.917	530	5.887	3337.8	3749.85	9.315	960	8.723	4156.58	4767.19	10.241
100	3.041	2644.2	2857.09	7.96	540	5.953	3355.38	3772.05	9.339	970	8.789	4177.09	4792.32	10.26
110	3.108	2659.17	2876.72	8.002	550	6.019	3373.03	3794.32	9.364	980	8.855	4197.67	4817.51	10.278
120	3.175	2674.16	2896.38	8.043	560	6.084	3390.75	3816.66	9.388	990	8.921	4218.3	4842.76	10.297
130	3.241	2689.19	2916.07	8.084	570	6.150	3408.53	3839.06	9.412	1000	8.987	4239	4868.07	10.316
140	3.308	2704.26	2935.8	8.123	580	6.216	3426.39	3861.54	9.436	1010	9.053	4259.76	4893.45	10.334
150	3.374	2719.38	2955.56	8.162	590	6.282	3444.33	3884.09	9.46	1020	9.119	4280.58	4918.89	10.353
160	3.441	2734.54	2975.37	8.2	600	6.348	3462.33	3906.71	9.484	1030	9.185	4301.46	4944.38	10.371
170	3.507	2749.74	2995.22	8.238	610	6.414	3480.4	3929.4	9.507	1040	9.251	4322.4	4969.94	10.389
180	3.573	2765	3015.13	8.275	620	6.480	3498.54	3952.17	9.53	1050	9.317	4343.4	4995.55	10.408
190	3.640	2780.31	3035.08	8.311	630	6.546	3516.76	3975	9.553	1060	9.383	4364.46	5021.23	10.426
200	3.706	2795.68	3055.08	8.347	640	6.612	3535.04	3997.9	9.576	1070	9.448	4385.57	5046.96	10.444
210	3.772	2811.1	3075.14	8.383	650	6.678	3553.4	4020.88	9.599	1080	9.514	4406.75	5072.75	10.462
220	3.838	2826.57	3095.25	8.417	660	6.744	3571.82	4043.92	9.622	1090	9.580	4427.98	5098.6	10.479
230	3.905	2842.11	3115.42	8.452	670	6.810	3590.32	4067.04	9.644	1100	9.646	4449.27	5124.51	10.497
240	3.971	2857.7	3135.65	8.485	680	6.876	3608.89	4090.22	9.667	1110	9.712	4470.62	5150.47	10.515
250	4.037	2873.35	3155.93	8.519	690	6.942	3627.53	4113.48	9.689	1120	9.778	4492.02	5176.49	10.533
260	4.103	2889.06	3176.27	8.552	700	7.008	3646.23	4136.8	9.711	1130	9.844	4513.48	5202.56	10.55
270	4.169	2904.83	3196.68	8.584	710	7.074	3665.01	4160.2	9.733	1140	9.910	4534.99	5228.69	10.568
280	4.235	2920.66	3217.14	8.616	720	7.140	3683.86	4183.66	9.755	1150	9.976	4556.56	5254.88	10.585
290	4.302	2936.56	3237.66	8.648	730	7.206	3702.78	4207.2	9.776	1160	10.042	4578.19	5281.12	10.602
300	4.368	2952.51	3258.25	8.679	740	7.272	3721.77	4230.8	9.798	1170	10.108	4599.87	5307.42	10.619
310	4.434	2968.53	3278.89	8.71	750	7.338	3740.82	4254.48	9.819	1180	10.174	4621.6	5333.76	10.637
320	4.500	2984.62	3299.6	8.741	760	7.404	3759.95	4278.22	9.84	1190	10.240	4643.39	5360.17	10.654
330	4.566	3000.77	3320.38	8.771	770	7.470	3779.14	4302.03	9.861	1200	10.306	4665.23	5386.62	10.671
340	4.632	3016.98	3341.22	8.801	780	7.536	3798.41	4325.91	9.882	1210	10.372	4687.12	5413.13	10.687
350	4.698	3033.26	3362.12	8.83	790	7.602	3817.74	4349.86	9.903	1220	10.438	4709.06	5439.69	10.704
360	4.764	3049.6	3383.09	8.859	800	7.668	3837.14	4373.88	9.924	1230	10.504	4731.06	5466.31	10.721
370	4.830	3066.01	3404.12	8.888	810	7.734	3856.61	4397.97	9.945	1240	10.569	4753.11	5492.97	10.738
380	4.896	3082.49	3425.22	8.917	820	7.800	3876.14	4422.12	9.965	1250	10.635	4775.21	5519.68	10.754
390	4.962	3099.03	3446.39	8.945	830	7.866	3895.75	4446.34	9.985	1260	10.701	4797.36	5546.45	10.771
400	5.028	3115.64	3467.62	8.973	840	7.932	3915.42	4470.63	10.006	1270	10.767	4819.56	5573.27	10.787
410	5.094	3132.32	3488.92	9.001	850	7.998	3935.15	4494.98	10.026	1280	10.833	4841.81	5600.13	10.804
420	5.160	3149.06	3510.29	9.028	860	8.064	3954.96	4519.4	10.046	1290	10.899	4864.11	5627.05	10.82
430	5.226	3165.87	3531.72	9.056	870	8.129	3974.83	4543.89	10.066	1300	10.965	4886.46	5654.01	10.836
440	5.292	3182.75	3553.22	9.083										

Table 3(A13) Superheated Steam Properties (Pressure Base) Pressure = 0.8 bar and saturation temperature, Tsat = 93.485°C

t (°C)	$v\left(\frac{m^3}{kg}\right)$	$u\left(\frac{kJ}{kg}\right)$	$h\left(\frac{kJ}{kg}\right)$	$s\left(\frac{kJ}{kg\,K}\right)$
Sat Vapour (g)	2.087	2498.22	2665.21	7.43401
10	2.148	2513.74	2685.56	7.489
20	2.208	2529.05	2705.68	7.542
30	2.268	2544.22	2725.64	7.592
40	2.327	2559.29	2745.47	7.642
50	2.387	2574.31	2765.23	7.69
60	2.446	2589.29	2784.93	7.737
70	2.504	2604.25	2804.6	7.782
80	2.563	2619.21	2824.26	7.827
90	2.622	2634.17	2843.91	7.87
100	2.680	2649.15	2863.57	7.913
110	2.739	2664.15	2883.25	7.955
120	2.797	2679.18	2902.95	7.995
130	2.855	2694.25	2922.69	8.036
140	2.914	2709.36	2942.45	8.075
150	2.972	2724.5	2962.26	8.114
160	3.030	2739.7	2982.1	8.152
170	3.088	2754.94	3001.99	8.189
180	3.146	2770.23	3021.93	8.226
190	3.204	2785.57	3041.92	8.262
200	3.262	2800.96	3061.95	8.298
210	3.320	2816.41	3082.04	8.333
220	3.378	2831.92	3102.19	8.368
230	3.436	2847.48	3122.39	8.402
240	3.494	2863.1	3142.64	8.435
250	3.552	2878.78	3162.96	8.469
260	3.610	2894.52	3183.33	8.501
270	3.668	2910.32	3203.76	8.534
280	3.726	2926.18	3224.25	8.566
290	3.784	2942.1	3244.8	8.597
300	3.842	2958.08	3265.42	8.628
310	3.900	2974.13	3286.09	8.659
320	3.957	2990.24	3306.83	8.69
330	4.015	3006.42	3327.64	8.72
340	4.073	3022.66	3348.5	8.749
350	4.131	3038.96	3369.43	8.779
360	4.189	3055.33	3390.43	8.808
370	4.247	3071.77	3411.49	8.837
380	4.304	3088.27	3432.62	8.865
390	4.362	3104.84	3453.81	8.893
400	4.420	3121.48	3475.07	8.921
410	4.478	3138.18	3496.4	8.949
420	4.536	3154.95	3517.79	8.976
430	4.593	3171.79	3539.26	9.003
440	4.651	3188.7	3560.79	9.03

t (°C)	$v\left(\frac{m^3}{kg}\right)$	$u\left(\frac{kJ}{kg}\right)$	$h\left(\frac{kJ}{kg}\right)$	$s\left(\frac{kJ}{kg\,K}\right)$
Sat Vapour (g)	2.087	2498.22	2665.21	7.43401
450	4.709	3205.67	3582.39	9.057
460	4.767	3222.72	3604.05	9.083
470	4.824	3239.83	3625.79	9.109
480	4.882	3257.02	3647.59	9.135
490	4.940	3274.27	3669.47	9.161
500	4.998	3291.59	3691.41	9.186
510	5.056	3308.99	3713.43	9.212
520	5.113	3326.45	3735.51	9.237
530	5.171	3343.98	3757.66	9.262
540	5.229	3361.59	3779.89	9.286
550	5.287	3379.26	3802.18	9.311
560	5.344	3397.01	3824.55	9.335
570	5.402	3414.82	3846.98	9.359
580	5.460	3432.71	3869.49	9.383
590	5.517	3450.67	3892.06	9.407
600	5.575	3468.7	3914.71	9.43
610	5.633	3486.79	3937.43	9.454
620	5.691	3504.96	3960.22	9.477
630	5.748	3523.2	3983.08	9.5
640	5.806	3541.51	4006.01	9.523
650	5.864	3559.9	4029.01	9.546
660	5.922	3578.35	4052.08	9.568
670	5.979	3596.87	4075.22	9.591
680	6.037	3615.47	4098.43	9.613
690	6.095	3634.13	4121.71	9.635
700	6.153	3652.86	4145.06	9.657
710	6.210	3671.67	4168.48	9.679
720	6.268	3690.54	4191.98	9.701
730	6.326	3709.48	4215.54	9.722
740	6.383	3728.5	4239.17	9.744
750	6.441	3747.58	4262.87	9.765
760	6.499	3766.73	4286.64	9.786
770	6.557	3785.95	4310.47	9.807
780	6.614	3805.24	4334.38	9.828
790	6.672	3824.59	4358.35	9.849
800	6.730	3844.02	4382.4	9.87
810	6.787	3863.51	4406.51	9.89
820	6.845	3883.07	4430.68	9.911
830	6.903	3902.7	4454.93	9.931
840	6.961	3922.4	4479.24	9.951
850	7.018	3942.16	4503.62	9.971
860	7.076	3961.99	4528.06	9.991
870	7.134	3981.88	4552.57	10.011

t (°C)	$v\left(\frac{m^3}{kg}\right)$	$u\left(\frac{kJ}{kg}\right)$	$h\left(\frac{kJ}{kg}\right)$	$s\left(\frac{kJ}{kg\,K}\right)$
Sat Vapour (g)	2.087	2498.22	2665.21	7.43401
880	7.191	4001.84	4577.15	10.031
890	7.249	4021.86	4601.79	10.051
900	7.307	4041.95	4626.5	10.07
910	7.365	4062.11	4651.27	10.09
920	7.422	4082.33	4676.11	10.109
930	7.480	4102.61	4701.01	10.128
940	7.538	4122.96	4725.97	10.148
950	7.595	4143.37	4751	10.167
960	7.653	4163.84	4776.09	10.186
970	7.711	4184.37	4801.24	10.205
980	7.769	4204.97	4826.45	10.223
990	7.826	4225.63	4851.73	10.242
1000	7.884	4246.35	4877.06	10.261
1010	7.942	4267.13	4902.46	10.279
1020	7.999	4287.97	4927.92	10.298
1030	8.057	4308.87	4953.44	10.316
1040	8.115	4329.83	4979.01	10.334
1050	8.172	4350.85	5004.65	10.352
1060	8.230	4371.93	5030.34	10.37
1070	8.288	4393.07	5056.1	10.388
1080	8.346	4414.27	5081.91	10.406
1090	8.403	4435.52	5107.78	10.424
1100	8.461	4456.83	5133.71	10.442
1110	8.519	4478.2	5159.69	10.46
1120	8.576	4499.62	5185.73	10.477
1130	8.634	4521.1	5211.82	10.495
1140	8.692	4542.64	5237.98	10.512
1150	8.750	4564.22	5264.18	10.529
1160	8.807	4585.87	5290.44	10.547
1170	8.865	4607.57	5316.76	10.564
1180	8.923	4629.32	5343.13	10.581
1190	8.980	4651.13	5369.55	10.598
1200	9.038	4672.99	5396.02	10.615
1210	9.096	4694.9	5422.55	10.632
1220	9.153	4716.86	5449.13	10.649
1230	9.211	4738.88	5475.76	10.665
1240	9.269	4760.94	5502.44	10.682
1250	9.327	4783.06	5529.18	10.699
1260	9.384	4805.23	5555.96	10.715
1270	9.442	4827.45	5582.8	10.732
1280	9.500	4849.71	5609.68	10.748
1290	9.557	4872.03	5636.61	10.764
1300	9.615	4894.4	5663.59	10.78

Table 3(A14) Superheated Steam Properties (Pressure Base) Pressure = 0.9 bar and saturation temperature, Tsat = 96.687°C

t (°C)	v $\left(\frac{m^3}{kg}\right)$	u $\left(\frac{kJ}{kg}\right)$	h $\left(\frac{kJ}{kg}\right)$	s $\left(\frac{kJ}{kg\,K}\right)$
Sat Vapour (g)	*1.870*	*2502.08*	*2670.34*	*7.39435*
10	1.924	2517.69	2690.81	7.449
20	1.977	2533.08	2711.02	7.501
30	2.031	2548.32	2731.06	7.552
40	2.084	2563.45	2750.97	7.601
50	2.136	2578.53	2770.79	7.649
60	2.189	2593.56	2790.55	7.696
70	2.241	2608.57	2810.28	7.741
80	2.294	2623.57	2829.99	7.785
90	2.346	2638.57	2849.69	7.829
100	2.398	2653.59	2869.39	7.871
110	2.450	2668.63	2889.12	7.913
120	2.502	2683.7	2908.86	7.953
130	2.554	2698.8	2928.63	7.993
140	2.606	2713.93	2948.43	8.033
150	2.657	2729.11	2968.27	8.071
160	2.709	2744.34	2988.15	8.109
170	2.761	2759.61	3008.07	8.146
180	2.812	2774.93	3028.04	8.183
190	2.864	2790.3	3048.06	8.219
200	2.916	2805.72	3068.13	8.254
210	2.967	2821.2	3088.25	8.289
220	3.019	2836.73	3108.42	8.324
230	3.070	2852.32	3128.65	8.358
240	3.122	2867.96	3148.94	8.392
250	3.173	2883.67	3169.28	8.425
260	3.225	2899.43	3189.68	8.457
270	3.276	2915.26	3210.14	8.489
280	3.328	2931.14	3230.66	8.521
290	3.379	2947.09	3251.23	8.553
300	3.431	2963.1	3271.88	8.584
310	3.482	2979.17	3292.58	8.614
320	3.534	2995.31	3313.34	8.645
330	3.585	3011.51	3334.17	8.675
340	3.637	3027.77	3355.06	8.704
350	3.688	3044.1	3376.02	8.734
360	3.739	3060.5	3397.04	8.763
370	3.791	3076.96	3418.13	8.791
380	3.842	3093.48	3439.28	8.82
390	3.894	3110.08	3460.5	8.848
400	3.945	3126.74	3481.79	8.876
410	3.996	3143.46	3503.14	8.903
420	4.048	3160.26	3524.56	8.931
430	4.099	3177.12	3546.05	8.958
440	4.151	3194.06	3567.6	8.984

t (°C)	v $\left(\frac{m^3}{kg}\right)$	u $\left(\frac{kJ}{kg}\right)$	h $\left(\frac{kJ}{kg}\right)$	s $\left(\frac{kJ}{kg\,K}\right)$
Sat Vapour (g)	*1.870*	*2502.08*	*2670.34*	*7.39435*
450	4.202	3211.06	3589.22	9.011
460	4.253	3228.13	3610.92	9.037
470	4.305	3245.26	3632.68	9.063
480	4.356	3262.47	3654.51	9.089
490	4.407	3279.75	3676.41	9.115
500	4.459	3297.1	3698.37	9.14
510	4.510	3314.51	3720.41	9.165
520	4.561	3332	3742.52	9.19
530	4.613	3349.56	3764.7	9.215
540	4.664	3367.19	3786.95	9.24
550	4.715	3384.88	3809.27	9.264
560	4.767	3402.65	3831.66	9.288
570	4.818	3420.49	3854.11	9.312
580	4.869	3438.4	3876.64	9.336
590	4.921	3456.38	3899.25	9.36
600	4.972	3474.44	3921.92	9.383
610	5.023	3492.56	3944.66	9.407
620	5.075	3510.75	3967.47	9.43
630	5.126	3529.02	3990.35	9.453
640	5.177	3547.35	4013.31	9.476
650	5.229	3565.76	4036.33	9.498
660	5.280	3584.23	4059.43	9.521
670	5.331	3602.78	4082.59	9.543
680	5.383	3621.39	4105.83	9.566
690	5.434	3640.08	4129.13	9.588
700	5.485	3658.84	4152.51	9.61
710	5.537	3677.67	4175.95	9.631
720	5.588	3696.56	4199.47	9.653
730	5.639	3715.53	4223.05	9.675
740	5.691	3734.56	4246.71	9.696
750	5.742	3753.67	4270.43	9.717
760	5.793	3772.84	4294.22	9.738
770	5.844	3792.09	4318.08	9.759
780	5.896	3811.4	4342.01	9.78
790	5.947	3830.78	4366.01	9.801
800	5.998	3850.22	4390.07	9.822
810	6.050	3869.74	4414.2	9.842
820	6.101	3889.32	4438.4	9.863
830	6.152	3908.97	4462.67	9.883
840	6.204	3928.69	4487.01	9.903
850	6.255	3948.47	4511.41	9.923
860	6.306	3968.32	4535.87	9.943
870	6.357	3988.24	4560.41	9.963

t (°C)	v $\left(\frac{m^3}{kg}\right)$	u $\left(\frac{kJ}{kg}\right)$	h $\left(\frac{kJ}{kg}\right)$	s $\left(\frac{kJ}{kg\,K}\right)$
Sat Vapour (g)	*1.870*	*2502.08*	*2670.34*	*7.39435*
880	6.409	4008.22	4585	9.983
890	6.460	4028.26	4609.67	10.003
900	6.511	4048.37	4634.39	10.022
910	6.563	4068.55	4659.19	10.042
920	6.614	4088.79	4684.04	10.061
930	6.665	4109.09	4708.96	10.08
940	6.717	4129.46	4733.95	10.099
950	6.768	4149.89	4759	10.118
960	6.819	4170.38	4784.11	10.137
970	6.870	4190.94	4809.28	10.156
980	6.922	4211.56	4834.51	10.175
990	6.973	4232.24	4859.81	10.194
1000	7.024	4252.97	4885.16	10.212
1010	7.076	4273.78	4910.58	10.231
1020	7.127	4294.64	4936.06	10.249
1030	7.178	4315.56	4961.6	10.267
1040	7.230	4336.54	4987.19	10.286
1050	7.281	4357.58	5012.85	10.304
1060	7.332	4378.68	5038.56	10.322
1070	7.383	4399.83	5064.34	10.34
1080	7.435	4421.05	5090.17	10.358
1090	7.486	4442.32	5116.05	10.375
1100	7.537	4463.65	5142	10.393
1110	7.589	4485.03	5168	10.411
1120	7.640	4506.47	5194.06	10.428
1130	7.691	4527.97	5220.17	10.446
1140	7.742	4549.52	5246.34	10.463
1150	7.794	4571.13	5272.57	10.481
1160	7.845	4592.79	5298.84	10.498
1170	7.896	4614.51	5325.18	10.515
1180	7.948	4636.28	5351.56	10.532
1190	7.999	4658.1	5378	10.549
1200	8.050	4679.98	5404.49	10.566
1210	8.102	4701.91	5431.04	10.583
1220	8.153	4723.89	5457.64	10.6
1230	8.204	4745.92	5484.28	10.616
1240	8.255	4768	5510.98	10.633
1250	8.307	4790.14	5537.73	10.65
1260	8.358	4812.32	5564.53	10.666
1270	8.409	4834.56	5591.38	10.682
1280	8.461	4856.84	5618.28	10.699
1290	8.512	4879.17	5645.23	10.715
1300	8.563	4901.55	5672.23	10.731

Table 3(A15) Superheated Steam Properties (Pressure Base) Pressure = 1.0 bar and saturation temperature, Tsat = 99.606°C

t (°C)	v $\left(\frac{m^3}{kg}\right)$	u $\left(\frac{kJ}{kg}\right)$	h $\left(\frac{kJ}{kg}\right)$	s $\left(\frac{kJ}{kg\,K}\right)$
Sat Vapour (g)	1.694	2505.56	2674.98	7.35893
10	1.743	2521.26	2695.55	7.413
20	1.791	2536.72	2715.85	7.466
30	1.839	2552.03	2735.96	7.516
40	1.887	2567.23	2755.94	7.565
50	1.935	2582.35	2775.83	7.613
60	1.982	2597.43	2795.65	7.659
70	2.030	2612.48	2815.43	7.704
80	2.077	2627.52	2835.19	7.749
90	2.124	2642.57	2854.93	7.792
100	2.171	2657.62	2874.68	7.834
110	2.218	2672.7	2894.44	7.875
120	2.264	2687.8	2914.22	7.916
130	2.311	2702.93	2934.03	7.956
140	2.358	2718.1	2953.87	7.995
150	2.404	2733.31	2973.74	8.033
160	2.451	2748.56	2993.65	8.071
170	2.498	2763.86	3013.61	8.108
180	2.544	2779.2	3033.61	8.144
190	2.591	2794.6	3053.65	8.18
200	2.637	2810.05	3073.75	8.216
210	2.684	2825.55	3093.9	8.251
220	2.730	2841.11	3114.1	8.285
230	2.776	2856.72	3134.36	8.319
240	2.823	2872.39	3154.67	8.352
250	2.869	2888.12	3175.03	8.385
260	2.916	2903.91	3195.46	8.418
270	2.962	2919.76	3215.94	8.45
280	3.008	2935.67	3236.49	8.482
290	3.055	2951.64	3257.09	8.513
300	3.101	2967.67	3277.76	8.544
310	3.147	2983.76	3298.48	8.575
320	3.194	2999.92	3319.27	8.605
330	3.240	3016.15	3340.13	8.635
340	3.286	3032.43	3361.04	8.664
350	3.332	3048.79	3382.02	8.693
360	3.379	3065.2	3403.07	8.722
370	3.425	3081.69	3424.18	8.751
380	3.471	3098.23	3445.35	8.779
390	3.518	3114.85	3466.6	8.807
400	3.564	3131.53	3487.91	8.835
410	3.610	3148.28	3509.28	8.863
420	3.656	3165.1	3530.72	8.89
430	3.703	3181.99	3552.23	8.917
440	3.749	3198.94	3573.81	8.943

t (°C)	v $\left(\frac{m^3}{kg}\right)$	u $\left(\frac{kJ}{kg}\right)$	h $\left(\frac{kJ}{kg}\right)$	s $\left(\frac{kJ}{kg\,K}\right)$
Sat Vapour (g)	1.694	2505.56	2674.98	7.35893
450	3.795	3215.96	3595.46	8.97
460	3.841	3233.05	3617.17	8.996
470	3.887	3250.22	3638.96	9.022
480	3.934	3267.45	3660.81	9.048
490	3.980	3284.75	3682.73	9.074
500	4.026	3302.11	3704.72	9.099
510	4.072	3319.55	3726.78	9.124
520	4.119	3337.06	3748.91	9.149
530	4.165	3354.64	3771.11	9.174
540	4.211	3372.29	3793.39	9.198
550	4.257	3390.01	3815.73	9.223
560	4.303	3407.8	3838.14	9.247
570	4.350	3425.66	3860.62	9.271
580	4.396	3443.6	3883.17	9.294
590	4.442	3461.6	3905.8	9.318
600	4.488	3479.67	3928.49	9.342
610	4.534	3497.82	3951.25	9.365
620	4.581	3516.03	3974.09	9.388
630	4.627	3534.32	3996.99	9.411
640	4.673	3552.67	4019.97	9.434
650	4.719	3571.1	4043.01	9.456
660	4.765	3589.6	4066.13	9.479
670	4.812	3608.17	4089.32	9.501
680	4.858	3626.8	4112.57	9.523
690	4.904	3645.51	4135.9	9.545
700	4.950	3664.29	4159.3	9.567
710	4.996	3683.14	4182.76	9.589
720	5.043	3702.06	4206.3	9.611
730	5.089	3721.04	4229.91	9.632
740	5.135	3740.1	4253.58	9.654
750	5.181	3759.23	4277.33	9.675
760	5.227	3778.42	4301.14	9.696
770	5.273	3797.68	4325.02	9.717
780	5.320	3817.01	4348.97	9.738
790	5.366	3836.41	4372.99	9.759
800	5.412	3855.88	4397.07	9.779
810	5.458	3875.42	4421.23	9.8
820	5.504	3895.02	4445.45	9.82
830	5.550	3914.69	4469.73	9.84
840	5.597	3934.43	4494.09	9.861
850	5.643	3954.23	4518.51	9.881
860	5.689	3974.1	4542.99	9.901
870	5.735	3994.03	4567.55	9.92

t (°C)	v $\left(\frac{m^3}{kg}\right)$	u $\left(\frac{kJ}{kg}\right)$	h $\left(\frac{kJ}{kg}\right)$	s $\left(\frac{kJ}{kg\,K}\right)$
Sat Vapour (g)	1.694	2505.56	2674.98	7.35893
880	5.781	4014.04	4592.17	9.94
890	5.828	4034.1	4616.85	9.96
900	5.874	4054.23	4641.6	9.979
910	5.920	4074.43	4666.41	9.999
920	5.966	4094.69	4691.28	10.018
930	6.012	4115.01	4716.22	10.037
940	6.058	4135.39	4741.23	10.056
950	6.105	4155.84	4766.29	10.075
960	6.151	4176.35	4791.42	10.094
970	6.197	4196.93	4816.61	10.113
980	6.243	4217.56	4841.86	10.132
990	6.289	4238.26	4867.18	10.15
1000	6.335	4259.02	4892.55	10.169
1010	6.382	4279.84	4917.99	10.187
1020	6.428	4300.72	4943.48	10.206
1030	6.474	4321.65	4969.04	10.224
1040	6.520	4342.65	4994.65	10.242
1050	6.566	4363.71	5020.33	10.26
1060	6.612	4384.83	5046.06	10.278
1070	6.659	4406	5071.85	10.296
1080	6.705	4427.23	5097.7	10.314
1090	6.751	4448.52	5123.6	10.332
1100	6.797	4469.86	5149.56	10.35
1110	6.843	4491.27	5175.58	10.367
1120	6.889	4512.72	5201.66	10.385
1130	6.936	4534.24	5227.79	10.402
1140	6.982	4555.81	5253.97	10.42
1150	7.028	4577.43	5280.21	10.437
1160	7.074	4599.11	5306.51	10.454
1170	7.120	4620.84	5332.86	10.471
1180	7.166	4642.63	5359.26	10.488
1190	7.213	4664.47	5385.71	10.505
1200	7.259	4686.36	5412.22	10.522
1210	7.305	4708.3	5438.78	10.539
1220	7.351	4730.3	5465.39	10.556
1230	7.397	4752.35	5492.06	10.573
1240	7.443	4774.44	5518.77	10.589
1250	7.489	4796.59	5545.54	10.606
1260	7.536	4818.79	5572.35	10.622
1270	7.582	4841.04	5599.22	10.639
1280	7.628	4863.34	5626.13	10.655
1290	7.674	4885.69	5653.09	10.671
1300	7.720	4908.08	5680.11	10.687

Table 3(A16) Superheated Steam Properties (Pressure Base) Pressure = 1.2 bar and saturation temperature, Tsat = 104.784°C

t (°C)	v $\left(\frac{m^3}{kg}\right)$	u $\left(\frac{kJ}{kg}\right)$	h $\left(\frac{kJ}{kg}\right)$	s $\left(\frac{kJ}{kg\,K}\right)$
Sat Vapour (g)	1.429	2511.66	2683.08	7.298
10	1.469	2527.52	2703.85	7.352
20	1.510	2543.12	2724.32	7.404
30	1.550	2558.55	2744.58	7.454
40	1.590	2573.86	2764.70	7.503
50	1.630	2589.08	2784.70	7.550
60	1.670	2604.25	2804.63	7.596
70	1.709	2619.39	2824.51	7.641
80	1.749	2634.51	2844.35	7.685
90	1.788	2649.62	2864.18	7.728
100	1.827	2664.75	2884.01	7.770
110	1.866	2679.88	2903.85	7.811
120	1.905	2695.05	2923.70	7.851
130	1.944	2710.24	2943.57	7.891
140	1.983	2725.46	2963.47	7.930
150	2.022	2740.72	2983.41	7.968
160	2.061	2756.03	3003.38	8.005
170	2.100	2771.38	3023.39	8.042
180	2.139	2786.77	3043.45	8.078
190	2.178	2802.22	3063.55	8.114
200	2.217	2817.71	3083.70	8.149
210	2.255	2833.26	3103.90	8.184
220	2.294	2848.86	3124.15	8.218
230	2.333	2864.52	3144.45	8.252
240	2.371	2880.24	3164.81	8.285
250	2.410	2896.01	3185.23	8.318
260	2.449	2911.84	3205.70	8.350
270	2.487	2927.73	3226.23	8.382
280	2.526	2943.68	3246.82	8.413
290	2.565	2959.70	3267.47	8.445
300	2.603	2975.77	3288.18	8.475
310	2.642	2991.91	3308.95	8.506
320	2.681	3008.11	3329.78	8.536
330	2.719	3024.37	3350.68	8.566
340	2.758	3040.70	3371.64	8.595
350	2.796	3057.09	3392.66	8.624
360	2.835	3073.55	3413.75	8.653
370	2.874	3090.07	3434.90	8.681
380	2.912	3106.66	3456.12	8.709
390	2.951	3123.32	3477.41	8.737
400	2.989	3140.04	3498.76	8.765
410	3.028	3156.83	3520.17	8.792
420	3.066	3173.69	3541.66	8.819
430	3.105	3190.61	3563.21	8.846
440	3.143	3207.61	3584.83	8.873

t (°C)	v $\left(\frac{m^3}{kg}\right)$	u $\left(\frac{kJ}{kg}\right)$	h $\left(\frac{kJ}{kg}\right)$	s $\left(\frac{kJ}{kg\,K}\right)$
Sat Vapour (g)	1.429	2511.66	2683.08	7.298
450	3.182	3224.67	3606.51	8.899
460	3.221	3241.80	3628.27	8.925
470	3.259	3259.00	3650.10	8.951
480	3.298	3276.27	3671.99	8.977
490	3.336	3293.61	3693.95	9.002
500	3.375	3311.02	3715.98	9.028
510	3.413	3328.50	3738.08	9.053
520	3.452	3346.05	3760.26	9.077
530	3.490	3363.66	3782.50	9.102
540	3.529	3381.35	3804.81	9.127
550	3.567	3399.11	3827.19	9.151
560	3.606	3416.94	3849.64	9.175
570	3.644	3434.84	3872.16	9.199
580	3.683	3452.81	3894.75	9.222
590	3.721	3470.86	3917.42	9.246
600	3.760	3488.97	3940.15	9.269
610	3.798	3507.15	3962.95	9.293
620	3.837	3525.40	3985.83	9.316
630	3.875	3543.73	4008.77	9.338
640	3.914	3562.12	4031.79	9.361
650	3.952	3580.59	4054.87	9.384
660	3.991	3599.12	4078.03	9.406
670	4.029	3617.73	4101.25	9.428
680	4.068	3636.41	4124.55	9.451
690	4.106	3655.15	4147.91	9.473
700	4.145	3673.97	4171.35	9.494
710	4.183	3692.85	4194.85	9.516
720	4.222	3711.81	4218.43	9.538
730	4.260	3730.83	4242.07	9.559
740	4.299	3749.93	4265.78	9.580
750	4.337	3769.09	4289.57	9.602
760	4.376	3788.32	4313.42	9.623
770	4.414	3807.62	4337.33	9.644
780	4.453	3826.99	4361.32	9.664
790	4.491	3846.43	4385.38	9.685
800	4.530	3865.93	4409.50	9.706
810	4.568	3885.50	4433.69	9.726
820	4.607	3905.14	4457.94	9.746
830	4.645	3924.84	4482.27	9.767
840	4.684	3944.62	4506.66	9.787
850	4.722	3964.46	4531.11	9.807
860	4.761	3984.36	4555.64	9.027
870	4.799	4004.33	4580.22	9.846

t (°C)	v $\left(\frac{m^3}{kg}\right)$	u $\left(\frac{kJ}{kg}\right)$	h $\left(\frac{kJ}{kg}\right)$	s $\left(\frac{kJ}{kg\,K}\right)$
Sat Vapour (g)	1.429	2511.66	2683.08	7.298
880	4.838	4024.36	4604.88	9.866
890	4.876	4044.46	4629.60	9.886
900	4.915	4064.63	4654.38	9.905
910	4.953	4084.86	4679.22	9.924
920	4.992	4105.15	4704.13	9.944
930	5.030	4125.51	4729.11	9.963
940	5.068	4145.93	4754.14	9.982
950	5.107	4166.41	4779.24	10.001
960	5.145	4186.95	4804.41	10.020
970	5.184	4207.56	4829.63	10.039
980	5.222	4228.23	4854.92	10.057
990	5.261	4248.96	4880.26	10.076
1000	5.299	4269.75	4905.67	10.094
1010	5.338	4290.60	4931.14	10.113
1020	5.376	4311.51	4956.66	10.131
1030	5.415	4332.48	4982.25	10.149
1040	5.453	4353.51	5007.90	10.168
1050	5.492	4374.60	5033.60	10.186
1060	5.530	4395.74	5059.36	10.204
1070	5.569	4416.95	5085.19	10.221
1080	5.607	4438.21	5111.06	10.239
1090	5.646	4459.53	5137.00	10.257
1100	5.684	4480.90	5162.99	10.275
1110	5.723	4502.33	5189.04	10.292
1120	5.761	4523.82	5215.14	10.310
1130	5.799	4545.36	5241.30	10.327
1140	5.838	4566.96	5267.52	10.344
1150	5.876	4588.62	5293.79	10.362
1160	5.915	4610.32	5320.11	10.379
1170	5.953	4632.08	5346.49	10.396
1180	5.992	4653.90	5372.92	10.413
1190	6.030	4675.76	5399.40	10.430
1200	6.069	4697.68	5425.94	10.447
1210	6.107	4719.65	5452.52	10.464
1220	6.146	4741.68	5479.16	10.480
1230	6.184	4763.75	5505.85	10.497
1240	6.223	4785.87	5532.60	10.514
1250	6.261	4808.05	5559.39	10.530
1260	6.300	4830.28	5586.23	10.547
1270	6.338	4852.55	5613.12	10.563
1280	6.377	4874.87	5640.06	10.579
1290	6.415	4897.25	5667.05	10.595
1300	6.453	4919.67	5694.09	10.612

Table 3(A17) Superheated Steam Properties (Pressure Base) Pressure = 1.4 bar and saturation temperature, Tsat = 109.292°C

t (°C)	$v\left(\frac{m^3}{kg}\right)$	$u\left(\frac{kJ}{kg}\right)$	$h\left(\frac{kJ}{kg}\right)$	$s\left(\frac{kJ}{kg\,K}\right)$	t (°C)	$v\left(\frac{m^3}{kg}\right)$	$u\left(\frac{kJ}{kg}\right)$	$h\left(\frac{kJ}{kg}\right)$	$s\left(\frac{kJ}{kg\,K}\right)$	t (°C)	$v\left(\frac{m^3}{kg}\right)$	$u\left(\frac{kJ}{kg}\right)$	$h\left(\frac{kJ}{kg}\right)$	$s\left(\frac{kJ}{kg\,K}\right)$
Sat Vapour (g)	1.237	2516.87	2690.02	7.246	Sat Vapour (g)	1.237	2516.87	2690.02	7.246	Sat Vapour (g)	1.237	2516.87	2690.02	7.246
10	1.272	2532.89	2710.97	7.300	450	2.742	3232.25	3616.14	8.840	880	4.161	4033.37	4615.95	9.804
20	1.307	2548.62	2731.60	7.352	460	2.775	3249.42	3637.93	8.866	890	4.194	4053.50	4640.70	9.823
30	1.342	2564.16	2752.00	7.402	470	2.808	3266.65	3659.79	8.892	900	4.227	4073.69	4665.51	9.843
40	1.376	2579.57	2772.23	7.451	480	2.841	3283.96	3681.72	8.917	910	4.260	4093.95	4690.39	9.862
50	1.410	2594.89	2792.34	7.498	490	2.874	3301.33	3703.72	8.943	920	4.293	4114.27	4715.33	9.881
60	1.444	2610.14	2812.37	7.544	500	2.907	3318.77	3725.79	8.968	930	4.326	4134.66	4740.33	9.900
70	1.478	2625.35	2832.34	7.588	510	2.940	3336.29	3747.92	8.993	940	4.359	4155.11	4765.40	9.919
80	1.512	2640.54	2852.27	7.632	520	2.973	3353.87	3770.13	9.017	950	4.392	4175.62	4790.53	9.938
90	1.546	2655.72	2872.18	7.674	530	3.006	3371.52	3792.41	9.042	960	4.425	4196.19	4815.72	9.957
100	1.580	2670.91	2892.08	7.716	540	3.039	3389.25	3814.75	9.066	970	4.458	4216.83	4840.97	9.976
110	1.613	2686.11	2911.98	7.757	550	3.072	3407.04	3837.17	9.090	980	4.491	4237.52	4866.29	9.995
120	1.647	2701.32	2931.90	7.797	560	3.105	3424.91	3859.66	9.114	990	4.524	4258.28	4891.66	10.013
130	1.680	2716.57	2951.83	7.836	570	3.138	3442.84	3882.21	9.138	1000	4.557	4279.10	4917.10	10.032
140	1.714	2731.84	2971.79	7.875	580	3.171	3460.85	3904.84	9.162	1010	4.590	4299.98	4942.59	10.050
150	1.747	2747.15	2991.78	7.913	590	3.204	3478.92	3927.54	9.185	1020	4.623	4320.91	4968.15	10.068
160	1.781	2762.51	3011.81	7.950	600	3.237	3497.07	3950.31	9.209	1030	4.656	4341.91	4993.76	10.086
170	1.814	2777.90	3031.87	7.987	610	3.270	3515.28	3973.15	9.232	1040	4.689	4362.97	5019.44	10.105
180	1.847	2793.34	3051.98	8.023	620	3.303	3533.57	3996.05	9.255	1050	4.722	4384.08	5045.17	10.123
190	1.881	2808.83	3072.13	8.058	630	3.336	3551.93	4019.03	9.278	1060	4.755	4405.26	5070.96	10.141
200	1.914	2824.37	3092.33	8.093	640	3.369	3570.36	4042.08	9.300	1070	4.788	4426.49	5096.81	10.158
210	1.947	2839.96	3112.57	8.128	650	3.402	3588.86	4065.20	9.323	1080	4.821	4447.77	5122.71	10.176
220	1.980	2855.60	3132.87	8.162	660	3.435	3607.42	4088.39	9.345	1090	4.854	4469.12	5148.67	10.194
230	2.014	2871.30	3153.22	8.195	670	3.468	3626.06	4111.65	9.367	1100	4.887	4490.52	5174.69	10.211
240	2.047	2887.06	3173.62	8.228	680	3.501	3644.77	4134.98	9.389	1110	4.920	4511.98	5200.76	10.229
250	2.080	2902.87	3194.08	8.261	690	3.534	3663.55	4158.38	9.411	1120	4.953	4533.49	5226.89	10.246
260	2.113	2918.74	3214.60	8.293	700	3.567	3682.40	4181.85	9.433	1130	4.986	4555.06	5253.08	10.264
270	2.146	2934.67	3235.17	8.325	710	3.600	3701.32	4205.39	9.455	1140	5.019	4576.68	5279.32	10.281
280	2.180	2950.66	3255.80	8.356	720	3.633	3720.31	4228.99	9.476	1150	5.052	4598.36	5305.61	10.298
290	2.213	2966.71	3276.49	8.387	730	3.666	3739.36	4252.67	9.498	1160	5.085	4620.09	5331.96	10.315
300	2.246	2982.82	3297.24	8.418	740	3.699	3758.49	4276.42	9.519	1170	5.118	4641.88	5358.36	10.332
310	2.279	2998.99	3318.05	8.448	750	3.732	3777.69	4300.23	9.540	1180	5.151	4663.71	5384.82	10.349
320	2.312	3015.23	3338.92	8.478	760	3.765	3796.95	4324.11	9.561	1190	5.184	4685.61	5411.33	10.366
330	2.345	3031.53	3359.86	8.507	770	3.798	3816.28	4348.06	9.582	1200	5.217	4707.55	5437.89	10.383
340	2.378	3047.89	3380.85	8.537	780	3.831	3835.68	4372.08	9.603	1210	5.250	4729.54	5464.50	10.400
350	2.411	3064.32	3401.92	8.566	790	3.864	3855.15	4396.17	9.623	1220	5.283	4751.59	5491.16	10.417
360	2.444	3080.81	3423.04	8.594	800	3.897	3874.68	4420.32	9.644	1230	5.316	4773.69	5517.87	10.433
370	2.478	3097.37	3444.23	8.623	810	3.930	3894.29	4444.55	9.664	1240	5.349	4795.84	5544.64	10.450
380	2.511	3114.00	3465.49	8.651	820	3.963	3913.96	4468.83	9.684	1250	5.382	4818.03	5571.45	10.466
390	2.544	3130.69	3486.81	8.679	830	3.996	3933.69	4493.19	9.705	1260	5.415	4840.28	5598.32	10.483
400	2.577	3147.45	3508.20	8.706	840	4.029	3953.50	4517.61	9.725	1270	5.448	4862.58	5625.23	10.499
410	2.610	3164.27	3529.65	8.733	850	4.062	3973.37	4542.10	9.745	1280	5.480	4884.93	5652.20	10.515
420	2.643	3181.17	3551.17	8.760	860	4.095	3993.30	4566.65	9.764	1290	5.513	4907.32	5679.21	10.532
430	2.676	3198.13	3572.76	8.787	870	4.128	4013.30	4591.27	9.784	1300	5.546	4929.76	5706.27	10.548
440	2.709	3215.16	3594.41	8.813										

Table 3(A18) Superheated Steam Properties (Pressure Base) Pressure = 1.6 bar and saturation temperature, Tsat = 113.298°C

t (°C)	v $\left(\frac{m^3}{kg}\right)$	u $\left(\frac{kJ}{kg}\right)$	h $\left(\frac{kJ}{kg}\right)$	s $\left(\frac{kJ}{kg\,K}\right)$	t (°C)	v $\left(\frac{m^3}{kg}\right)$	u $\left(\frac{kJ}{kg}\right)$	h $\left(\frac{kJ}{kg}\right)$	s $\left(\frac{kJ}{kg\,K}\right)$	t (°C)	v $\left(\frac{m^3}{kg}\right)$	u $\left(\frac{kJ}{kg}\right)$	h $\left(\frac{kJ}{kg}\right)$	s $\left(\frac{kJ}{kg\,K}\right)$
Sat Vapour (g)	*1.091*	*2521.43*	*2696.07*	*7.201*	*Sat Vapour (g)*	*1.091*	*2521.43*	*2696.07*	*7.201*	*Sat Vapour (g)*	*1.091*	*2521.43*	*2696.07*	*7.201*
10	1.123	2537.59	2717.20	7.255	450	2.411	3238.99	3624.69	8.788	880	3.653	4041.37	4625.80	9.750
20	1.153	2553.45	2737.97	7.307	460	2.440	3256.19	3646.51	8.814	890	3.682	4061.53	4650.58	9.769
30	1.184	2569.09	2758.50	7.357	470	2.468	3273.45	3668.40	8.840	900	3.710	4081.75	4675.42	9.789
40	1.214	2584.60	2778.85	7.405	480	2.497	3290.79	3690.37	8.866	910	3.739	4102.03	4700.32	9.808
50	1.244	2600.00	2799.06	7.452	490	2.526	3308.19	3712.40	8.891	920	3.768	4122.38	4725.29	9.827
60	1.274	2615.33	2819.19	7.498	500	2.555	3325.67	3734.50	8.916	930	3.797	4142.79	4750.31	9.846
70	1.304	2630.61	2839.24	7.542	510	2.584	3343.21	3756.67	8.941	940	3.826	4163.27	4775.41	9.865
80	1.334	2645.87	2859.25	7.586	520	2.613	3360.83	3778.91	8.965	950	3.855	4183.81	4800.56	9.884
90	1.363	2661.11	2879.23	7.628	530	2.642	3378.51	3801.21	8.990	960	3.884	4204.40	4825.78	9.903
100	1.393	2676.35	2899.20	7.670	540	2.671	3396.26	3823.59	9.014	970	3.912	4225.07	4851.06	9.922
110	1.422	2691.60	2919.17	7.710	550	2.700	3414.09	3846.04	9.038	980	3.941	4245.79	4876.40	9.940
120	1.452	2706.87	2939.14	7.750	560	2.729	3431.98	3868.56	9.062	990	3.970	4266.57	4901.80	9.959
130	1.481	2722.17	2959.14	7.789	570	2.757	3449.95	3891.15	9.086	1000	3.999	4287.41	4927.26	9.977
140	1.510	2737.49	2979.15	7.827	580	2.786	3467.98	3913.81	9.110	1010	4.028	4308.31	4952.78	9.996
150	1.540	2752.85	2999.19	7.865	590	2.815	3486.09	3936.54	9.133	1020	4.057	4329.28	4978.36	10.014
160	1.569	2768.24	3019.27	7.902	600	2.844	3504.27	3959.33	9.156	1030	4.086	4350.30	5004.00	10.032
170	1.598	2783.68	3039.38	7.939	610	2.873	3522.51	3982.20	9.179	1040	4.114	4371.38	5029.69	10.050
180	1.627	2799.16	3059.54	7.975	620	2.902	3540.83	4005.14	9.202	1050	4.143	4392.52	5055.45	10.068
190	1.657	2814.69	3079.73	8.010	630	2.931	3559.22	4028.15	9.225	1060	4.172	4413.71	5081.26	10.086
200	1.686	2830.27	3099.97	8.045	640	2.960	3577.68	4051.23	9.247	1070	4.201	4434.97	5107.14	10.104
210	1.715	2845.90	3120.26	8.079	650	2.989	3596.21	4074.38	9.270	1080	4.230	4456.28	5133.06	10.122
220	1.744	2861.58	3140.60	8.113	660	3.017	3614.81	4097.60	9.292	1090	4.259	4477.65	5159.05	10.139
230	1.773	2877.31	3160.99	8.146	670	3.046	3633.47	4120.89	9.314	1100	4.288	4499.07	5185.09	10.157
240	1.802	2893.10	3181.43	8.179	680	3.075	3652.21	4144.25	9.336	1110	4.316	4520.55	5211.19	10.174
250	1.831	2908.95	3201.93	8.212	690	3.104	3671.02	4167.68	9.358	1120	4.345	4542.09	5237.34	10.192
260	1.860	2924.86	3222.48	8.244	700	3.133	3689.90	4191.18	9.380	1130	4.374	4563.68	5263.55	10.209
270	1.889	2940.82	3243.09	8.275	710	3.162	3708.85	4214.75	9.402	1140	4.403	4585.33	5289.81	10.226
280	1.918	2956.84	3263.76	8.307	720	3.191	3727.86	4238.39	9.423	1150	4.432	4607.03	5316.13	10.244
290	1.947	2972.93	3284.49	8.337	730	3.220	3746.95	4262.09	9.444	1160	4.461	4628.78	5342.50	10.261
300	1.976	2989.07	3305.27	8.368	740	3.249	3766.11	4285.87	9.466	1170	4.490	4650.59	5368.92	10.278
310	2.005	3005.28	3326.12	8.398	750	3.277	3785.33	4309.71	9.487	1180	4.518	4672.45	5395.40	10.295
320	2.034	3021.55	3347.03	8.428	760	3.306	3804.62	4333.62	9.508	1190	4.547	4694.36	5421.93	10.312
330	2.063	3037.88	3368.00	8.457	770	3.335	3823.98	4357.60	9.528	1200	4.576	4716.32	5448.51	10.328
340	2.092	3054.28	3389.03	8.486	780	3.364	3843.41	4381.65	9.549	1210	4.605	4738.34	5475.14	10.345
350	2.121	3070.74	3410.13	8.515	790	3.393	3862.91	4405.77	9.570	1220	4.634	4760.41	5501.83	10.362
360	2.150	3087.27	3431.29	8.544	800	3.422	3882.47	4429.95	9.590	1230	4.663	4782.52	5528.56	10.378
370	2.179	3103.86	3452.51	8.572	810	3.451	3902.10	4454.20	9.611	1240	4.692	4804.69	5555.35	10.395
380	2.208	3120.51	3473.80	8.600	820	3.479	3921.80	4478.52	9.631	1250	4.720	4826.91	5582.18	10.411
390	2.237	3137.24	3495.16	8.628	830	3.508	3941.56	4502.90	9.651	1260	4.749	4849.18	5609.07	10.428
400	2.266	3154.03	3516.58	8.655	840	3.537	3961.39	4527.35	9.671	1270	4.778	4871.50	5636.00	10.444
410	2.295	3170.88	3538.07	8.682	850	3.566	3981.29	4551.86	9.691	1280	4.807	4893.86	5662.98	10.460
420	2.324	3187.81	3559.62	8.709	860	3.595	4001.25	4576.44	9.711	1290	4.836	4916.28	5690.02	10.476
430	2.353	3204.80	3581.24	8.736	870	3.624	4021.28	4601.09	9.730	1300	4.865	4938.74	5717.10	10.492
440	2.382	3221.86	3602.93	8.762										

Table 3(A19) Superheated Steam Properties (Pressure Base) Pressure =1.8 bar and saturation temperature, Tsat = 116.912°C

t (°C)	v $\left(\frac{m^3}{kg}\right)$	u $\left(\frac{kJ}{kg}\right)$	h $\left(\frac{kJ}{kg}\right)$	s $\left(\frac{kJ}{kg\,K}\right)$	t (°C)	v $\left(\frac{m^3}{kg}\right)$	u $\left(\frac{kJ}{kg}\right)$	h $\left(\frac{kJ}{kg}\right)$	s $\left(\frac{kJ}{kg\,K}\right)$	t (°C)	v $\left(\frac{m^3}{kg}\right)$	u $\left(\frac{kJ}{kg}\right)$	h $\left(\frac{kJ}{kg}\right)$	s $\left(\frac{kJ}{kg\,K}\right)$
Sat Vapour (g)	0.978	2525.47	2701.44	7.162	Sat Vapour (g)	0.978	2525.47	2701.44	7.162	Sat Vapour (g)	0.978	2525.47	2701.44	7.162
10	1.005	2541.78	2722.74	7.216	450	2.152	3245.07	3632.40	8.743	880	3.256	4048.60	4634.69	9.703
20	1.033	2557.74	2743.65	7.268	460	2.178	3262.29	3654.25	8.769	890	3.282	4068.78	4659.49	9.722
30	1.060	2573.49	2764.30	7.317	470	2.203	3279.59	3676.17	8.795	900	3.307	4089.02	4684.35	9.741
40	1.087	2589.09	2784.76	7.366	480	2.229	3296.95	3698.16	8.820	910	3.333	4109.33	4709.28	9.761
50	1.114	2604.57	2805.07	7.412	490	2.255	3314.39	3720.22	8.845	920	3,359	4129.70	4734.27	9.780
60	1.141	2619.97	2825.28	7.458	500	2.280	3331.89	3742.35	8.870	930	3.384	4150.14	4759.32	9.799
70	1.167	2635.32	2845.42	7.502	510	2.306	3349.46	3764.55	8.895	940	3.410	4170.64	4784.44	9.818
80	1.194	2650.64	2865.50	7.545	520	2.332	3367.10	3786.82	8.920	950	3.436	4191.20	4809.62	9.837
90	1.220	2665.94	2885.55	7.587	530	2.357	3384.81	3809.16	8.944	960	3.461	4211.82	4834.86	9.855
100	1.246	2681.24	2905.58	7.629	540	2.383	3402.60	3831.57	8.968	970	3.487	4232.50	4860.16	9.874
110	1.273	2696.54	2925.61	7.669	550	2.409	3420.45	3854.04	8.993	980	3.513	4253.25	4885.52	9.893
120	1.299	2711.86	2945.64	7.709	560	2.435	3438.37	3876.59	9.016	990	3.538	4274.05	4910.94	9.911
130	1.325	2727.20	2965.69	7.748	570	2.460	3456.36	3899.21	9.040	1000	3.564	4294.92	4936.43	9.930
140	1.351	2742.56	2985.76	7.786	580	2.486	3474.43	3921.89	9.064	1010	3.590	4315.84	4961.97	9.948
150	1.377	2757.96	3005.85	7.823	590	2.512	3492.56	3944.65	9.087	1020	3.615	4336.83	4987.57	9.966
160	1.403	2773.40	3025.97	7.860	600	2.537	3510.77	3967.48	9.110	1030	3.641	4357.87	5013.23	9.984
170	1.429	2788.88	3046.13	7.897	610	2.563	3529.04	3990.38	9.133	1040	3.667	4378.97	5038.95	10.002
180	1.455	2804.40	3066.33	7.932	620	2.589	3547.38	4013.34	9.156	1050	3.692	4400.13	5064.73	10.020
190	1.481	2819.96	3086.56	7.968	630	2.614	3565.80	4036.38	9.179	1060	3.718	4421.35	5090.57	10.038
200	1.507	2835.58	3106.85	8.002	640	2.640	3584.29	4059.49	9.201	1070	3.744	4442.63	5116.46	10.056
210	1.533	2851.24	3127.18	8.036	650	2.666	3602.84	4082.67	9.224	1080	3.769	4463.96	5142.41	10.074
220	1.559	2866.96	3147.55	8.070	660	2.691	3621.47	4105.92	9.246	1090	3.795	4485.35	5168.42	10.091
230	1.585	2882.72	3167.98	8.103	670	2.717	3640.16	4129.23	9.268	1100	3.820	4506.79	5194.48	10.109
240	1.611	2898.55	3188.46	8.136	680	2.743	3658.93	4152.62	9.290	1110	3.846	4528.29	5220.60	10.126
250	1.636	2914.43	3209.00	8.168	690	2.768	3677.76	4176.08	9.312	1120	3.872	4549.85	5246.77	10.144
260	1.662	2930.37	3229.58	8.200	700	2.794	3696.67	4199.60	9.333	1130	3.897	4571.46	5273.00	10.161
270	1.688	2946.36	3250.23	8.232	710	2.820	3715.64	4223.20	9.355	1140	3.923	4593.13	5299.28	10.178
280	1.714	2962.42	3270.93	8.263	720	2.845	3734.69	4246.86	9.376	1150	3.949	4614.85	5325.62	10.195
290	1.740	2978.53	3291.69	8.294	730	2.871	3753.80	4270.59	9.398	1160	3.974	4636.62	5352.01	10.212
300	1.766	2994.71	3312.51	8.324	740	2.897	3772.98	4294.40	9.419	1170	4.000	4658.45	5378.45	10.229
310	1.791	3010.95	3333.39	8.354	750	2.922	3792.23	4318.27	9.440	1180	4.026	4680.33	5404.95	10.246
320	1.817	3027.25	3354.33	8.384	760	2.948	3811.55	4342.20	9.461	1190	4.051	4702.26	5431.50	10.263
330	1.843	3043.61	3375.34	8.413	770	2.974	3830.93	4366.21	9.482	1200	4.077	4724.24	5458.10	10.280
340	1.869	3060.04	3396.40	8.442	780	2.999	3850.39	4390.29	9.502	1210	4.103	4746.28	5484.75	10.297
350	1.894	3076.53	3417.53	8.471	790	3.025	3869.91	4414.43	9.523	1220	4.128	4768.36	5511.45	10.313
360	1.920	3093.08	3438.72	8.500	800	3.051	3889.50	4438.64	9.543	1230	4.154	4790.50	5538.21	10.330
370	1.946	3109.70	3459.98	8.528	810	3.076	3909.15	4462.91	9.563	1240	4.180	4812.69	5565.01	10.346
380	1.972	3126.39	3481.30	8.556	820	3.102	3928.88	4487.25	9.584	1250	4.205	4834.92	5591.87	10.363
390	1.997	3143.14	3502.68	8.583	830	3.128	3948.67	4511.66	9.604	1260	4.231	4857.21	5618.77	10.379
400	2.023	3159.96	3524.14	8.610	840	3.153	3968.52	4536.14	9.624	1270	4.257	4879.55	5645.72	10.395
410	2.049	3176.85	3545.65	8.638	850	3.179	3988.44	4560.68	9.644	1280	4.282	4901.93	5672.72	10.412
420	2.075	3193.80	3567.24	8.664	860	3.205	4008.43	4585.28	9.663	1290	4.308	4924.36	5699.77	10.428
430	2.100	3210.82	3588.89	8.691	870	3.230	4028.48	4609.95	9.683	1300	4.333	4946.84	5726.87	10.444
440	2.126	3227.91	3610.61	8.717										

Table 3(A20) Superheated Steam Properties (Pressure Base) Pressure = 2.0 bar and saturation temperature, Tsat = 120.211°C

t (°C)	v (m³/kg)	u (kJ/kg)	h (kJ/kg)	s (kJ/kg K)	t (°C)	v (m³/kg)	u (kJ/kg)	h (kJ/kg)	s (kJ/kg K)	t (°C)	v (m³/kg)	u (kJ/kg)	h (kJ/kg)	s (kJ/kg K)
Sat Vapour (g)	0.886	2529.11	2706.26	7.127	Sat Vapour (g)	0.886	2529.11	2706.26	7.127	Sat Vapour (g)	0.886	2529.11	2706.26	7.127
10	0.911	2545.54	2727.72	7.181	450	1.944	3250.61	3639.43	8.703	880	2.938	4055.20	4642.80	9.660
20	0.936	2561.62	2748.78	7.232	460	1.967	3267.87	3661.31	8.729	890	2.961	4075.40	4667.63	9.680
30	0.960	2577.47	2769.54	7.282	470	1.990	3285.19	3683.26	8.754	900	2.984	4095.67	4692.51	9.699
40	0.985	2593.15	2790.10	7.330	480	2.014	3302.58	3705.28	8.780	910	3.007	4116.00	4717.46	9.718
50	1.009	2608.70	2810.51	7.377	490	2.037	3320.04	3727.37	8.805	920	3.030	4136.39	4742.47	9.737
60	1.033	2624.18	2830.80	7.422	500	2.060	3337.57	3749.52	8.830	930	3.054	4156.85	4767.55	9.756
70	1.057	2639.59	2851.01	7.466	510	2.083	3355.16	3771.75	8.855	940	3.077	4177.37	4792.69	9.775
80	1.081	2654.97	2871.16	7.509	520	2.106	3372.83	3794.04	8.879	950	3.100	4197.95	4817.89	9.794
90	1.105	2670.32	2891.28	7.551	530	2.129	3390.57	3816.41	8.904	960	3.123	4218.59	4843.15	9.813
100	1.129	2685.67	2911.38	7.592	540	2.152	3408.38	3838.84	8.928	970	3.146	4239.30	4868.47	9.832
110	1.152	2701.02	2931.46	7.633	550	2.175	3426.26	3861.35	8.952	980	3.169	4260.06	4893.85	9.850
120	1.176	2716.38	2951.55	7.672	560	2.199	3444.20	3883.92	8.975	990	3.192	4280.89	4919.30	9.869
130	1.199	2731.77	2971.65	7.711	570	2.222	3462.22	3906.56	8.999	1000	3.215	4301.77	4944.80	9.887
140	1.223	2747.18	2991.76	7.749	580	2.245	3480.31	3929.28	9.023	1010	3.238	4322.72	4970.37	9.905
150	1.246	2762.61	3011.90	7.786	590	2.268	3498.47	3952.06	9.046	1020	3.261	4343.72	4995.99	9.923
160	1.270	2778.09	3032.07	7.823	600	2.291	3516.70	3974.91	9.069	1030	3.284	4364.79	5021.67	9.942
170	1.293	2793.60	3052.27	7.859	610	2.314	3535.00	3997.84	9.092	1040	3.308	4385.91	5047.41	9.960
180	1.317	2809.16	3072.50	7.895	620	2.337	3553.37	4020.83	9.115	1050	3.331	4407.09	5073.21	9.977
190	1.340	2824.76	3092.78	7.930	630	2.360	3571.81	4043.89	9.137	1060	3.354	4428.33	5099.06	9.995
200	1.363	2840.41	3113.10	7.964	640	2.384	3590.32	4067.03	9.160	1070	3.377	4449.62	5124.98	10.013
210	1.387	2856.10	3133.47	7.998	650	2.407	3608.90	4090.23	9.182	1080	3.400	4470.97	5150.95	10.031
220	1.410	2871.85	3153.88	8.032	660	2.430	3627.55	4113.50	9.204	1090	3.423	4492.38	5176.97	10.048
230	1.433	2887.65	3174.35	8.065	670	2.453	3646.27	4136.85	9.226	1100	3.446	4513.84	5203.05	10.066
240	1.457	2903.51	3194.86	8.098	680	2.476	3665.06	4160.26	9.248	1110	3.469	4535.36	5229.19	10.083
250	1.480	2919.42	3215.43	8.130	690	2.499	3683.92	4183.74	9.270	1120	3.492	4556.94	5255.38	10.101
260	1.503	2935.39	3236.05	8.162	700	2.522	3702.85	4207.29	9.292	1130	3.515	4578.57	5281.63	10.118
270	1.527	2951.41	3256.73	8.193	710	2.545	3721.85	4230.91	9.313	1140	3.538	4600.25	5307.93	10.135
280	1.550	2967.50	3277.46	8.224	720	2.568	3740.92	4254.60	9.335	1150	3.561	4621.99	5334.29	10.152
290	1.573	2983.64	3298.26	8.255	730	2.592	3760.05	4278.36	9.356	1160	3.585	4643.78	5360.70	10.169
300	1.596	2999.85	3319.11	8.285	740	2.615	3779.26	4302.18	9.377	1170	3.608	4665.63	5387.16	10.186
310	1.620	3016.11	3340.02	8.315	750	2.638	3798.53	4326.08	9.398	1180	3.631	4687.52	5413.67	10.203
320	1.643	3032.44	3360.99	8.345	760	2.661	3817.87	4350.04	9.419	1190	3.654	4709.47	5440.24	10.220
330	1.666	3048.83	3382.02	8.374	770	2.684	3837.28	4374.07	9.440	1200	3.677	4731.47	5466.86	10.237
340	1.689	3065.29	3403.12	8.403	780	2.707	3856.76	4398.17	9.460	1210	3.700	4753.53	5493.53	10.254
350	1.712	3081.81	3424.28	8.432	790	2.730	3876.31	4422.33	9.481	1220	3.723	4775.63	5520.25	10.270
360	1.736	3098.39	3445.50	8.460	800	2.753	3895.92	4446.57	9.501	1230	3.746	4797.78	5547.02	10.287
370	1.759	3115.04	3466.78	8.488	810	2.776	3915.60	4470.87	9.521	1240	3.769	4819.99	5573.84	10.303
380	1.782	3131.75	3488.13	8.516	820	2.799	3935.34	4495.23	9.542	1250	3.792	4842.24	5600.71	10.320
390	1.805	3148.53	3509.55	8.543	830	2.823	3955.15	4519.66	9.562	1260	3.815	4864.54	5627.63	10.336
400	1.828	3165.38	3531.03	8.571	840	2.846	3975.03	4544.16	9.582	1270	3.839	4886.90	5654.60	10.352
410	1.851	3182.29	3552.57	8.598	850	2.869	3994.98	4568.72	9.601	1280	3.862	4909.30	5681.62	10.368
420	1.875	3199.27	3574.19	8.624	860	2.892	4014.99	4593.35	9.621	1290	3.885	4931.75	5708.68	10.384
430	1.898	3216.32	3595.87	8.651	870	2.915	4035.06	4618.05	9.641	1300	3.908	4954.24	5735.79	10.401
440	1.921	3233.43	3617.61	8.677										

Table 3(A21) Superheated Steam Properties (Pressure Base) Pressure = 3.0 bar and saturation temperature, Tsat = 133.523°C

t (°C)	v $\left(\frac{m^3}{kg}\right)$	u $\left(\frac{kJ}{kg}\right)$	h $\left(\frac{kJ}{kg}\right)$	s $\left(\frac{kJ}{kg\,K}\right)$
Sat Vapour (g)	0.606	2543.17	2724.91	6.992
10	0.623	2560.22	2747.12	7.046
20	0.640	2576.79	2768.78	7.097
30	0.657	2593.06	2790.07	7.146
40	0.673	2609.12	2811.09	7.194
50	0.690	2625.02	2831.91	7.240
60	0.706	2640.80	2852.58	7.285
70	0.722	2656.49	2873.13	7.328
80	0.738	2672.13	2893.60	7.371
90	0.754	2687.73	2914.01	7.412
100	0.770	2703.30	2934.37	7.453
110	0.786	2718.86	2954.71	7.493
120	0.802	2734.42	2975.04	7.532
130	0.818	2750.00	2995.36	7.570
140	0.834	2765.59	3015.69	7.607
150	0.849	2781.20	3036.03	7.644
160	0.865	2796.84	3056.39	7.681
170	0.881	2812.51	3076.78	7.716
180	0.897	2828.22	3097.20	7.751
190	0.912	2843.98	3117.65	7.786
200	0.928	2859.77	3138.13	7.820
210	0.944	2875.61	3158.66	7.854
220	0.959	2891.50	3179.24	7.887
230	0.975	2907.43	3199.85	7.919
240	0.990	2923.42	3220.52	7.951
250	1.006	2939.46	3241.23	7.983
260	1.021	2955.56	3262.00	8.015
270	1.037	2971.71	3282.82	8.046
280	1.053	2987.92	3303.69	8.076
290	1.068	3004.18	3324.62	8.107
300	1.084	3020.51	3345.60	8.136
310	1.099	3036.90	3366.65	8.166
320	1.115	3053.34	3387.75	8.195
330	1.130	3069.85	3408.91	8.224
340	1.146	3086.42	3430.13	8.253
350	1.161	3103.06	3451.41	8.281
360	1.177	3119.75	3472.76	8.309
370	1.192	3136.51	3494.17	8.337
380	1.208	3153.34	3515.64	8.364
390	1.223	3170.23	3537.17	8.392
400	1.239	3187.19	3558.77	8.418
410	1.254	3204.21	3580.44	8.445
420	1.270	3221.30	3602.17	8.472
430	1.285	3238.46	3623.96	8.498
440	1.300	3255.69	3645.83	8.524

t (°C)	v $\left(\frac{m^3}{kg}\right)$	u $\left(\frac{kJ}{kg}\right)$	h $\left(\frac{kJ}{kg}\right)$	s $\left(\frac{kJ}{kg\,K}\right)$
Sat Vapour (g)	0.606	2543.17	2724.91	6.992
450	1.316	3272.98	3667.76	8.550
460	1.331	3290.34	3689.75	8.575
470	1.347	3307.77	3711.82	8.600
480	1.362	3325.27	3733.95	8.625
490	1.378	3342.83	3756.15	8.650
500	1.393	3360.47	3778.42	8.675
510	1.409	3378.17	3800.76	8.700
520	1.424	3395.95	3823.17	8.724
530	1.439	3413.79	3845.64	8.748
540	1.455	3431.70	3868.18	8.772
550	1.470	3449.69	3890.80	8.796
560	1.486	3467.74	3913.48	8.819
570	1.501	3485.86	3936.23	8.843
580	1.517	3504.06	3959.06	8.866
590	1.532	3522.32	3981.95	8.889
600	1.548	3540.65	4004.91	8.912
610	1.563	3559.05	4027.94	8.935
620	1.578	3577.53	4051.04	8.957
630	1.594	3596.07	4074.21	8.980
640	1.609	3614.68	4097.44	9.002
650	1.625	3633.36	4120.75	9.024
660	1.640	3652.11	4144.13	9.046
670	1.655	3670.94	4167.58	9.068
680	1.671	3689.83	4191.09	9.090
690	1.686	3708.79	4214.68	9.112
700	1.702	3727.81	4238.33	9.133
710	1.717	3746.91	4262.05	9.154
720	1.733	3766.08	4285.84	9.176
730	1.748	3785.31	4309.70	9.197
740	1.763	3804.61	4333.62	9.218
750	1.779	3823.98	4357.62	9.238
760	1.794	3843.42	4381.68	9.259
770	1.810	3862.93	4405.81	9.280
780	1.825	3882.50	4430.00	9.300
790	1.840	3902.14	4454.27	9.321
800	1.856	3921.85	4478.59	9.341
810	1.871	3941.62	4502.99	9.361
820	1.887	3961.46	4527.45	9.381
830	1.902	3981.36	4551.98	9.401
840	1.917	4001.33	4576.57	9.421
850	1.933	4021.37	4601.22	9.440
860	1.948	4041.47	4625.95	9.460
870	1.964	4061.63	4650.73	9.479

t (°C)	v $\left(\frac{m^3}{kg}\right)$	u $\left(\frac{kJ}{kg}\right)$	h $\left(\frac{kJ}{kg}\right)$	s $\left(\frac{kJ}{kg\,K}\right)$
Sat Vapour (g)	0.606	2543.17	2724.91	6.992
880	1.979	4081.86	4675.58	9.499
890	1.994	4102.15	4700.49	9.518
900	2.010	4122.51	4725.47	9.537
910	2.025	4142.93	4750.51	9.556
920	2.041	4163.41	4775.61	9.575
930	2.056	4183.95	4800.77	9.594
940	2.071	4204.56	4826.00	9.613
950	2.087	4225.22	4851.29	9.632
960	2.102	4245.95	4876.63	9.650
970	2.118	4266.74	4902.04	9.669
980	2.133	4287.59	4927.51	9.687
990	2.148	4308.50	4953.04	9.706
1000	2.164	4329.46	4978.63	9.724
1010	2.179	4350.49	5004.27	9.742
1020	2.195	4371.58	5029.98	9.760
1030	2.210	4392.72	5055.74	9.778
1040	2.225	4413.92	5081.56	9.796
1050	2.241	4435.18	5107.44	9.814
1060	2.256	4456.50	5133.38	9.832
1070	2.272	4477.87	5159.37	9.849
1080	2.287	4499.30	5185.42	9.867
1090	2.302	4520.79	5211.52	9.884
1100	2.318	4542.33	5237.68	9.902
1110	2.333	4563.92	5263.90	9.919
1120	2.349	4585.57	5290.17	9.936
1130	2.364	4607.28	5316.49	9.954
1140	2.379	4629.04	5342.86	9.971
1150	2.395	4650.85	5369.29	9.988
1160	2.410	4672.71	5395.78	10.005
1170	2.426	4694.63	5422.31	10.022
1180	2.441	4716.60	5448.90	10.038
1190	2.456	4738.62	5475.54	10.055
1200	2.472	4760.69	5502.23	10.072
1210	2.487	4782.81	5528.97	10.088
1220	2.503	4804.98	5555.76	10.105
1230	2.518	4827.20	5582.60	10.121
1240	2.533	4849.48	5609.49	10.138
1250	2.549	4871.80	5636.43	10.154
1260	2.564	4894.17	5663.41	10.170
1270	2.580	4916.59	5690.45	10.186
1280	2.595	4939.05	5717.53	10.203
1290	2.610	4961.57	5744.67	10.219
1300	2.626	4984.13	5771.84	10.235

Table 3(A22) Superheated Steam Properties (Pressure Base) Pressure = 4.0 bar and saturation temperature, Tsat = 143.609°C

t (°C)	$v\left(\frac{m^3}{kg}\right)$	$u\left(\frac{kJ}{kg}\right)$	$h\left(\frac{kJ}{kg}\right)$	$s\left(\frac{kJ}{kg\,K}\right)$	t (°C)	$v\left(\frac{m^3}{kg}\right)$	$u\left(\frac{kJ}{kg}\right)$	$h\left(\frac{kJ}{kg}\right)$	$s\left(\frac{kJ}{kg\,K}\right)$	t (°C)	$v\left(\frac{m^3}{kg}\right)$	$u\left(\frac{kJ}{kg}\right)$	$h\left(\frac{kJ}{kg}\right)$	$s\left(\frac{kJ}{kg\,K}\right)$
Sat Vapour (g)	0.462	2553.11	2738.08	6.896	Sat Vapour (g)	0.462	2553.11	2738.08	6.896	Sat Vapour (g)	0.462	2553.11	2738.08	6.896
10	0.476	2570.70	2760.95	6.950	450	0.998	3289.89	3689.15	8.442	880	1.496	4102.08	4700.43	9.385
20	0.489	2587.72	2783.15	7.001	460	1.010	3307.33	3711.24	8.467	890	1.507	4122.44	4725.42	9.405
30	0.501	2604.36	2804.90	7.050	470	1.021	3324.85	3733.40	8.492	900	1.519	4142.87	4750.46	9.424
40	0.514	2620.74	2826.32	7.098	480	1.033	3342.43	3755.62	8.517	910	1.531	4163.35	4775.57	9.443
50	0.526	2636.93	2847.50	7.144	490	1.045	3360.08	3777.91	8.542	920	1.542	4183.90	4800.74	9.462
60	0.539	2652.98	2868.50	7.188	500	1.056	3377.80	3800.27	8.566	930	1.554	4204.51	4825.97	9.480
70	0.551	2668.92	2889.34	7.232	510	1.068	3395.59	3822.69	8.591	940	1.565	4225.18	4851.26	9.499
80	0.563	2684.77	2910.08	7.274	520	1.079	3413.44	3845.19	8.615	950	1.577	4245.91	4876.61	9.518
90	0.575	2700.58	2930.74	7.315	530	1.091	3431.37	3867.75	8.639	960	1.588	4266.70	4902.03	9.536
100	0.588	2716.34	2951.34	7.355	540	1.103	3449.37	3890.38	8.663	970	1.600	4287.56	4927.50	9.555
110	0.600	2732.08	2971.90	7.395	550	1.114	3467.43	3913.08	8.686	980	1.611	4308.47	4953.04	9.573
120	0.612	2747.82	2992.43	7.433	560	1.126	3485.57	3935.85	8.710	990	1.623	4329.44	4978.63	9.591
130	0.623	2763.55	3012.94	7.471	570	1.137	3503.77	3958.68	8.733	1000	1.635	4350.47	5004.28	9.609
140	0.635	2779.29	3033.46	7.508	580	1.149	3522.04	3981.59	8.756	1010	1.646	4371.56	5029.99	9.627
150	0.647	2795.05	3053.97	7.545	590	1.160	3540.39	4004.57	8.779	1020	1.658	4392.71	5055.76	9.645
160	0.659	2810.83	3074.50	7.581	600	1.172	3558.80	4027.61	8.802	1030	1.669	4413.91	5081.59	9.663
170	0.671	2826.64	3095.04	7.616	610	1.184	3577.28	4050.72	8.824	1040	1.681	4435.18	5107.47	9.681
180	0.683	2842.48	3115.61	7.651	620	1.195	3595.83	4073.91	8.847	1050	1.692	4456.50	5133.41	9.699
190	0.695	2858.35	3136.21	7.685	630	1.207	3614.45	4097.16	8.869	1060	1.704	4477.87	5159.41	9.717
200	0.706	2874.26	3156.84	7.719	640	1.218	3633.14	4120.48	8.891	1070	1.715	4499.31	5185.46	9.734
210	0.718	2890.22	3177.50	7.752	650	1.230	3651.90	4143.87	8.913	1080	1.727	4520.79	5211.57	9.752
220	0.730	2906.22	3198.21	7.785	660	1.241	3670.73	4167.33	8.935	1090	1.738	4542.34	5237.73	9.769
230	0.742	2922.27	3218.95	7.817	670	1.253	3689.63	4190.85	8.957	1100	1.750	4563.94	5263.95	9.786
240	0.753	2938.37	3239.74	7.849	680	1.265	3708.60	4214.45	8.979	1110	1.762	4585.59	5290.22	9.804
250	0.765	2954.51	3260.58	7.880	690	1.276	3727.63	4238.11	9.000	1120	1.773	4607.30	5316.55	9.821
260	0.777	2970.71	3281.46	7.911	700	1.288	3746.74	4261.84	9.021	1130	1.785	4629.06	5342.93	9.838
270	0.789	2986.97	3302.40	7.942	710	1.299	3765.91	4285.64	9.043	1140	1.796	4650.87	5369.36	9.855
280	0.800	3003.28	3323.38	7.972	720	1.311	3785.15	4309.51	9.064	1150	1.808	4672.74	5395.85	9.872
290	0.812	3019.64	3344.42	8.002	730	1.322	3804.46	4333.45	9.085	1160	1.819	4694.66	5422.39	9.889
300	0.824	3036.07	3365.52	8.032	740	1.334	3823.84	4357.45	9.106	1170	1.831	4716.63	5448.98	9.906
310	0.835	3052.55	3386.67	8.061	750	1.346	3843.28	4381.52	9.126	1180	1.842	4738.65	5475.62	9.922
320	0.847	3069.09	3407.88	8.090	760	1.357	3862.79	4405.66	9.147	1190	1.854	4760.72	5502.31	9.939
330	0.859	3085.69	3429.14	8.119	770	1.369	3882.37	4429.86	9.167	1200	1.866	4782.85	5529.05	9.956
340	0.870	3102.36	3450.47	8.147	780	1.380	3902.02	4454.13	9.188	1210	1.877	4805.02	5555.85	9.972
350	0.882	3119.08	3471.85	8.175	790	1.392	3921.73	4478.47	9.208	1220	1.889	4827.25	5582.69	9.989
360	0.894	3135.87	3493.30	8.203	800	1.403	3941.51	4502.87	9.228	1230	1.900	4849.52	5609.59	10.005
370	0.905	3152.72	3514.80	8.231	810	1.415	3961.35	4527.34	9.248	1240	1.912	4871.85	5636.53	10.021
380	0.917	3169.64	3536.37	8.258	820	1.427	3981.26	4551.87	9.268	1250	1.923	4894.22	5663.52	10.038
390	0.928	3186.62	3558.00	8.285	830	1.438	4001.24	4576.47	9.288	1260	1.935	4916.64	5690.56	10.054
400	0.940	3203.67	3579.70	8.312	840	1.450	4021.28	4601.14	9.308	1270	1.946	4939.11	5717.65	10.070
410	0.952	3220.78	3601.46	8.338	850	1.461	4041.38	4625.86	9.327	1280	1.958	4961.62	5744.78	10.086
420	0.963	3237.96	3623.28	8.364	860	1.473	4061.55	4650.66	9.347	1290	1.969	4984.19	5771.96	10.102
430	0.975	3255.20	3645.17	8.390	870	1.484	4081.79	4675.51	9.366	1300	1.981	5006.80	5799.19	10.118
440	0.987	3272.51	3667.13	8.416										

Table 3(A23) Superheated Steam Properties (Pressure Base) Pressure = 5.0 bar and saturation temperature, Tsat = 151.832°C

t (°C)	$v\left(\frac{m^3}{kg}\right)$	$u\left(\frac{kJ}{kg}\right)$	$h\left(\frac{kJ}{kg}\right)$	$s\left(\frac{kJ}{kg\,K}\right)$	t (°C)	$v\left(\frac{m^3}{kg}\right)$	$u\left(\frac{kJ}{kg}\right)$	$h\left(\frac{kJ}{kg}\right)$	$s\left(\frac{kJ}{kg\,K}\right)$	t (°C)	$v\left(\frac{m^3}{kg}\right)$	$u\left(\frac{kJ}{kg}\right)$	$h\left(\frac{kJ}{kg}\right)$	$s\left(\frac{kJ}{kg\,K}\right)$
Sat Vapour (g)	0.375	2560.72	2748.13	6.821	Sat Vapour (g)	0.375	2560.72	2748.13	6.821	Sat Vapour (g)	0.375	2560.72	2748.13	6.821
10	0.386	2578.81	2771.62	6.875	450	0.806	3303.64	3706.54	8.359	880	1.204	4118.58	4720.70	9.298
20	0.396	2596.21	2794.30	6.927	460	0.815	3321.16	3728.70	8.384	890	1.213	4139.00	4745.74	9.317
30	0.407	2613.20	2816.47	6.976	470	0.824	3338.74	3750.93	8.409	900	1.223	4159.47	4770.84	9.336
40	0.417	2629.88	2838.26	7.024	480	0.834	3356.40	3773.23	8.434	910	1.232	4180.01	4796.01	9.355
50	0.427	2646.33	2859.77	7.069	490	0.843	3374.12	3795.60	8.459	920	1.241	4200.62	4821.23	9.374
60	0.437	2662.61	2881.06	7.114	500	0.852	3391.91	3818.03	8.483	930	1.250	4221.28	4846.52	9.393
70	0.447	2678.77	2902.17	7.157	510	0.862	3409.76	3840.53	8.507	940	1.260	4242.00	4871.87	9.411
80	0.457	2694.82	2923.16	7.199	520	0.871	3427.69	3863.10	8.531	950	1.269	4262.79	4897.27	9.430
90	0.466	2710.81	2944.04	7.240	530	0.880	3445.68	3885.73	8.555	960	1.278	4283.63	4922.74	9.448
100	0.476	2726.75	2964.85	7.280	540	0.889	3463.75	3908.44	8.579	970	1.287	4304.54	4948.27	9.467
110	0.486	2742.65	2985.60	7.319	550	0.899	3481.88	3931.21	8.602	980	1.297	4325.50	4973.86	9.485
120	0.496	2758.53	3006.32	7.357	560	0.908	3500.08	3954.05	8.625	990	1.306	4346.53	4999.50	9.503
130	0.505	2774.41	3027.01	7.395	570	0.917	3518.35	3976.96	8.649	1000	1.315	4367.61	5025.21	9.521
140	0.515	2790.28	3047.68	7.432	580	0.926	3536.69	3999.94	8.671	1010	1.324	4388.75	5050.97	9.539
150	0.524	2806.17	3068.35	7.468	590	0.936	3555.10	4022.98	8.694	1020	1.334	4409.95	5076.79	9.557
160	0.534	2822.07	3089.02	7.504	600	0.945	3573.58	4046.10	8.717	1030	1.343	4431.20	5102.67	9.575
170	0.543	2838.00	3109.71	7.539	610	0.954	3592.13	4069.28	8.739	1040	1.352	4452.52	5128.60	9.593
180	0.553	2853.95	3130.41	7.573	620	0.964	3610.75	4092.53	8.762	1050	1.361	4473.88	5154.59	9.610
190	0.562	2869.93	3151.14	7.607	630	0.973	3629.44	4115.85	8.784	1060	1.371	4495.31	5180.64	9.628
200	0.572	2885.95	3171.89	7.641	640	0.982	3648.19	4139.24	8.806	1070	1.380	4516.79	5206.74	9.645
210	0.581	2902.01	3192.68	7.674	650	0.991	3667.02	4162.70	8.828	1080	1.389	4538.33	5232.90	9.663
220	0.591	2918.11	3213.50	7.706	660	1.001	3685.91	4186.22	8.850	1090	1.398	4559.92	5259.11	9.680
230	0.600	2934.26	3234.36	7.738	670	1.010	3704.87	4209.81	8.871	1100	1.408	4581.56	5285.38	9.698
240	0.610	2950.45	3255.26	7.770	680	1.019	3723.90	4233.48	8.893	1110	1.417	4603.26	5311.70	9.715
250	0.619	2966.69	3276.20	7.801	690	1.028	3743.00	4257.21	8.914	1120	1.426	4625.02	5338.07	9.732
260	0.628	2982.98	3297.19	7.832	700	1.038	3762.17	4281.00	8.936	1130	1.435	4646.82	5364.50	9.749
270	0.638	2999.32	3318.22	7.863	710	1.047	3781.40	4304.87	8.957	1140	1.445	4668.68	5390.98	9.766
280	0.647	3015.71	3339.31	7.893	720	1.056	3800.71	4328.80	8.978	1150	1.454	4690.60	5417.51	9.783
290	0.657	3032.17	3360.45	7.923	730	1.065	3820.08	4352.80	8.999	1160	1.463	4712.56	5444.10	9.800
300	0.666	3048.67	3381.64	7.952	740	1.075	3839.52	4376.87	9.019	1170	1.472	4734.58	5470.73	9.816
310	0.675	3065.24	3402.88	7.981	750	1.084	3859.02	4401.00	9.040	1180	1.482	4756.64	5497.42	9.833
320	0.685	3081.86	3424.18	8.010	760	1.093	3878.59	4425.20	9.060	1190	1.491	4778.76	5524.16	9.850
330	0.694	3098.54	3445.53	8.038	770	1.102	3898.23	4449.47	9.081	1200	1.500	4800.93	5550.94	9.866
340	0.703	3115.29	3466.95	8.066	780	1.112	3917.94	4473.80	9.101	1210	1.509	4823.15	5577.78	9.883
350	0.713	3132.09	3488.42	8.094	790	1.121	3937.71	4498.20	9.121	1220	1.519	4845.41	5604.67	9.899
360	0.722	3148.96	3509.95	8.122	800	1.130	3957.55	4522.66	9.141	1230	1.528	4867.73	5631.61	9.915
370	0.731	3165.88	3531.54	8.149	810	1.139	3977.45	4547.19	9.161	1240	1.537	4890.10	5658.59	9.932
380	0.741	3182.88	3553.19	8.176	820	1.149	3997.42	4571.79	9.181	1250	1.546	4912.51	5685.62	9.948
390	0.750	3199.93	3574.91	8.203	830	1.158	4017.45	4596.45	9.201	1260	1.555	4934.97	5712.70	9.964
400	0.759	3217.05	3596.69	8.230	840	1.167	4037.55	4621.17	9.220	1270	1.565	4957.48	5739.83	9.980
410	0.769	3234.24	3618.53	8.256	850	1.176	4057.71	4645.96	9.240	1280	1.574	4980.04	5767.01	9.996
420	0.778	3251.49	3640.43	8.282	860	1.186	4077.94	4670.81	9.259	1290	1.583	5002.64	5794.23	10.012
430	0.787	3268.81	3662.40	8.308	870	1.195	4098.23	4695.72	9.279	1300	1.592	5025.29	5821.50	10.028
440	0.796	3286.19	3684.44	8.334										

Table 3(A24) Superheated Steam Properties (Pressure Base) Pressure = 6.0 bar and saturation temperature, Tsat = 158.827°C

t (°C)	v $\left(\frac{m^3}{kg}\right)$	u $\left(\frac{kJ}{kg}\right)$	h $\left(\frac{kJ}{kg}\right)$	s $\left(\frac{kJ}{kg\,K}\right)$	t (°C)	v $\left(\frac{m^3}{kg}\right)$	u $\left(\frac{kJ}{kg}\right)$	h $\left(\frac{kJ}{kg}\right)$	s $\left(\frac{kJ}{kg\,K}\right)$	t (°C)	v $\left(\frac{m^3}{kg}\right)$	u $\left(\frac{kJ}{kg}\right)$	h $\left(\frac{kJ}{kg}\right)$	s $\left(\frac{kJ}{kg\,K}\right)$
Sat Vapour (g)	0.316	2566.80	2756.16	6.759	Sat Vapour (g)	0.316	2566.80	2756.16	6.759	Sat Vapour (g)	0.316	2566.80	2756.16	6.759
10	0.325	2585.36	2780.23	6.814	450	0.677	3315.31	3721.28	8.292	880	1.009	4132.62	4737.94	9.227
20	0.334	2603.13	2803.36	6.866	460	0.684	3332.89	3743.51	8.317	890	1.017	4153.08	4763.03	9.246
30	0.342	2620.43	2825.92	6.916	470	0.692	3350.54	3765.81	8.342	900	1.024	4173.61	4788.18	9.265
40	0.351	2637.38	2848.05	6.963	480	0.700	3368.25	3788.18	8.366	910	1.032	4194.19	4813.39	9.284
50	0.360	2654.08	2869.86	7.009	490	0.708	3386.03	3810.61	8.391	920	1.040	4214.84	4838.67	9.303
60	0.368	2670.58	2891.42	7.053	500	0.715	3403.88	3833.11	8.415	930	1.047	4235.55	4864.00	9.321
70	0.376	2686.93	2912.79	7.096	510	0.723	3421.80	3855.67	8.439	940	1.055	4256.32	4889.40	9.340
80	0.385	2703.17	2934.00	7.138	520	0.731	3439.79	3878.30	8.463	950	1.063	4277.15	4914.85	9.358
90	0.393	2719.33	2955.09	7.179	530	0.739	3457.84	3901.00	8.487	960	1.071	4298.04	4940.37	9.377
100	0.401	2735.42	2976.09	7.218	540	0.740	3475.97	3923.77	8.510	970	1.078	4318.99	4965.94	9.395
110	0.409	2751.47	2997.03	7.257	550	0.754	3494.16	3946.61	8.534	980	1.086	4340.00	4991.57	9.413
120	0.417	2767.49	3017.91	7.296	560	0.762	3512.42	3969.51	8.557	990	1.094	4361.07	5017.27	9.431
130	0.425	2783.49	3038.75	7.333	570	0.770	3530.75	3992.48	8.580	1000	1.101	4382.19	5043.01	9.449
140	0.433	2799.49	3059.58	7.370	580	0.777	3549.15	4015.52	8.603	1010	1.109	4403.38	5068.82	9.467
150	0.441	2815.49	3080.39	7.406	590	0.785	3567.61	4038.63	8.626	1020	1.117	4424.62	5094.69	9.485
160	0.449	2831.51	3101.20	7.441	600	0.793	3586.15	4061.80	8.648	1030	1.124	4445.92	5120.61	9.503
170	0.457	2847.54	3122.01	7.476	610	0.800	3604.76	4085.04	8.671	1040	1.132	4467.27	5146.58	9.521
180	0.465	2863.59	3142.84	7.510	620	0.808	3623.43	4108.36	8.693	1050	1.140	4488.68	5172.62	9.538
190	0.473	2879.68	3163.68	7.544	630	0.816	3642.17	4131.73	8.715	1060	1.148	4510.15	5198.71	9.556
200	0.481	2895.79	3184.55	7.577	640	0.824	3660.98	4155.18	8.737	1070	1.155	4531.67	5224.85	9.573
210	0.489	2911.94	3205.45	7.610	650	0.831	3679.86	4178.70	8.759	1080	1.163	4553.25	5251.05	9.591
220	0.497	2928.13	3226.37	7.643	660	0.839	3698.81	4202.28	8.781	1090	1.171	4574.88	5277.31	9.608
230	0.505	2944.36	3247.33	7.675	670	0.847	3717.83	4225.93	8.802	1100	1.178	4596.57	5303.61	9.625
240	0.513	2960.64	3268.33	7.706	680	0.855	3736.92	4249.65	8.823	1110	1.186	4618.31	5329.98	9.642
250	0.521	2976.96	3289.37	7.737	690	0.862	3756.07	4273.44	8.845	1120	1.194	4640.10	5356.39	9.660
260	0.529	2993.33	3310.45	7.768	700	0.870	3775.29	4297.29	8.866	1130	1.202	4661.95	5382.86	9.677
270	0.536	3009.75	3331.58	7.798	710	0.878	3794.58	4321.21	8.887	1140	1.209	4683.85	5409.38	9.693
280	0.544	3026.22	3352.75	7.828	720	0.885	3813.94	4345.20	8.908	1150	1.217	4705.80	5435.95	9.710
290	0.552	3042.75	3373.98	7.858	730	0.893	3833.36	4369.26	8.929	1160	1.225	4727.80	5462.57	9.727
300	0.560	3059.33	3395.25	7.887	740	0.901	3852.85	4393.38	8.949	1170	1.232	4749.85	5489.25	9.744
310	0.568	3075.97	3416.58	7.916	750	0.909	3872.41	4417.57	8.970	1180	1.240	4771.96	5515.97	9.760
320	0.575	3092.67	3437.96	7.944	760	0.916	3892.03	4441.82	8.990	1190	1.248	4794.11	5542.75	9.777
330	0.583	3109.42	3459.39	7.973	770	0.924	3911.72	4466.14	9.011	1200	1.255	4816.32	5569.58	9.794
340	0.591	3126.23	3480.89	8.001	780	0.932	3931.48	4490.53	9.031	1210	1.263	4838.57	5596.45	9.810
350	0.599	3143.11	3502.43	8.029	790	0.939	3951.30	4514.98	9.051	1220	1.271	4860.88	5623.37	9.826
360	0.607	3160.04	3524.04	8.056	800	0.947	3971.19	4539.50	9.071	1230	1.279	4883.23	5650.35	9.843
370	0.614	3177.04	3545.71	8.083	810	0.955	3991.15	4564.08	9.091	1240	1.286	4905.63	5677.37	9.859
380	0.622	3194.09	3567.44	8.110	820	0.963	4011.16	4588.73	9.111	1250	1.294	4928.08	5704.44	9.875
390	0.630	3211.22	3589.23	8.137	830	0.970	4031.25	4613.44	9.130	1260	1.302	4950.58	5731.56	9.891
400	0.638	3228.40	3611.08	8.163	840	0.978	4051.39	4638.21	9.150	1270	1.309	4973.12	5758.72	9.907
410	0.646	3245.65	3632.99	8.189	850	0.986	4071.60	4663.05	9.169	1280	1.317	4995.71	5785.93	9.923
420	0.653	3262.97	3654.97	8.215	860	0.993	4091.88	4687.95	9.189	1290	1.325	5018.35	5813.19	9.939
430	0.661	3280.35	3677.01	8.241	870	1.001	4112.22	4712.92	9.208	1300	1.332	5041.03	5840.49	9.955
440	0.669	3297.80	3699.11	8.267										

Table 3(A25) Superheated Steam Properties (Pressure Base) Pressure = 7.0 bar and saturation temperature, Tsat = 164.946°C

t (°C)	v $\left(\frac{m^3}{kg}\right)$	u $\left(\frac{kJ}{kg}\right)$	h $\left(\frac{kJ}{kg}\right)$	s $\left(\frac{kJ}{kg\,K}\right)$	t (°C)	v $\left(\frac{m^3}{kg}\right)$	u $\left(\frac{kJ}{kg}\right)$	h $\left(\frac{kJ}{kg}\right)$	s $\left(\frac{kJ}{kg\,K}\right)$	t (°C)	v $\left(\frac{m^3}{kg}\right)$	u $\left(\frac{kJ}{kg}\right)$	h $\left(\frac{kJ}{kg}\right)$	s $\left(\frac{kJ}{kg\,K}\right)$
Sat Vapour (g)	0.273	2571.82	2762.77	6.707	Sat Vapour (g)	0.273	2571.82	2762.77	6.707	Sat Vapour (g)	0.273	2571.82	2762.77	6.707
10	0.281	2590.82	2787.38	6.763	450	0.584	3325.49	3734.13	8.235	880	0.869	4144.90	4753.02	9.167
20	0.289	2608.94	2810.95	6.815	460	0.590	3343.13	3756.43	8.260	890	0.875	4165.40	4778.16	9.186
30	0.296	2626.52	2833.87	6.864	470	0.597	3360.83	3778.79	8.285	900	0.882	4185.97	4803.35	9.205
40	0.304	2643.73	2856.32	6.912	480	0.604	3378.60	3801.21	8.310	910	0.889	4206.60	4828.61	9.224
50	0.311	2660.66	2878.41	6.957	490	0.610	3396.44	3823.70	8.334	920	0.895	4227.29	4853.92	9.243
60	0.318	2677.37	2900.23	7.002	500	0.617	3414.34	3846.26	8.358	930	0.902	4248.04	4879.30	9.261
70	0.326	2693.91	2921.83	7.045	510	0.624	3432.31	3868.88	8.382	940	0.908	4268.85	4904.74	9.280
80	0.333	2710.32	2943.25	7.086	520	0.630	3450.35	3891.57	8.406	950	0.915	4289.72	4930.23	9.298
90	0.340	2726.63	2964.54	7.127	530	0.637	3468.46	3914.33	8.430	960	0.922	4310.65	4955.79	9.317
100	0.347	2742.87	2985.73	7.167	540	0.644	3486.64	3937.16	8.453	970	0.928	4331.64	4981.40	9.335
110	0.354	2759.05	3006.83	7.206	550	0.650	3504.88	3960.05	8.476	980	0.935	4352.69	5007.08	9.353
120	0.361	2775.20	3027.87	7.244	560	0.657	3523.19	3983.01	8.500	990	0.941	4373.79	5032.81	9.371
130	0.368	2791.32	3048.86	7.281	570	0.664	3541.58	4006.03	8.522	1000	0.948	4394.96	5058.60	9.389
140	0.375	2807.43	3069.82	7.318	580	0.670	3560.03	4029.13	8.545	1010	0.955	4416.18	5084.44	9.407
150	0.382	2823.54	3090.77	7.353	590	0.677	3578.54	4052.29	8.568	1020	0.961	4437.46	5110.35	9.425
160	0.389	2839.66	3111.70	7.389	600	0.683	3597.13	4075.52	8.590	1030	0.968	4458.79	5136.31	9.443
170	0.395	2855.79	3132.63	7.423	610	0.690	3615.79	4098.82	8.613	1040	0.974	4480.18	5162.32	9.460
180	0.402	2871.94	3153.57	7.458	620	0.697	3634.51	4122.18	8.635	1050	0.981	4501.63	5188.39	9.478
190	0.409	2888.11	3174.53	7.491	630	0.703	3653.30	4145.61	8.657	1060	0.988	4523.14	5214.52	9.495
200	0.416	2904.31	3195.50	7.524	640	0.710	3672.17	4169.11	8.679	1070	0.994	4544.69	5240.70	9.513
210	0.423	2920.55	3216.50	7.557	650	0.717	3691.10	4192.68	8.701	1080	1.001	4566.31	5266.94	9.530
220	0.430	2936.82	3237.52	7.589	660	0.723	3710.09	4216.32	8.722	1090	1.008	4587.98	5293.23	9.547
230	0.436	2953.13	3258.57	7.621	670	0.730	3729.16	4240.02	8.744	1100	1.014	4609.70	5319.57	9.565
240	0.443	2969.48	3279.66	7.652	680	0.736	3748.29	4263.79	8.765	1110	1.021	4631.47	5345.97	9.582
250	0.450	2985.88	3300.79	7.683	690	0.743	3767.49	4287.63	8.786	1120	1.027	4653.30	5372.42	9.599
260	0.457	3002.32	3321.96	7.714	700	0.750	3786.76	4311.53	8.807	1130	1.034	4675.18	5398.92	9.616
270	0.463	3018.82	3343.17	7.744	710	0.756	3806.10	4335.50	8.828	1140	1.041	4697.11	5425.48	9.633
280	0.470	3035.36	3364.42	7.774	720	0.763	3825.50	4359.54	8.849	1150	1.047	4719.10	5452.09	9.649
290	0.477	3051.95	3385.73	7.803	730	0.770	3844.97	4383.65	8.870	1160	1.054	4741.13	5478.74	9.666
300	0.484	3068.60	3407.08	7.832	740	0.776	3864.51	4407.82	8.891	1170	1.060	4763.22	5505.45	9.683
310	0.490	3085.31	3428.48	7.861	750	0.783	3884.12	4432.05	8.911	1180	1.067	4785.36	5532.21	9.699
320	0.497	3102.07	3449.94	7.890	760	0.789	3903.79	4456.36	8.931	1190	1.074	4807.54	5559.02	9.716
330	0.504	3118.89	3471.45	7.918	770	0.796	3923.52	4480.73	8.952	1200	1.080	4829.78	5585.88	9.732
340	0.510	3135.76	3493.01	7.946	780	0.803	3943.33	4505.16	8.972	1210	1.087	4852.07	5612.79	9.749
350	0.517	3152.70	3514.63	7.973	790	0.809	3963.19	4529.66	8.992	1220	1.093	4874.40	5639.75	9.765
360	0.524	3169.69	3536.31	8.001	800	0.816	3983.13	4554.22	9.012	1230	1.100	4896.79	5666.75	9.781
370	0.530	3186.75	3558.05	8.028	810	0.822	4003.12	4578.85	9.032	1240	1.107	4919.22	5693.80	9.797
380	0.537	3203.87	3579.84	8.054	820	0.829	4023.19	4603.54	9.051	1250	1.113	4941.70	5720.91	9.814
390	0.544	3221.05	3601.70	8.081	830	0.836	4043.31	4628.30	9.071	1260	1.120	4964.23	5748.05	9.830
400	0.550	3238.30	3623.61	8.107	840	0.842	4063.50	4653.12	9.090	1270	1.126	4986.80	5775.25	9.846
410	0.557	3255.61	3645.59	8.133	850	0.849	4083.76	4678.00	9.110	1280	1.133	5009.43	5802.49	9.861
420	0.564	3272.98	3667.63	8.159	860	0.856	4104.08	4702.95	9.129	1290	1.140	5032.09	5829.78	9.877
430	0.570	3290.42	3689.73	8.185	870	0.862	4124.46	4727.95	9.148	1300	1.146	5054.81	5857.11	9.893
440	0.577	3307.92	3711.90	8.210										

Table 3(A26) Superheated Steam Properties (Pressure Base) Pressure = 8.0 bar and saturation temperature, Tsat = 170.407°C

t (°C)	v $\left(\frac{m^3}{kg}\right)$	u $\left(\frac{kJ}{kg}\right)$	h $\left(\frac{kJ}{kg}\right)$	s $\left(\frac{kJ}{kg\ K}\right)$	t (°C)	v $\left(\frac{m^3}{kg}\right)$	u $\left(\frac{kJ}{kg}\right)$	h $\left(\frac{kJ}{kg}\right)$	s $\left(\frac{kJ}{kg\ K}\right)$	t (°C)	v $\left(\frac{m^3}{kg}\right)$	u $\left(\frac{kJ}{kg}\right)$	h $\left(\frac{kJ}{kg}\right)$	s $\left(\frac{kJ}{kg\ K}\right)$
Sat Vapour (g)	0.240	2576.04	2768.32	6.662	Sat Vapour (g)	0.240	2576.04	2768.32	6.662	Sat Vapour (g)	0.240	2576.04	2768.32	6.662
10	0.247	2595.45	2793.44	6.718	450	0.514	3334.55	3745.56	8.187	880	0.763	4155.85	4766.48	9.116
20	0.254	2613.90	2817.42	6.770	460	0.520	3352.24	3767.91	8.212	890	0.769	4176.40	4791.65	9.135
30	0.261	2631.77	2840.69	6.820	470	0.525	3369.99	3790.33	8.236	900	0.775	4197.00	4816.88	9.154
40	0.268	2649.22	2863.45	6.867	480	0.531	3387.81	3812.81	8.261	910	0.781	4217.67	4842.18	9.172
50	0.274	2666.36	2885.82	6.913	490	0.537	3405.70	3835.36	8.285	920	0.786	4238.39	4867.53	9.191
60	0.281	2683.27	2907.88	6.957	500	0.543	3423.65	3857.97	8.309	930	0.792	4259.18	4892.95	9.210
70	0.287	2699.98	2929.70	7.000	510	0.549	3441.67	3880.64	8.333	940	0.798	4280.03	4918.42	9.228
80	0.293	2716.55	2951.32	7.042	520	0.555	3459.76	3903.39	8.357	950	0.804	4300.93	4943.95	9.247
90	0.300	2733.01	2972.80	7.082	530	0.560	3477.92	3926.20	8.380	960	0.810	4321.90	4969.55	9.265
100	0.306	2749.39	2994.15	7.122	540	0.566	3496.14	3949.07	8.404	970	0.815	4342.92	4995.20	9.283
110	0.312	2765.70	3015.41	7.161	550	0.572	3514.44	3972.02	8.427	980	0.821	4364.00	5020.91	9.301
120	0.318	2781.97	3036.60	7.199	560	0.578	3532.80	3995.03	8.450	990	0.827	4385.14	5046.67	9.319
130	0.324	2798.20	3057.73	7.236	570	0.584	3551.22	4018.10	8.473	1000	0.833	4406.34	5072.50	9.337
140	0.331	2814.42	3078.83	7.272	580	0.589	3569.72	4041.25	8.496	1010	0.838	4427.60	5098.38	9.355
150	0.337	2830.63	3099.89	7.308	590	0.595	3588.29	4064.46	8.518	1020	0.844	4448.91	5124.32	9.373
160	0.343	2846.85	3120.95	7.343	600	0.601	3606.92	4087.74	8.541	1030	0.850	4470.28	5150.31	9.391
170	0.349	2863.07	3141.99	7.378	610	0.607	3625.62	4111.09	8.563	1040	0.856	4491.71	5176.36	9.408
180	0.355	2879.30	3163.04	7.412	620	0.613	3644.39	4134.50	8.585	1050	0.862	4513.19	5202.47	9.426
190	0.361	2895.56	3184.09	7.446	630	0.618	3663.23	4157.98	8.607	1060	0.867	4534.72	5228.63	9.443
200	0.367	2911.84	3205.16	7.479	640	0.624	3682.13	4181.53	8.629	1070	0.873	4556.31	5254.85	9.461
210	0.373	2928.16	3226.25	7.511	650	0.630	3701.11	4205.14	8.651	1080	0.879	4577.96	5281.11	9.478
220	0.379	2944.50	3247.37	7.543	660	0.636	3720.15	4228.83	8.672	1090	0.885	4599.66	5307.44	9.495
230	0.385	2960.89	3268.51	7.575	670	0.642	3739.26	4252.58	8.693	1100	0.891	4621.41	5333.82	9.512
240	0.390	2977.31	3289.69	7.606	680	0.647	3758.44	4276.39	8.715	1110	0.896	4643.22	5360.25	9.529
250	0.396	2993.78	3310.90	7.637	690	0.653	3777.68	4300.28	8.736	1120	0.902	4665.08	5386.73	9.546
260	0.402	3010.29	3332.14	7.667	700	0.659	3796.99	4324.23	8.757	1130	0.908	4686.99	5413.26	9.563
270	0.408	3026.85	3353.43	7.697	710	0.665	3816.37	4348.24	8.778	1140	0.914	4708.95	5439.85	9.580
280	0.414	3043.46	3374.76	7.727	720	0.671	3835.82	4372.33	8.799	1150	0.919	4730.97	5466.49	9.597
290	0.420	3060.12	3396.14	7.756	730	0.676	3855.33	4396.47	8.819	1160	0.925	4753.03	5493.18	9.613
300	0.426	3076.83	3417.57	7.785	740	0.682	3874.91	4420.69	8.840	1170	0.931	4775.15	5519.92	9.630
310	0.432	3093.60	3439.04	7.814	750	0.688	3894.56	4444.97	8.860	1180	0.937	4797.32	5546.70	9.647
320	0.438	3110.42	3460.57	7.842	760	0.694	3914.27	4469.32	8.881	1190	0.943	4819.53	5573.54	9.663
330	0.444	3127.29	3482.14	7.870	770	0.700	3934.05	4493.73	8.901	1200	0.948	4841.80	5600.43	9.680
340	0.449	3144.23	3503.77	7.898	780	0.705	3953.89	4518.20	8.921	1210	0.954	4864.11	5627.37	9.696
350	0.455	3161.22	3525.46	7.926	790	0.711	3973.80	4542.75	8.941	1220	0.960	4886.48	5654.36	9.712
360	0.461	3178.27	3547.20	7.953	800	0.717	3993.77	4567.35	8.961	1230	0.966	4908.89	5681.39	9.728
370	0.467	3195.39	3569.00	7.980	810	0.723	4013.81	4592.02	8.981	1240	0.971	4931.35	5708.47	9.745
380	0.473	3212.56	3590.86	8.007	820	0.729	4033.91	4616.75	9.000	1250	0.977	4953.86	5735.60	9.761
390	0.479	3229.80	3612.78	8.033	830	0.734	4054.08	4641.55	9.020	1260	0.983	4976.41	5762.78	9.777
400	0.485	3247.10	3634.75	8.059	840	0.740	4074.31	4666.41	9.039	1270	0.989	4999.02	5790.00	9.793
410	0.490	3264.46	3656.79	8.085	850	0.746	4094.60	4691.34	9.058	1280	0.995	5021.66	5817.27	9.808
420	0.496	3281.89	3678.89	8.111	860	0.752	4114.95	4716.32	9.078	1290	1.000	5044.36	5844.59	9.824
430	0.502	3299.38	3701.05	8.136	870	0.757	4135.37	4741.37	9.097	1300	1.006	5067.10	5871.95	9.840
440	0.508	3316.93	3723.27	8.162										

Table 3(A27) Superheated Steam Properties (Pressure Base) Pressure = 9,0 bar and saturation temperature, Tsat = 175.35°C

t (°C)	$v\left(\frac{m^3}{kg}\right)$	$u\left(\frac{kJ}{kg}\right)$	$h\left(\frac{kJ}{kg}\right)$	$s\left(\frac{kJ}{kg\,K}\right)$	t (°C)	$v\left(\frac{m^3}{kg}\right)$	$u\left(\frac{kJ}{kg}\right)$	$h\left(\frac{kJ}{kg}\right)$	$s\left(\frac{kJ}{kg\,K}\right)$	t (°C)	$v\left(\frac{m^3}{kg}\right)$	$u\left(\frac{kJ}{kg}\right)$	$h\left(\frac{kJ}{kg}\right)$	$s\left(\frac{kJ}{kg\,K}\right)$
Sat Vapour (g)	0.215	2579.65	2773.05	6.621	Sat Vapour (g)	0.215	2579.65	2773.05	6.621	Sat Vapour (g)	0.215	2579.65	2773.05	6.621
10	0.221	2599.46	2798.67	6.678	450	0.459	3342.74	3755.88	8.144	880	0.681	4165.77	4778.66	9.071
20	0.228	2618.22	2823.04	6.730	460	0.464	3360.47	3778.28	8.169	890	0.686	4186.35	4803.87	9.090
30	0.234	2636.36	2846.65	6.780	470	0.469	3378.27	3800.75	8.194	900	0.691	4206.99	4829.13	9.108
40	0.240	2654.04	2869.70	6.828	480	0.475	3396.14	3823.28	8.218	910	0.696	4227.69	4854.46	9.127
50	0.245	2671.39	2892.32	6.874	490	0.480	3414.07	3845.88	8.242	920	0.702	4248.44	4879.85	9.146
60	0.251	2688.48	2914.62	6.918	500	0.485	3432.07	3868.54	8.266	930	0.707	4269.26	4905.30	9.164
70	0.257	2705.36	2936.65	6.961	510	0.490	3450.13	3891.27	8.290	940	0.712	4290.14	4930.81	9.183
80	0.263	2722.08	2958.47	7.003	520	0.495	3468.27	3914.06	8.314	950	0.717	4311.08	4956.38	9.201
90	0.268	2738.69	2980.12	7.043	530	0.501	3486.47	3936.92	8.337	960	0.722	4332.08	4982.00	9.219
100	0.274	2755.19	3001.63	7.083	540	0.506	3504.74	3959.84	8.360	970	0.727	4353.14	5007.69	9.238
110	0.279	2771.62	3023.04	7.122	550	0.511	3523.07	3982.83	8.384	980	0.732	4374.25	5033.43	9.256
120	0.285	2788.00	3044.37	7.159	560	0.516	3541.47	4005.89	8.407	990	0.738	4395.42	5059.23	9.274
130	0.290	2804.35	3065.64	7.196	570	0.521	3559.95	4029.01	8.429	1000	0.743	4416.65	5085.09	9.292
140	0.296	2820.67	3086.86	7.233	580	0.526	3578.49	4052.20	8.452	1010	0.748	4437.94	5111.00	9.309
150	0.301	2836.97	3108.04	7.269	590	0.532	3597.09	4075.46	8.475	1020	0.753	4459.28	5136.97	9.327
160	0.307	2853.28	3129.21	7.304	600	0.537	3615.77	4098.79	8.497	1030	0.758	4480.68	5163.00	9.345
170	0.312	2869.58	3150.36	7.338	610	0.542	3634.51	4122.18	8.519	1040	0.763	4502.14	5189.08	9.362
180	0.317	2885.90	3171.51	7.372	620	0.547	3653.32	4145.64	8.541	1050	0.768	4523.65	5215.21	9.380
190	0.323	2902.24	3192.66	7.405	630	0.552	3672.20	4169.16	8.563	1060	0.774	4545.21	5241.41	9.397
200	0.328	2918.60	3213.83	7.438	640	0.557	3691.15	4192.75	8.585	1070	0.779	4566.84	5267.65	9.415
210	0.333	2934.99	3235.00	7.471	650	0.562	3710.16	4216.41	8.607	1080	0.784	4588.51	5293.95	9.432
220	0.339	2951.41	3256.21	7.503	660	0.568	3729.25	4240.14	8.628	1090	0.789	4610.24	5320.31	9.449
230	0.344	2967.86	3277.43	7.534	670	0.573	3748.40	4263.93	8.649	1100	0.794	4632.02	5346.71	9.466
240	0.349	2984.35	3298.69	7.565	680	0.578	3767.61	4287.79	8.671	1110	0.799	4653.85	5373.17	9.483
250	0.355	3000.88	3319.97	7.596	690	0.583	3786.90	4311.72	8.692	1120	0.804	4675.74	5399.68	9.500
260	0.360	3017.46	3341.30	7.626	700	0.588	3806.25	4335.71	8.713	1130	0.810	4697.68	5426.25	9.517
270	0.365	3034.08	3362.66	7.656	710	0.593	3825.67	4359.77	8.734	1140	0.815	4719.67	5452.86	9.534
280	0.370	3050.75	3384.06	7.686	720	0.599	3845.15	4383.89	8.754	1150	0.820	4741.71	5479.53	9.551
290	0.376	3067.47	3405.51	7.715	730	0.604	3864.71	4408.08	8.775	1160	0.825	4763.81	5506.25	9.567
300	0.381	3084.24	3427.00	7.744	740	0.609	3884.32	4432.34	8.795	1170	0.830	4785.95	5533.01	9.584
310	0.386	3101.06	3448.54	7.773	750	0.614	3904.01	4456.66	8.816	1180	0.835	4808.14	5559.83	9.600
320	0.391	3117.94	3470.13	7.801	760	0.619	3923.76	4481.04	8.836	1190	0.840	4830.39	5586.70	9.617
330	0.397	3134.87	3491.77	7.829	770	0.624	3943.57	4505.49	8.856	1200	0.845	4852.68	5613.61	9.633
340	0.402	3151.86	3513.47	7.856	780	0.630	3963.45	4530.01	8.876	1210	0.851	4875.02	5640.58	9.650
350	0.407	3168.90	3535.22	7.884	790	0.635	3983.40	4554.59	8.896	1220	0.856	4897.41	5667.59	9.666
360	0.412	3186.01	3557.02	7.911	800	0.640	4003.41	4579.23	8.916	1230	0.861	4919.85	5694.65	9.682
370	0.417	3203.18	3578.88	7.938	810	0.645	4023.48	4603.94	8.936	1240	0.866	4942.34	5721.76	9.698
380	0.423	3220.40	3600.79	7.964	820	0.650	4043.62	4628.71	8.955	1250	0.871	4964.87	5748.91	9.714
390	0.428	3237.69	3622.77	7.991	830	0.655	4063.82	4653.55	8.975	1260	0.876	4987.45	5776.12	9.730
400	0.433	3255.04	3644.80	8.017	840	0.660	4084.08	4678.44	8.994	1270	0.881	5010.07	5803.37	9.746
410	0.438	3272.45	3666.89	8.043	850	0.666	4104.41	4703.41	9.013	1280	0.887	5032.75	5830.66	9.762
420	0.443	3289.92	3689.05	8.069	860	0.671	4124.80	4728.43	9.033	1290	0.892	5055.47	5858.00	9.778
430	0.449	3307.46	3711.26	8.094	870	0.676	4145.26	4753.51	9.052	1300	0.897	5078.23	5885.39	9.793
440	0.454	3325.07	3733.54	8.119										

Table 3(A28) Superheated Steam Properties (Pressure Base) Pressure = 10.0 bar and saturation temperature, Tsat = 179.878°C

t (°C)	v $\left(\frac{m^3}{kg}\right)$	u $\left(\frac{kJ}{kg}\right)$	h $\left(\frac{kJ}{kg}\right)$	s $\left(\frac{kJ}{kg\,K}\right)$	t (°C)	v $\left(\frac{m^3}{kg}\right)$	u $\left(\frac{kJ}{kg}\right)$	h $\left(\frac{kJ}{kg}\right)$	s $\left(\frac{kJ}{kg\,K}\right)$	t (°C)	v $\left(\frac{m^3}{kg}\right)$	u $\left(\frac{kJ}{kg}\right)$	h $\left(\frac{kJ}{kg}\right)$	s $\left(\frac{kJ}{kg\,K}\right)$
Sat Vapour (g)	*0.194*	*2582.76*	*2777.12*	*6.585*	*Sat Vapour (g)*	*0.194*	*2582.76*	*2777.12*	*6.585*	*Sat Vapour (g)*	*0.194*	*2582.76*	*2777.12*	*6.585*
10	0.200	2602.95	2803.22	6.642	450	0.415	3350.21	3765.30	8.106	880	0.615	4174.85	4789.81	9.031
20	0.206	2622.02	2827.98	6.695	460	0.420	3367.99	3787.75	8.131	890	0.620	4195.46	4815.05	9.049
30	0.211	2640.41	2851.91	6.745	470	0.424	3385.83	3810.27	8.155	900	0.624	4216.13	4840.35	9.068
40	0.217	2658.32	2875.24	6.793	480	0.429	3403.74	3832.85	8.180	910	0.629	4236.86	4865.71	9.087
50	0.222	2675.87	2898.11	6.839	490	0.434	3421.71	3855.49	8.204	920	0.633	4257.65	4891.13	9.105
60	0.227	2693.13	2920.63	6.883	500	0.438	3439.76	3878.20	8.228	930	0.638	4278.50	4916.61	9.124
70	0.233	2710.18	2942.86	6.926	510	0.443	3457.86	3900.97	8.252	940	0.643	4299.41	4942.15	9.142
80	0.238	2727.05	2964.87	6.968	520	0.448	3476.04	3923.81	8.275	950	0.647	4320.38	4967.75	9.161
90	0.243	2743.79	2986.69	7.008	530	0.452	3494.28	3946.71	8.299	960	0.652	4341.40	4993.41	9.179
100	0.248	2760.42	3008.36	7.048	540	0.457	3512.59	3969.68	8.322	970	0.657	4362.49	5019.12	9.197
110	0.253	2776.96	3029.92	7.086	550	0.462	3530.97	3992.72	8.345	980	0.661	4383.63	5044.90	9.215
120	0.258	2793.45	3051.38	7.124	560	0.466	3549.41	4015.82	8.368	990	0.666	4404.83	5070.73	9.233
130	0.263	2809.90	3072.77	7.161	570	0.471	3567.92	4038.99	8.391	1000	0.671	4426.09	5096.61	9.251
140	0.268	2826.31	3094.11	7.197	580	0.476	3586.50	4062.22	8.413	1010	0.675	4447.41	5122.56	9.269
150	0.273	2842.71	3115.41	7.233	590	0.480	3605.15	4085.52	8.436	1020	0.680	4468.78	5148.56	9.286
160	0.278	2859.10	3136.68	7.268	600	0.485	3623.86	4108.89	8.458	1030	0.684	4490.21	5174.61	9.304
170	0.282	2875.49	3157.94	7.302	610	0.490	3642.64	4132.32	8.480	1040	0.689	4511.69	5200.72	9.322
180	0.287	2891.89	3179.18	7.336	620	0.494	3661.49	4155.82	8.502	1050	0.694	4533.23	5226.89	9.339
190	0.292	2908.30	3200.43	7.370	630	0.499	3680.41	4179.39	8.524	1060	0.698	4554.82	5253.11	9.356
200	0.297	2924.73	3221.68	7.402	640	0.504	3699.40	4203.02	8.546	1070	0.703	4576.47	5279.38	9.374
210	0.302	2941.19	3242.94	7.435	650	0.508	3718.45	4226.72	8.567	1080	0.708	4598.17	5305.71	9.391
220	0.307	2957.68	3264.23	7.467	660	0.513	3737.57	4250.49	8.589	1090	0.712	4619.93	5332.09	9.408
230	0.311	2974.20	3285.53	7.498	670	0.518	3756.76	4274.32	8.610	1100	0.717	4641.73	5358.52	9.425
240	0.316	2990.75	3306.86	7.529	680	0.522	3776.01	4298.22	8.631	1110	0.721	4663.59	5385.01	9.442
250	0.321	3007.35	3328.22	7.560	690	0.527	3795.33	4322.18	8.652	1120	0.726	4685.51	5411.55	9.459
260	0.326	3023.98	3349.62	7.590	700	0.531	3814.72	4346.21	8.673	1130	0.731	4707.47	5438.14	9.476
270	0.330	3040.66	3371.05	7.620	710	0.536	3834.17	4370.31	8.694	1140	0.735	4729.49	5464.78	9.493
280	0.335	3057.39	3392.52	7.649	720	0.541	3853.69	4394.47	8.715	1150	0.740	4751.56	5491.47	9.509
290	0.340	3074.16	3414.04	7.678	730	0.545	3873.28	4418.70	8.735	1160	0.745	4773.67	5518.22	9.526
300	0.345	3090.99	3435.59	7.707	740	0.550	3892.94	4442.99	8.756	1170	0.749	4795.84	5545.01	9.543
310	0.349	3107.86	3457.20	7.736	750	0.555	3912.65	4467.35	8.776	1180	0.754	4818.06	5571.85	9.559
320	0.354	3124.79	3478.85	7.764	760	0.559	3932.44	4491.77	8.796	1190	0.758	4840.33	5598.74	9.576
330	0.359	3141.78	3500.55	7.792	770	0.564	3952.29	4516.26	8.816	1200	0.763	4862.64	5625.68	9.592
340	0.363	3158.82	3522.30	7.819	780	0.569	3972.20	4540.81	8.836	1210	0.768	4885.01	5652.67	9.608
350	0.368	3175.91	3544.11	7.847	790	0.573	3992.18	4565.43	8.856	1220	0.772	4907.42	5679.71	9.624
360	0.373	3193.07	3565.97	7.874	800	0.578	4012.22	4590.11	8.876	1230	0.777	4929.88	5706.79	9.641
370	0.378	3210.28	3587.88	7.900	810	0.583	4032.33	4614.85	8.896	1240	0.782	4952.39	5733.93	9.657
380	0.382	3227.55	3609.85	7.927	820	0.587	4052.50	4639.66	8.915	1250	0.786	4974.95	5761.11	9.673
390	0.387	3244.89	3631.88	7.953	830	0.592	4072.74	4664.53	8.935	1260	0.791	4997.55	5788.33	9.689
400	0.392	3262.28	3653.97	7.979	840	0.596	4093.04	4689.46	8.954	1270	0.795	5020.20	5815.60	9.704
410	0.396	3279.74	3676.11	8.005	850	0.601	4113.40	4714.45	8.973	1280	0.800	5042.90	5842.92	9.720
420	0.401	3297.26	3698.32	8.031	860	0.606	4133.82	4739.51	8.993	1290	0.805	5065.64	5870.29	9.736
430	0.406	3314.85	3720.58	8.056	870	0.610	4154.30	4764.63	9.012	1300	0.809	5088.42	5897.69	9.752
440	0.410	3332.50	3742.91	8.081										

STEAM TABLES
[With SI Units]

Performance analysis of boilers, steam turbines, plant process and other mechanical equipment connected with power generation requires the thermodynamic data and transport properties of water and steam which are essential for the design and performance specifications.

The Tables contain thermodynamic properties of water and steam which cover the whole practical range being practised in power plants and process industries. The salient features of this book is that the properties of steam have been given fora wide range of temperatures and pressures.

The subject matter is intended for practising engineers in power plants and process industries and students pursuing undergraduate and postgraduate courses in Engineering. It also helps the students preparing for AMIE and competitive examinations such as GATE, IES, PSU and Civil Services, etc.

B. B. Arora is working as faculty member in Delhi Technological University (formerly Delhi College of Engineering). He has around 30 years of teaching, research and industrial experience. He has taught undergraduate, postgraduate and research scholars and has guided a number of dissertations and theses and published numerous research papers in national and international journals of repute. He has successfully completed many consultancy and industrial research projects.

Akhilesh Arora is working as faculty member in Delhi Technological University (formerly Delhi College of Engineering). He has more than 20 years of teaching, research and industrial experience. He has taught undergraduate, postgraduate and research scholars. He has guided a number of postgraduate dissertations and published numerous research papers in national and international journals of repute.

Vipin is working as faculty member in Delhi Technological University (formerly Delhi College of Engineering). He has more than 20 years of teaching, research and industrial experience. He has taught undergraduate, postgraduate and research scholars. He has guided a number of dissertations and theses and published numerous research papers in national and international journals of repute.

I.K. International Publishing House Pvt. Ltd.

S-25, Green Park Extension, Uphaar Cinema Market
New Delhi-110016, India • E-mail: info@ikinternational.com

9 789384 588533

www. ikbooks.com